Safety &

Health

on the

Internet

Third Edition

Ralph B. Stuart III
Christopher Moore

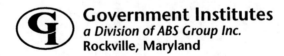

Government Institutes
a Division of ABS Group Inc.
Rockville, Maryland

Government Institutes, a Division of ABS Group Inc.
4 Research Place, Rockville, Maryland 20850, USA
Phone: (301) 921-2300
Fax: (301) 921-0373
Email: giinfo@govinst.com
Internet: http://www.govinst.com

03 02 01 00 99   5   4   3   2   1

ISBN: 0-86587-669-X

Printed in the United States of America

Contents

Chapter 6: Emerging Issues on the Internet 111

Chapter 7: A Web Safety and Health Directory 125

Chapter 8: Discussion Directory 453

Appendix 1: An Internet Glossary 513

Appendix 2: Email Reference Material 521

Index 537

About the Authors

Ralph Stuart, CIH has been in the Risk Management Department at the University of Vermont since 1985. In 1989, he started the SAFETY email list. This list has developed into the largest email list for safety professionals on the Internet. He has also helped organize the Vermont SIRI Web site at **http://hazard.com** (also **http://siri.org**), which contains the largest collection of safety data available on the Internet. In 1996, he was given the Chemical Health and Safety Award from the Chemical Health and Safety Division of the American Chemical Society for this work.

Chris Moore, MLS, has been working with, and teaching people to use, health and safety information systems since joining the Canadian Centre for Occupational Health and Safety (CCOHS) in 1986. He has primary responsibility for the content of CCOHS's World Wide Web site (**http://www.ccohs.ca**), including its popular directory of health and safety resources, which is accessed by thousands of people every month. He is the "list owner" for HS-Canada, an email list for health and safety discussions of interest to people working in Canada. He has written numerous articles on health and safety information systems and writes a regular "SafetyNet" column in *Canadian Occupational Safety*. He has given numerous courses on using the Internet as a health and safety information resource in Canada, the United States, Australia, New Zealand, and South Africa.

1 Introduction to the Internet

"Where can I get an MSDS for acetone?"

"What does the OSHA regulation on confined spaces say?"

"What does 'CIH' mean?"

Increasingly, the place to turn for answers to these and other commonly asked safety questions is the Internet. As it has grown through the 1990s, the Internet has proven to be a fertile ground for many different kinds of information, and the amount of information available there is increasing daily. What started off as the province of computer hobbyists and university research scientists has broadened into an important information tool for other professionals and the general public.

As a research tool, there is none faster than the World Wide Web. However, as anyone who has "surfed the Web" knows, there is a lot of material to sift through before you find the information you want. And the very structure of the Web is designed to make it easy to distract you from what you are looking for. Making the best use of the Internet requires information and practice, as does any other professional tool.

The Internet also shines as a professional networking tool. In 1980, there were 100 million emails sent each year, compared with 135 billion pieces of first class mail. In 1997, the two media were about equal, at 190 billion pieces each. So, the Internet has grown to become as important as paper for general communications. This chapter provides some basic information about the Internet to provide a context for the rest of this book.

What is the Internet?

The Internet in general, and the World Wide Web in particular, have been the subjects of much discussion, much of it sensational and some of it misleading. Although commercial issues have dominated much of the public discussion of the Internet as it has matured, the uses of most interest to safety professionals do not require familiarity with the latest Internet innovations. As with any computer resource, using the Internet involves many details, but the basics can be described with a few relatively simple concepts.

The Physical Internet

Physically, the Internet is a group of computers and computer networks that are physically connected and speak the same language. The physical connections can be through commercial phone lines, over high-speed cables specifically designed for digital communications, or over wireless connections. The value of these connections is that they allow a wide variety of computers and other electronic equipment to exchange data, in the form of commands, files, and raw data from electronic instruments.

An Internet connection can be made with a personal computer such as a Windows or Macintosh machine, a mainframe computer, or an electronic instrument such as video camera. This flexibility creates a versatile and open-ended medium, which allows a wide variety of people to actively participate. For these reasons, the Internet has become the method of choice for public access to electronic information.

The common language used on the Internet is the Terminal Control Protocol/Internet Protocol (TCP/IP). Fortunately, as the Internet has developed into a tool for the general public, the amount of jargon associated with using TCP/IP and the Internet has decreased. The computer concepts and tools used to gather information from the Internet are described in Chapter 2, and the more common jargon is defined in the Glossary.

Using the Internet

While the physical and technical basis for exchanging files between computers is interesting to some, what practical use is it? There are three specific functions that the Internet serves for users of safety information.

First, the Internet provides the ability to access a wide variety of technical information that is available in the form of electronic data files on the World Wide Web. Government agencies, large corporations, colleges and universities, and manufacturers of safety equipment all use the Internet as ways to make information available to the people they serve, and to the world at large. This is because making information available over the Internet can be much cheaper than using traditional paper or phone media.

Second, the Internet makes possible discussions of professional interest among people who are working in similar jobs. Many of the issues that safety and health professionals deal with on a day to day basis are unique to particular settings and often involve concerns outside the expertise of the general practitioner. The ability to consult in a general way with colleagues who may have faced similar situations can be an invaluable tool for a safety professional faced with an unfamiliar situation.

Third, as the Internet has become accepted as a public form of communication, new avenues have arisen for getting health and safety information out to the audiences that safety and health professionals serve. Web-based training, distribution of safety information, and responding to email are ways in which the Internet can help safety professional increase their outreach to their audiences.

While the first aspect, the Internet as a reference library, is important, the interactive discussions between a large number of professional peers on an ongoing basis is the unique aspect of the Internet as an information tool. Professional experience with the Internet as a safety marketing tool is still developing, but it presents exciting possibilities for reaching diverse and far flung audiences.

Culture of the Internet

It is important to remember that the Internet is not controlled by a single organization or authority. The Internet protocol itself is a language that computers and computer networks use. Consequently, the Internet has an organizational nature similar to any language: there are many local authorities, but no global power. This decentralized nature results in a culture on the Internet that is:

Diverse: People use the Internet in many different ways for many different purposes. They use many different kinds of computers using different sorts of information tools. It is important to avoid making any assumptions about the people you "meet" on the Internet, to whatever extent that is possible. It is also best to not take anything on the Internet at face value (that is without confirming it through other information), because the source of the information is generally ambiguous.

Disorganized at the casual-user level: The Internet can be very confusing for people who are new to it or people who do not maintain regular contact with it. This does not mean that it cannot be used casually; it means that more patience is required for casual use, or on the bright side, habitual use increases its ease of use.

Inventive: There are many different ways of getting a particular thing done on the Internet. If you find yourself wondering if you can use the Internet in a particular way, it is likely that other people are wondering the same thing and are developing a way to do it themselves. Often, they are quite willing to share these techniques. This results in the ever increasing number of tools available to use on the Net and many pleasant surprises.

Brief History of the Internet

As with any cultural phenomenon, it is important to have some awareness of the history of the Internet in order to use it successfully. The Internet evolved out of the integration of a group of independent computer networks, beginning in the late 1960s. These included academically oriented networks, military networks, commercial networks, and networks established by hobbyists. The final merger of these networks was not completed until early in the 1990s. The reason it took this long was that exchanging data easily between computers requires more hardware and software resources than were generally available until the powerful desktop machines of the 1990s became widely available.

Vestiges of these different origins can be seen in some stylistic points of the Internet. For example, the most developed mailing lists are associated with academic institutions that were formerly part of BITNET, because that is where the most powerful mailing list manager was available. The Internet newsgroups evolved in hobbyist-oriented groups, so their greatest strengths are in technical computer information. In addition, these groups have pioneered the development of "open source" software—programs (usually low level programs such as operating systems) which are distributed for free in a way that allows others to modify them for their own use. The most famous example of open source software is the Linux operating system, which is used by many Internet hosts.

Because of its flexibility and the low cost of connecting to it, the Internet has grown rapidly through the 1990s. Estimates of the number of computers hooked to the Internet and/or the number of users of the Internet quickly become obsolete, because these numbers have been growing exponentially for several years. However, in 1999 a survey posted at **http://www.nua.ie/** estimated that across the world 160 million people were connected. This figure was broken out geographically as 1 million in Africa, 27 million in Asia/Pacific, 37 million in Europe, 1 million in the Middle East, 89 million in Canada and the United States, and 5 million in South America. It is safe to say that the Internet will continue to grow rapidly and that computer access to networked information will become a feature of everyday life for many people.

The Internet and the computer tools that use it are continuing to evolve, driven by many different forces. The power of commonly available computers is increasing regularly, so many of the tools that made data transmission more efficient and conserved computer memory or hard disk space are less important. As more people use the Internet, software needs to become more flexible and user-friendly. As the resources devoted to the Internet become larger, some way must be found to pay for them. This situation creates a rapidly shifting set of Internet tools and some of the tools which were vital in 1995 are no longer of much interest.

Connecting to the Internet

Connecting to the Internet need not be expensive or technically challenging. For many people, Internet connections are now provided at their workplace. In those cases, the technical details are usually managed by the computer department of the organization. However, because the Internet is so useful for a wide variety of things, many people are interested in home access to the Internet. This section describes some of the considerations in arranging for this access.

The computer and connection requirements depend on how you want to use the Internet. Sending email and browsing text file libraries on the World Wide Web do not require more than a standard phone line and an Internet account; providing information to the Web requires more. This section describes some of the options available in making this connection.

Hardware Requirements

Because of the flexibility of the Internet protocol, the hardware specifications for a computer to be used to connect to the Internet are minimal. Deciding what kind of computer you need means deciding how fast and how convenient you want your Internet access to be. As with automobiles, the details of the computer setup best for you depends on your individual preferences and available options. The best way to explore those options is to contact a local computer user's group, computer store or Internet service provider. This chapter describes some of the generic considerations involved.

At the low end of the scale, any computer that can run a terminal emulation program can be used to access the Internet information services based on text systems. Most publicly available information fits into that category. Machines such as a computer that can run early DOS or other early 1980s personal computer operating systems can provide this kind of connection. However, to use the full graphics capabilities of modern programs such as Netscape, or to handle and archive a large amount of email, you will need more modern machines that can handle Windows or the Macintosh operating system. In general, it is likely that if you currently have a computer, you can use it to access the Internet at some level.

The critical component in determining the power of your connection to the Internet is the method that your computer uses to communicate with the Internet. This connection can be done either with a modem or with a more specialized piece of communications equipment. It is important to acquire the fastest communication connection that your computer can use. Currently, 56 kilobaud modems are commonly available for general use. Higher speeds are only available over phone lines specifically designed for improved data transfer. People who use the Internet regularly will find that upgrading their connection speed as much as possible will likely pay for itself very quickly in increased convenience and decreased connection charges.

Software Requirements

The software that your computer can run will depend primarily on the type of connection to the Internet you have and the type and processing power of your computer. For example, if you are using an IBM PC XT with a 2400 baud modem (typical of the first generation of computers to have a hard disk, circa 1985), you will be able to access email and the text of some Web sites. If you have a modern machine (such as one with a Pentium Processor, 32 megabytes of RAM and a 56 K modem), you will be able to use Internet sites which provide animation, sound, and Java programs and manipulate and archive your email automatically.

Connection Options

There are a number of ways to get your computer talking to others on the Internet. The way you choose will depend on a variety of factors: your location, your needs, and your budget. All of the following methods of access require that your computer have a modem, and that it be connected to a telephone line. An ordinary telephone line is fine; cellular phone connections will also work in situations where wireless access is important.

The most basic decision you need to make is the type of connection that you will use. Your options are terminal access or PPP (Point to Point Protocol). If you access the Internet through a Freenet or a university/college, you may be limited to terminal access. If you connect to the Internet through a commercial Internet Service Provider, you will almost certainly be using PPP.

Terminal access

If you have terminal access, it means that you dial into another computer, usually running Unix as the operating system, and access the Internet from that computer. You are limited to the programs and tools provided on the computer you are connected to. You can access all the types of resources described in this book—email, Usenet Newsgroups, and World Wide Web servers—but usually just the text.

The advantage to terminal access is that you can use any type of computer running any standard telecommunications package and since there are no graphics, sound, or video being transmitted, you can access the information more quickly. As an example, if you have an old IBM PC XT compatible with a 2400 bps modem, you can effectively access Internet resources using a terminal connection.

The disadvantage to terminal access is that text-based Internet software is generally harder to use, because many World Wide Web servers rely heavily on graphical material to convey their information.

Direct access

Direct access means that you dial into a network that is part of the Internet. For the time you are connected, your personal computer is directly connected to the Internet, using a protocol called PPP (Point to Point Protocol). With this type of connection, you can use any Windows or Macintosh Internet clients that your hardware can manage. Direct access to the Internet is generally preferable to terminal access.

The advantages to this type of connection is that these "client" programs are generally easier to use than those accessible through a terminal connection, and they provide access to multimedia material (graphics, sound, and video) that is otherwise not accessible.

The potential disadvantages are that you need a more powerful computer (at least a 386, preferably a Pentium on the Windows side or a PowerPC Macintosh) and at least a 14.4K bps modem to take advantage of this type of access. At the time of writing, 33.6K and 56K modems are common although obtaining a 56K connection can be a problem for many household phone lines. Even with a 56K modem, accessing graphic-intensive World Wide Web pages can be tedious.

Types of Service Providers

Commercial Internet service providers - These are companies that specialize in providing Internet access to individuals and organizations. These companies usually provide PPP access to their clients. They may or may not provide other services as well. Most Internet service providers have a variety of plans, depending on the level of Internet access you need. A few hours a month of individual access in North America can be had for as little as $10-$15 (prices quoted are 1999 U.S. Dollars) per month. Typical rates for unlimited monthly dial access are in the $20 per month range. Unlimited corporate access through a dedicated connection can be had for a few hundred dollars per month. With increased competition, prices are dropping quickly. Shop around.

Online services - Services such as America Online, Prodigy and CompuServe provide full Internet access, along with access to their own services. They all provide direct access to the Internet or an equivalent.

Freenets - Freenets are community-based computer networks that are springing up in urban centres all over North America. There are also Freenets in Australia and New Zealand. Their main focus is community information, but they also provide access to Internet resources. Usually, as part of your membership with a Freenet, you receive an Internet email address and access to Usenet Newsgroups. Different Freenets provide different combinations of Telnet, FTP, Gopher and World Wide Web access. Most provide terminal access only. As the name implies, Freenets do not usually require that you pay a membership fee, though they do encourage donations. Most Freenets only offer full access to local residents.

Newer Connection Options

Although modem and telephone access is by far the most common way for an individual to connect to the Internet, there are several other possibilities being developed to increase the speed of Internet access at retail prices. Not all alternatives will be available in all areas.

Some cable television companies are now offering Internet access through their coaxial cable networks. The bottom line with cable access is that it provides Internet access at speeds up to 40 times faster than those achievable with a 28.8K modem, but at a somewhat higher price. A monthly charge $39 for unlimited access is common. Two other approaches to Internet access that are available in some places are Internet access to your computer via satellite and "Web TV," which uses your television as a Web browser.

Rules of Thumb for Using the Internet

The Internet changes rapidly, so it is difficult to prescribe specific rules for using it effectively. However, keeping certain rules of thumb in mind while exploring and using the Internet will prove useful. Remember that even habitual users of the Internet find themselves lost occasionally; they get used to the feeling and learn that it means that is it time to find help.

A. A good way to learn the Internet is to use it to pursue hobby information.

For example, if you are a baseball fan, you can find mailing lists and Web sites with statistics, schedules and discussions of major league, minor league, and college baseball. Similarly, music fans can find databases and discussions on nearly any kind of music. The knowledge you pick up in pursuit of hobby information will be directly transferable to other interests. This strategy uses preexisting motivation and knowledge to get you past the technical aspects of learning to use the Internet.

B. Learning the Internet requires a willingness to experiment.

It is unlikely that you will permanently break anything, either on your computer or on someone else's, by using the Internet. The most serious consequence of using the Internet is the possibility of acquiring a computer virus that affects your machine. This is likely only if you download and use program files from an unscreened file library. The same steps you take to protect your computer from any virus (backing up your data, screening programs with virus protection software, etc.) will protect you from Internet computer viruses.

As you experiment, computer problems will occur and error messages are the result. Some of the error messages you see will be because of a mistake that you made and some will be because of mistakes other people make. Ignore those error messages that don't seem to affect you and you don't understand. If the mistake is someone else's and you understand the error, send gentle, brief information to the errormaker. Users helping users is an important part of the Internet. If a persistent error hampers your use of the Net (either because it stops your path or because it is so annoying), you need to find a way to work around that error. This means going to the next rule of thumb.

C. Look for help from the Internet itself.

Files and discussions about using the Internet are easily available on the Internet. You are quite likely to find help about using the Internet from the Internet itself, or from the help files that come with the software you use to access the Internet. It is good netiquette (points of etiquette specific to the Internet) to look for appropriate files (for example, Frequently Asked Questions or FAQs) on the Internet first, but don't be afraid to ask questions when you have found what seems to be an appropriate discussion area. While you may get negative feedback from some people if your question is "too basic," you are also likely to get a response pointing you in the right direction.

D. On the Internet, nobody knows if you're a dog, but everybody knows if you're a buffoon.

A famous cartoon shows two dogs working on a computer terminal, with one saying, "The nice thing about the Internet is that nobody knows if you're a dog." Because nobody can judge you on your physical attributes on the Internet, your netiquette habits are very important. Gaining a reputation for being polite and useful on the Internet pays back many times over.

The Future

The Internet is changing rapidly. The options described above are likely to continue to improve, in the sense of becoming cheaper and easier to use. Connections options are increasing and becoming more convenient and cheaper. It is now possible to routinely monitor your email while travelling, use the resources of the World Wide Web over wireless connections, and connect to your personal information from any computer on the Internet. These options all require a good understanding of how the Internet works and some planning.

The best way to find out how to use the Internet in your particular situation is to find someone who is already using the Internet and ask their advice. A few well-placed questions can save hours of frustration.

2 Internet Information Tools

The information available on the Internet is continually growing and changing, so looking for a specific piece of data on it can be a challenge. However, the types of information that are available on the Net are fairly straightforward and the information tools that are used to collect it are consistent in their approach to the task.

These facts, plus the speed and low cost of searching the Internet, make it a useful tool for professional research and communication, and one that is improving. This section describes Internet information tools in a general way. Remember that the specific tools that are available to you will depend on the type of computer you are using and how you are connected to the Internet. As with everything else on the Internet, the details of these tools change regularly, so be prepared for surprises, usually pleasant.

Internet Information

There are two basic forms of information available on the Internet. The first is the collection of file libraries containing data, usually text or graphics, on the World Wide Web. The second type of information is informal discussions between people of similar interests and of varying expertise. Which type of information will be more valuable to you will depend on your personal needs.

Gathering information from the Internet involves the use of various software programs, here called information tools. Different kinds of software can be used to perform the same job, depending on the specifics of your Internet connection and computer. This section briefly describes some of these tools to help you select the one most appropriate to your needs.

Clients and Servers

Two important terms to keep in mind when you are working with Internet software are *client* and *server*. Client programs are written to interact with specific server programs and can receive data from any server program of the same type. In general, client software is on the machine you are using and server software is on the machine hosting the information.

For example, there are Web clients and Web servers. Because they use the HTTP protocol (Hyper Text Transfer Protocol, the standard file format for presenting information on the World Wide Web) to communicate, the transmissions between the two programs are minimized and network traffic is decreased. HTTP is designed to be machine independent, so the computers can communicate with each other regardless of the hardware available at each end. It is up to the local software on each end to interpret the information received for the hardware available at the local machine.

An important effect of the client-server structure is that a continuous connection is not usually established between the client and the server. The connections are open only as long as a particular transmission is continuing. This allows the server to serve many different clients at the same time.

Node Names, Email Addresses, and URLs

Like most computer-related phenomenon, the Internet has jargon associated with it. Fortunately, it is not necessary to master it all to begin using the Internet. However, there are three pieces of jargon that you need to understand for most uses of the Internet. (Other common terms used on the Internet are explained in the Glossary.) These are **node names, email addresses**, and **URLs**.

Node names are the names of computers that are connected to the Internet. These come in two versions: numeric and alphabetic. The numeric versions, which every computer on the Internet has, comes in four segments, such as 132.198.205.29. However, these numeric versions are hard for people to remember and easy to mistype. Therefore, there are computers, called Domain Name Servers (DNS), which convert the numeric names into alphabetic names such as esf.uvm.edu. These are the names which are most commonly used. It should be noted that not every computer on the Internet is given an alphabetic version of the node name, and a single alphabetic node name can represent more than one computer. However, the basic concept that a node name represents a particular Internet site is valid.

Email addresses are used to send notes to particular individuals on the Internet. Email notes are usually ASCII text files, but can include binary files as well. Email addresses are of the form: **user@node-name**. "User" is the user-id (user identification) assigned at that computer to the person who will receive the file; "node name" is the computer that the file will be sent to. For example, email sent to **rstuart@esf.uvm.edu** goes to the **esf.uvm.edu** computer and is held for the person with the user-id (and the password for that id) **rstuart**.

When a computer receives an email message, it keeps it in the file assigned to that user-id until the person picks it up with his email software. Note that not all computers on the Internet are set up to receive email. An Internet computer must be configured as an email server in order to know what to do with email it receives.

Email addresses can represent computer programs (such as LISTSERV) or functions within an organization (such as the sales department) as well as particular people. Usually, you can figure out what type of entity will be receiving the message from the user-id. Note that a particular individual's user-id can change as computer systems are reconfigured. It is a good idea to be alert for these changes.

URL stands for Uniform Resource Locator, which is the address of a particular file on the Internet. These come in the form: **tag://node-name/path**. "Tag" is usually "http," although there are a variety of other tags available on the Internet. The tag indicates what kind of file is being referenced. "Node-

More on Node Names

The alphabetic version of node names have specific structures that provide information about the computer they refer to. They are generally composed of three or more parts, separated by periods, usually called "dots."

The letters between the dots correspond to domains and subdomains, with the primary domain on the right. In the United States, the primary domain consists of three letters which indicate what type of organization the node is located within. The major U.S. domains include ".edu," which indicates an academic site, ".com" for commercial sites, ".mil" for military sites, ".gov" for government sites, and ".net" and ".org" for miscellaneous sites. Outside the United States, and in some U.S. locations, the primary domain is a two-letter country code. For example, ".ca" indicates Canada, ".au" indicates Australia and ".uk" indicate Great Britain. The United States country code, ".us" is used by some American sites, particularly those representing governments. Country codes in node names may be preceded by two letter versions of the American type domain names described above, but this is not the rule.

In the node name before the type or country code are one or more alphabetic strings. The string preceding the last code usually provides an indication of the institution hosting the node. Strings before that are usually set by the operators of a particular computer and can be any unique strings. Often, these are whimsical names (*e.g.,* "calvin" and "hobbes" are two of the more popular).

An example of a node name is moose.uvm.edu. This node is at an American educational institution, the University of Vermont (nicknamed UVM). Moose is one of a group of computers, which also includes elk and gnu (together they form the UVM zoo, or zoo.uvm.edu). An example of a non-U.S. name is ccohs.ca. This is a Canadian site, at the Canadian Centre for Occupational Health and Safety. ".ca" is the country code and ".ccohs" is the institution.

name" is the computer that the file resides on, and often includes "www" as a prefix to the node-name (for example, www.ccohs.ca). "Path" is an optional part of the URL; it is used to indicate the specific location of the file that you are after. If the path is not given, then you will be given the "home page" of the Web site hosted at the computer with that node name. For example, typing in **http://www.ccohs.ca** produces the home page of the Canadian Centre for Health and Safety site, while typing in **http://www.ccohs.ca/ resources/** takes you directly to their Internet resources page.

File Libraries

One of the main attractions of the Internet is the availability of public file libraries. These are put together for various purposes by academic institutions, commercial and noncommercial organizations, and government agencies. These libraries are interconnected to form the "World Wide Web" (WWW). You visit these libraries by using software tools called Web browsers (Web browsers are the client software for Web servers).

The most popular Web browsers are Netscape Navigator, Microsoft Internet Explorer, and Lynx. Web browsers allow you to seamlessly download a variety of files from the file libraries, including multimedia files such as graphics, animation and sound files. You select the files to download by pointing at a link on the screen and clicking. Some Internet file libraries predate the development of Web browsers and do not use multimedia files. These older libraries are usually limited to text files, however they can still be explored with Web browsers. Note that Lynx (which is the Web client available on most Unix systems) can only read text files. This makes Lynx much faster than other browsers for using this information, although many sites are not organized in a way that allows Lynx to organize them easily.

Some of these libraries can also be used with other types of software, particularly FTP (File Transfer Protocol) software. FTP clients display only text information on the screen. However, graphics or program files downloaded by FTP can be converted on most computers to graphics. For this reason, many people use FTP to transmit large files that are meant to be read by a Web browser. FTP can also be used to move a file from your machine to someone else's.

More on URLs

Many Web browsers will assume that the beginning of a URL is "http://www." and that the ending of the node name is ".com". So, just typing in the primary subdomain of the Web site is enough to reach its home page. (Many advertisements in other media such as print and television take advantage of this by simply announcing "See our Web site at www.abcinc.com" or see us on the Web.) This practice makes it relatively easy to guess the URL for a strong brand identity. For information on a brand of cars, you can type "ford" into your browser; for electronic products, "sony" will take you to this brand's site.

For this reason, the ".com" domain is extremely popular for people hoping to attract traffic to their site. Many non-U.S. companies and organizations have acquired ".com" endings to make it easier for their customers to use their site. For example, the Canadian Centre for Occupational Health and Safety has Web addresses of **http://www.ccohs.ca** and **http://www.ccohs.com,** which both point to the same Web page. Thus just typing "ccohs" into the address bar of your Web browser will find that site.

On the other hand, many organizations that wish to avoid identification as a commercial enterprise avoid the ".com" domain. Membership organizations such as the American Chemical Society are usually found in the ".org" domain. The ACS site is at **http://www.acs.org**. Again guessing the URL site of an organization is usually successful if you add "www." to the front of its acronym and ".org" to the back.

There is some concern that the ".com" name domain is "filling up." People worry that many of the prime subdomain names are taken and have proposed breaking out more specific domains. However, ".com" is so deeply engrained into both software and people's

What are Web Sites?

The World Wide Web is linked together by a common language, known as HyperText Mark-up Language. A "Web site" is organized as a collection of pages. Most pages consist of a text file and one or more graphics files. Capabilities such as conducting searches for particular words within a large group of text files, or the ability to send messages to the owner of the site are also included on many Web sites.

The opening page of a particular Web site is known as the "home page." The home page is usually found at the address that you use to access a particular organization's site. A good home page serves as a table of contents to the information that is provided at a particular Web site. Complex Web sites can have hundreds or even thousands of pages, so a good home page is important in making the site useful.

You can also connect directly to any one of the other pages at a site without going through the home page. Entering at pages other than the home page can be confusing if you are not familiar with the site. Well-designed Web pages take this possibility into account and provide you with clues as to how to get to the home page on each of the other pages.

Establishing a Web site is an inexpensive proposition, so the number and types of pages are proliferating rapidly. Some are designed professionally, while others are homemade. Professional pages usually have more graphics and interactive features. However, the actual value of the information available at a particular site is hard to determine from looking at the graphics used at the site. While some Web developers focus on the graphic contents of the page, others focus on the information they want to make available. Exploring the site is the only way to develop a sense of what information is actually available.

Discussion Groups

The other major type of Internet information is interactive discussions organized around particular subjects. These take place in discussion groups, which operate over either email lists or newsgroups. The information in these discussions is less formal than the file libraries, and thus it is a different sort of information than Web sites contain.

Email

Electronic mail (email) is one of the few Internet tools that conveys its function in its name. Email is the ability to exchange notes with other people connected to the computer network and is the most commonly used feature of the Internet.

To use email, you need an email address and email software. These are usually provided by the same source as your Internet connection. As a medium, email shares many features with familiar means of communication such as telephones and face-to-face conversations. However, it has attributes that make it distinct from these media. In particular, it is much faster and usually more convenient than paper mail for exchanging information. It is important to understand these differences in order to use email effectively.

Nature of Email

Email is asynchronous: The most distinctive feature of email is that it is a fast asynchronous form of communication that is cheap, convenient, and fast. *Asynchronous communication* means that two people do not have to be using the medium at the same time; people exchange information at their convenience rather than trying to find a simultaneous free spot in their schedules. This feature also allows participants to spend time thinking about the issues under discussion without interrupting the flow of the conversation.

Email shares this attribute with other media, such as letters or phonemail, but its speed is distinctive—detailed interactive discussions can take place over a two- or three-day period without changing anybody's schedule. These discussions are easily retrievable and may be restarted at a later time. Email is also very flexible in terms of audience. It can facilitate discussions among thousands of people as easily as between two people, using the same tools and in the same amount of time. This flexibility is an important strength of the medium.

Email is conversational: Because of the rapid delivery of email from one person to another, it works well as a conversational medium. This means that when you write email notes, you should try to minimize their length and expect follow-up notes to continue the discussion.

Brevity is important for several reasons. First, there is a limit to the length of a message that can be comfortably read on a computer screen. More than a couple of screenfuls of a note requires the recipient to pay as much attention to the mechanics of paging through the note as its content.

Secondly, because email strips many of the interactive cues people use to interpret messages (tone of voice, body language), it is easy to take a conversation in the wrong direction. It is important to make email notes short enough to give people a chance to provide feedback about what you're saying. Sending a long note can overwhelm the recipient and make an answer difficult or impossible.

Email is public: Email is tricky. It can feel very personal and many private conversations take place happily by email. Unfortunately, this can be deceptive. Email should always be considered a public medium, because the machines that an email note passes through on the way to its destination are under the control of other people.

It is unlikely that someone will take the time to randomly read email notes passing through their machine, just because of the sheer volume of email going by. On the other hand, if a particular notes gets misdirected, it is quite likely that a note will end up in unexpected places. Also, it is possible for software to be written that could detect and capture a credit card number or Internet account username and password that is part of an email message. For this reason, sensitive information should be encrypted if it is sent by email.

Thus, email should always be considered to be a public medium and every note which you send should include the thought: "What if someone I know sees this?" This consideration generally does not limit the usefulness of email for professional purposes, but it can affect the tone of a professional discussion by making people more cautious in stating opinions.

Email Lists

While email conversations with other individuals can be productive, the major difference between email and other communication methods is the email mailing list. These lists are run by automated software which receives a note and then redistributes it to a list of interested people. The lists provide kinds of interaction that can not be achieved over the phone or at conferences and provide for sharing information and expertise in a uniquely convenient and timely way. However, they do require some practice to be used effectively.

Computer Side of Mailing Lists

While an email list is a relatively simple concept, they can be confusing for people inexperienced in their use. Mailing lists are run by mailing list management (MLM) software packages. The options available to you for controlling your subscription to a list will depend on the features of the software running the list. For example, most MLMs allow you to "digest" your subscription—combine each day's postings into a single file. This is particularly valuable for active lists in which you have only marginal interest.

One of the reasons that mailing lists can be confusing for human users of the system is that there are several different mailing list managers operating on the Internet. Popular MLMs include Listserv, and majordomo. There are subtle but critical differences in the syntax of the commands for the same functions in the various MLMs. Therefore, it is important to keep the introductory file you get when signing on a particular list, in order to know which program is managing that list and which commands apply.

There is usually a human mailing list owner for each list. Questions about using the list should be directed to that person rather than the list as a whole. Fortunately, the commands needed to use the most common mailing list managers have been summarized in a file, available on the World Wide Web at **http://lawwww.cwru.edu/cwrulaw/faculty/milles/mailser.html.** You can also get specific information about the commands of that mailing list management software running on a particular computer by sending a message to that software, *e.g.*, listserv@list.uvm.edu. The body of the message should be "help."

The general principles of most mailing list managers are similar: there are generally three email addresses involved in using the list. The first is the address which processes commands to the MLM regarding your subscription. "Subscribe," "unsubscribe," and similar commands are sent to this address. These commands are handled automatically by computer software, as long as they have the correct syntax.

The second important address involved in working with an email list is the address of the list owner, who is the human being who can help you with problems when the syntax that you've sent to the first address doesn't work. These people have varying amounts of time to devote to such administrative tasks, so the timeliness of their responses is not as reliable as the computer software's.

The third address is the one which results in your note being redistributed to the list itself. Take care that notes that you send to this address are meant for the list as a whole, because their redistribution is usually automatic and immediate, so that the list owner can not help you stop a missent message from going out to a large number of people.

As an example, the three email addresses associated with the HS-Canada mailing list are requests@ccohs.ca (the MLM), owner-hs-canada@ccohs.ca (the human owner) and hs-canada@ccohs.ca (the list itself).

Human Side of Mailing Lists (Netiquette)

While interacting with the MLM software may be confusing, with a little practice, it is a straightforward process. However, working with the other people on the mailing list will require continued attention. Depending on the nature of the audience for a particular mailing list, different standards of email manners apply. These standards may change over time if the list becomes busier.

Remember that email is a delicate medium, and the tone of email correspondence is very important in getting a useful response. Since there is no body language or tone of voice to help convey meaning in email, communicating effectively by email requires a certain amount of practice. This is especially true on email lists. Electronic mail is about communication with other people. When you compose an email message, read it over before sending it and ask yourself what your reaction would be if you received it. Time spent making email clearer is time well spent.

It is important to avoid assumptions about your correspondents from what you see in their email notes. There are many different hardware/software combinations connected to the Internet, and things that you can do naturally with your email (such as, attaching a signature to your posting or spell-checking a note) can be nearly impossible for someone on another system to accomplish. If misspellings or lack of signatures make it difficult to understand a message, making a gentle inquiry to the author is the best approach.

Email lists are most useful when the question you're trying to answer is general, or when you have no idea where to look for an answer. The usual goal of asking a question to a list should be to get a reference to a good place to look for the answer. The power of the Internet is that you can pose your inquiry to a large number of people simultaneously and get responses quickly, usually within 24 hours.

There are many summaries of good netiquette written and available both on the Internet and in books about the Internet. One of the best sources of information on this subject can be found at Arlene Rinaldi's Netiquette Web site at **http://www.fau.edu/netiquette/net**.

Considerations in Using Mailing Lists

When you first become involved in using electronic discussions over the Internet, it is easy to become very (perhaps, over-) enthusiastic about them. Being able to talk to people who understand and are interested in your job may be a new experience. However, it is also easy to find yourself spending more time than you intended on the computer as the discussions on various lists develop in ways that you didn't expect. This section describes some of the issues you should consider to assure that your email time is well spent.

Joining email lists

Joining an email list is straightforward. You send the appropriate "subscribe" command to the address of the mailing list management program that runs the list. However, choosing which lists to subscribe to takes practice. It is difficult to participate effectively in a busy email list on an occasional basis. The speed of email discussions is such that they can begin and end within 48 hours. Tracking discussions and moving them into directions that you find useful means reading list traffic at least every other day. Browsing a list on a weekly basis will enable you to keep track of what information is available in the list's archives. However, it is unlikely that people will respond to messages about issues that came up two weeks ago unless a new angle is given to the question. Less active lists do not require as much effort to follow, but they are less likely to generate useful discussions.

Things to consider when deciding whether to subscribe to or participate in a particular list are the subject focus of the list, the volume of the list (number of messages per day), the tone of the list and the "signal-to-noise ratio" of the list. In order to judge these things, it is usually necessary to sign on to a list and monitor it for a week or more before making a decision about your continued participation or jumping into the discussion. There are no negative consequences associated with subscribing to and unsubscribing from mailing lists. Almost all of them are free, so the only cost to you is your time and effort to gain a sense of the value of a particular list to your interests. If your need for information is too urgent to allow this "warm-up period," it is a good idea to search the list archives to see if your need is likely to be met by that particular list.

Some mailing lists are not set up to provide forums for discussions. Rather they are one-way distribution channels for announcements and notices. For example, the U.S. EPA has a group of mailing lists for people interested in receiving Federal Register notices on the day they are published. These notices are the only things that appear on the list. This distinction will be clear from the nature of the traffic you get from the list. A list of discussion groups related to professional health and safety issues is provided in Chapter 7.

Managing a lot of email

Because there is little or no charge to most people for sending email, email is abundant and often overwhelming. To manage it easily, you need good email software. Ideally, your software should have easy use of several important functions: an address book for commonly used email addresses; the ability to automatically append a signature file with contact information; easy replies and forwarding for messages you receive; sorting (or filtering) your messages based on the sender or title line; and the ability to search the content of messages you have sent and received. Software that has most of these features include Eudora Pro and Netscape Messenger.

Mailers that possess all or most of these functions have been developed for most platforms, so if using email is a chore for you, it will probably pay to investigate other software which can be used with your current platform. Managing 150 notes in a half-hour is possible with good email software, while with other mailers, 20 messages can take that long. Consider the limitations of your email software when choosing which email lists to subscribe to.

Newsgroups

Newsgroups are similar to email lists in concept and perform a similar function—electronic messages are made available to a large number of interested people. However, newsgroups are not as readily accessible to as many people on the Internet, so specialized newsgroups, such as ones specifically related to occupational health and safety, are not as common as electronic mailing lists.

In addition, there are cultural differences between mailing lists and newsgroups that are likely to make the answers to a particular question different when posted in one forum rather than the other. Newsgroups tend to attract more members of the general public, which creates a higher noise level in the technical groups. These cultural differences result from the nature of the two media.

Newsgroups were developed as an alternative to mailing lists when network managers realized that a similar function could be served more efficiently (in terms of network resources) if a single copy of a message was kept in a place where people would be able to browse and read it. The messages are sorted into groups based on their subject to form "newsgroups." To read newsgroups, you use newsreader software to access a newsserver site. Newsreaders emphasize handling a large number of messages as efficiently as possible. They present a list of the title lines available and you select ones of interest. If you want to keep a particular posting, you need to tell the newsreader to email you a copy.

Although the intent of newsgroups and email lists are similar, the technical differences between them create important functional differences. Newsgroups *feel* more public than email lists, so they attract people who are interested in addressing a mass audience. Because newsgroup users tend to be casual visitors to the group and have less history with each other, flaming (see glossary) happens more quickly in newsgroups and questions tend to be repeated more often.

On the other hand, newsgroups are convenient places for casual users to find answers to common questions. Many high traffic newsgroups develop Frequently Asked Question (FAQ) files which are posted regularly to answer questions that are repeatedly asked. These FAQ files are usually available on the Web as well. A list of health and safety FAQs is available at the CCOHS site at **http://www.ccohs.ca/resources/faqs.html.**

3 The Internet as a Research Tool

One of the biggest challenges facing health and safety professionals is finding information when they need it. While a well-stocked paper library can go a long way towards answering this need, changing regulations and new uses for hazardous chemicals make paper an unreliable and often expensive medium for researching the latest aspects of hazardous material use. Happily, the Internet has developed into an important research tool over the last few years and now provides a legitimate alternative to an extensive library of books, manuals, and regulations.

This chapter first describes some of the tools that you can use to become familiar with the types of information available on the Net. Then it describes a strategy for answering specific questions that arise in daily health and safety work.

Finding Information on the Internet

Just as being familiar with the contents of a paper library makes searching for information there easier, being familiar with the Internet makes searching for particular information easier. So it is a good idea to spend some time building your net-surfing skills before you need to answer a particular question.

Web Directories

Exploring the Internet has been made easier by the development of Web directories. These are Web sites that list a large number of general interest Internet sites, organized into subject categories. One of the most popular of these directories is Yahoo (**http://www.yahoo.com**). Yahoo's home page lists a variety of subject areas (*e.g.*, Arts, Business and Economy, Computers and the Internet, Education, Entertainment, Government, Health, News, Recreation and Sports, Reference, Regional, Science, Social Science, Society and Culture). Clicking on one of these titles leads to subcategories and then to a list of Web sites that provide information about these subject areas. By using Yahoo as a starting point, you can get a sense of the capabilities of the Web and the type of information available there.

For example, if you are looking to see which U.S. government agencies that regulate hazardous materials have information available on the Web, you can select the U.S. Government category on the Yahoo home page. Under the "Independent Agencies" (*i.e.*, not part of a cabinet-level department) item, you can find 152 different EPA Web pages, established by different offices within the agency. These offices include the Environmental Response Team, the Chemical Emergency Preparedness and Prevention Office and the Emergency Response Notification System, as well as the various EPA regions. This is many more than is evident from visiting the EPA home page at **http://www.epa.gov**.

Search Engines

Many times, rather than just browsing the web, you are interested in information on a specific subject. Using a subject-based directory such as Yahoo is not the most efficient way of approaching this sort of search on the Internet. Rather, you want to be able to specify particular words that you want to find and have the computer find files that contain those words for you. This requires the use of a Web search engine. One of the most powerful search engines currently available on the Net is AltaVista, found at **http://www.altavista.com/**. Others that are valuable include Google (**http://www.google.com**) and Northern Light (**http://www.northernlight.com**). Web search engines are developing rapidly and competing with each other for users, so trying more than one may be useful. Each search engine has different features and which will work best for you will depend on what kind of information you're most interested in. A good description of the various search engines is available at the Internet Scout site at **http:// wwwscout.cs.wisc.edu/scout/toolkit/searching/index.html**

Conducting Web Searches

The difference between search engines and Web directories is best given by example. A search for the words *hazardous materials* with Yahoo produced 192 hits. AltaVista produced about 60,000. Yahoo found sites that specifically listed *hazardous materials* as keywords associated with their site— primarily commercial producers of hazardous materials handling equipment and training courses. AltaVista found many other files containing the words *hazardous materials*, including job descriptions, safety plans, and information about specific chemicals.

Although AltaVista can generate a large number of hits for a particular search, it is important to remember that AltaVista does not represent a "netwide" search. The rapidly changing nature of Web sites makes it impossible for even the most powerful computers to track them with complete accuracy. For example, even AltaVista does not cover all the Web sites that are available, and many of its references in its index are out of date. It can take as much as a month for AltaVista to become aware of a new Web site or detect a change in one that is included in its catalog. Again, careful selection of keywords and use of subject-specific indices will increase the usefulness of your searches.

Refining the Search

As the *hazardous materials* example demonstrates, simply putting in the first words that occur to you to search on can be rather inefficient. Fortunately, Internet search engines provide ways of refining your search so that it can be more selective and the results more useful. While the precise format of these refinements varies from search engine to search engine, the concepts they use are similar.

The first step in refining your search involves phrasing your question in such a way that you will get the answer you need. This is often rather easy (*Does OSHA have any regulations that specifically cover the use of this chemical?*), but other times this can be more difficult (*e.g., Is this workplace situation a confined space?*).

If you are having trouble coming up with a question that describes your need, it may be better to think up the name of a magazine article that would be just what you need. A four or five word phrase is a good place to start your search. It is important to think about possible other meanings for the words you select. For example, *safety* may refer to chemical concerns in your mind, while it refers to law enforcement issues in many other people's minds. An ambiguous word such as this is usually a poor choice to search for.

Subject Specific Indexes

It is best to test out your initial set of keywords by using it with one search engine with the idea of seeing how many useful responses it produces before using other search engines. In this phase of your research, it is better to start with a more subject-focused index.

Fortunately, in the occupational and environmental health and safety area there is an index specifically focused at this area. At **http://list.uvm.edu/ archives/safety.html**, there is an index to the SAFETY email list archives. This database consists of over 50,000 email messages discussing a wide variety of safety issues. As well as providing access to the discussions by SAFETY participants, this database provides a good way to refine your search strategy in a limited universe of subjects. The advantage of using this database is that the keywords are likely to be used in the same context that you are thinking of them, so the results are likely to be germane to your question.

By searching in this database, you can see if the phrase you chose is commonly used by other safety professionals to describe the situation you are thinking of. If the results are not related to your specific concerns, you can change the words you are using until the results are more appropriate. Once the keywords are refined in this search, they are likely to be more effective when searching larger databases like Yahoo and AltaVista.

general, more than 20 hits in a search indicate that the search needs to be further refined if you are looking for specific information. It can become quite time-consuming to check out more than 20 links unless you are in an exploratory mode. Various search engines provide different ways of manipulating the keywords you have decided to use. For example, large databases usually require that all of the keywords you enter be included in the file for it to be considered a relevant hit. This is because the wide variety of subjects they cover creates many hits on an individual word. Most Internet search engines require you to use an alternate screen to conduct a nonstandard search. The details of composing these searches are usually found on those screens.

Internet Search Strategy

Although the tools above can make searching the Internet for specific information easier, it is still easy to get distracted from the question you are trying to answer. In order to conduct research on the Internet efficiently, it is important to have a search strategy in mind while you are looking. This strategy may be as follows.

Refining the Question

Consider whether the Internet is the best place to look for the information you're interested in. Information that does not change often or is of wide application is probably available on paper, in a library. On the other hand, for very specific or very new information, the Internet can be an unmatched resource.

A. Decide what kind of information you're looking for.

Looking for a specific piece of data (*e.g.*, the flashpoint of acetone) is different from looking for a technical interpretation of that data (use of acetone requires adequate ventilation due to its flammability), which is different from looking for a rule of thumb (use acetone in a fume hood if you're using more than 500 milliliters). These different types of information will be found with different strategies in different places on the Net.

Organizing the Web

Web Directories and Search Engines

What is the difference between Web directories and search engines?

As the Internet has grown, various attempts to organize the information available on it have been developed. These efforts fall into two basic types: **Web directories** and **search engines**. These are different in both structure and purpose. Web directories provide hierarchical access to descriptions of Web sites or pages; they are designed to provide browsable lists of Web sites concerned with a specific subject. Search engines (described in more detail in the next sidebar of this chapter), provide the ability to search the actual text of Web pages. Web directories usually allow keyword searching as well. However, these searches are more limited than those provided by search engines; Web directories index just the titles and brief descriptions of the Web sites.

These two different functions are analogous to different sections of a book: Web directories act as a table of contents for the web, while search engines act as an index to the web. If you want to browse around and see what sites exist that contain information on a general subject, use a Web directory. If you want to search for Web content on a specific subject or combination of subjects, use a search engine.

Popular Web directories

One of the most popular general purpose directories is "Yahoo!" (**http://www.yahoo.com**). It is arranged into major subject categories, such as Computers and Internet, Government, Health, Science, etc. Choose a category such as Environmental Health, and you will see a list of sites containing environmental health information. If you are interested in getting more specific, subcategories such as Multiple Chemical Sensitivities and Toxicology are available. Yahoo! does allow keyword searching, but bear in mind that you are just searching on the names of the Web sites, and sometimes on short text descriptions of them. For this reason, Yahoo makes it convenient to pass your inquiry on to more extensive search engines.

In addition to the directory provided in this book, there are several web-based directories specific to health and safety. Two of the most popular are the "Health and Safety Internet Directory" maintained at the Canadian Centre for Occupational Health and Safety (**http://www.ccohs.ca/resources**) and OSHWeb, maintained at the Institute of Occupational Safety Engineering at the Tampere University of Technology in Finland (**http://oshweb.me.tut.fi**). These and other health and safety directories are described under "Pointers to Other Sites" in the directory in Chapter 7.

Formal databases: For specific pieces of data, formal databases are the best places to look. There is a variety of such sources, such as MSDS collections and databases containing government regulations. These databases are usually indexed to allow for keyword searches. Selecting keywords carefully will make your search more efficient.

Professional interpretations: For technical interpretations of raw data, the best places to look are in collections of policies and procedures that are available on line. Such collections are usually associated with information systems that companies and institutions put on-line for the convenience of their employees or customers. A good way to find these documents is to use the Internet search engines available. For this search, use keywords that apply as specifically as possible to your item of interest. Be prepared for many *false hits*—returns that are of no interest—for these searches. However, you are also likely to find several useful unexpected sources.

Informal information: Because informal knowledge requires technical expertise to apply appropriately, it is unlikely to be found in the formal information sources on the Net. However, the Internet has many informal information collections available in the archives of electronic mailing lists that operate on the Net. These are the first places to check for this type of information. The SAFETY archives (**http://list.uvm.edu/archives/safety.html**) are a good place to start such a search. Even if you do not find the information you are after there, you may find a reference to another Internet resource that has the information you are looking for.

B. Select keywords to search for.

The result of refining your question should be a set of keywords that you want to search for. You will use these keywords in performing searches at various Web sites that are likely to contain appropriate information. For example, if you are simply after the flashpoint of acetone, *flashpoint* and *acetone* are appropriate keywords. Whereas if you are concerned about ventilation requirements for using acetone, *flashpoint* is not likely to be helpful and *flammable liquid* may be a reasonable substitute for *acetone*.

Keywords need to be as specific as possible while allowing for variations in terminology that are likely to arise. Using keywords such as *safety* or *health* are likely to produce too many sites for most purposes. Most Web site indexes allow you to use logical connectors such as *and*, *or* and *not* when conducting your search. This can help you refine your keyword search until you have about 20-40 hits. Lists of hits longer than that are probably too long to effectively search and an indication that your keyword strategy should be refined.

Selecting a Web Site

Once you have a good idea of what kind of information you need to answer your question and what keywords are likely to be associated with that information, you are ready to start searching the file libraries on the Net. Start with a hotlist of Web sites that you are familiar with. If you are still building your hotlist and haven't visited a site that would have the type of information that you're looking for, consult the Web site listings in Chapter 7.

Asking a Discussion Group

If your search of the file libraries fails to produce the information you're after, or you're looking for more informal information than is available at Web sites, it is time to post a request for information to an appropriate email list or a newsgroup. To increase your chances of success when you ask a question of a list, be sure to follow netiquette (Internet etiquette) guidelines appropriate to that group.

First, check the archives of the group's discussions to see if it is the right group of which to ask the question and to be sure that it isn't a question that has been asked and answered repeatedly. When framing the question, be as specific as possible in asking the question, so those who read it can determine what type of answers are appropriate (*i.e.*, general pointers to the professional literature versus specific interpretations of your information).

Chapter 8 has a list of safety-related email lists to provide places to start your inquiry. If possible, monitor the traffic on the list for about a week before asking your question so that you can see what sorts of questions are appropriate for the list.

Checking Your Information

Always be sure to confirm information you've gotten from the Net before you act on it. Remember that the information available on the Internet was written based on someone else's assumptions, in ignorance of the details of your particular situation. There may be specific, critical differences between the situation that you face and that of the person writing the information. The effort involved in confirming net information may range from asking yourself *Does this make sense?,* to checking a paper reference source, to consulting with a professional with more expertise than yours.

Using Web Search Engines and Directories

General purpose search engines are often the easiest way to look for information on the World Wide Web. There are a number of them—AltaVista, Excite, HotBot, InfoSeek, Northern Light and WebCrawler (*see* Chapter 7 for the URLs)—to name just a few. However, their best use is not always obvious. Some of the considerations involved in using them are listed here.

Where Does the Data Come From?

In most cases, search engines use some form of "Web crawler" software to gather their information. Generically, Web crawlers are programs that travel the Internet, accumulating the content of World Wide Web sites in a central database. When a site is visited, a Web crawler will try to access and retrieve the text content from all of its pages. The content is stored in an indexed database which you can search through a fill-in form on the Web.

Using the Search Engine

You type in a word or a series of words. When you submit the form, the database search program accesses its index containing words from thousands or millions of Web pages. If there are some Web pages containing the word or words you are looking for the software will present the search results to you. You will see the title of the page, a few lines of text extracted from the page itself, and a link to the page on the remote Web server. This "hit list" allows you to browse through a list of the pages that have been retrieved and decide whether you want to visit any of the pages.

Interpreting the Results

Search engines use **relevance ranking** to present the results of their searches in what is hoped to be the most usable order. This means that the hits that appear to the search engine to be the most relevant search results (based on the ratio of the search words to the rest of the document) are presented first. As an example, if you were searching for confined spaces, summaries of pages containing both words in the title would usually appear first, followed by those where the words appeared frequently, and/or near the beginning of the text of the page. Sometimes this will give you the information you need. Sometimes it will not be specific enough.

If you are searching for information where the concept can be described in a phrase, most search engines will allow you to search for the phrase by enclosing it in quotes. If you perform a search on *confined spaces*, the search engine will only return links to pages where the two words appear together.

Refining the Search

Sometimes, using the basic strategy described above is not specific enough for you to find the information that you are looking for. Some search engines let you narrow your search using other criteria. For example, AltaVista allows you to narrow your search results using a variety of other criteria, such as title, host/domain name, dates, and use of boolean operators.

Title - Search for Web pages where your word, words, or phrase appear in the title of the page itself, rather than simply the body of the text.

Host/domain name - Specify that the information must come from a specific organization or type of organization.

Date range - Specify that the information must have been added to the index or updated between certain dates.

Boolean operators - Functions such as *and*, *or*, *not*, and *near,* which indicate that certain words must or must not appear in the retrieved pages in particular combinations.

Unfortunately, the syntax associated with these options varies from engine to engine. Fortunately, the Internet Scout Web page at **http://wwwscout.cs.wisc.edu** provides one-page summaries (called SideKicks) of the syntax of the most popular search engines.

What Are Meta Search Engines?

Meta search engines are services that access a number of Web search engines for you with a single inquiry, remove the duplicate results, and provide you with a list of sites that meet your search criteria. Among the meta search engines available are Inference, MetaCrawler, ProFusion and Savvy Search. They all work differently, and use various combinations of search engines. As with the search engines themselves, it is a good idea to work with them yourself in order to determine which ones suit your needs best.

Why Use General Search Engines If Meta Engines Are Available?

A meta search engine will give you a more all-inclusive search result than a single search engine. However, there is so much information available from regular search engines that the problem most people face is making their searches specific enough to retrieve usable results.

Meta search engines do not make this any easier, for two main reasons. First, depending on the topic, they may retrieve references to more pages than individual search engines. This is not necessarily a good thing! Second, since they need to work on a number of search engines, they aim at the "lowest common denominator" in terms of search features. They do not provide any of the additional capabilities found in individual search engines.

The authors' opinion is that, in the majority of cases, you may be better off starting with an individual search engine. If, however, your topic is relatively obscure, or if you want to be sure that you are retrieving all available information on a topic, a meta search engine can be useful to follow up a general search effort.

Top 50 Safety Web Sites

One of the most important steps in searching for information on the Internet is to start with a "hot list" of sites likely to have your information. Developing such a hot list is an individual exercise, depending on your particular interests. However, to help you get a sense of the types of information available on the Web and a place to start in exploring it, this section describes some of the more important health and safety sites on the Internet in some detail.

The purpose of this list of health and safety Web sites is to give you a good idea of the types of information available on the Internet. The sites chosen are the most interesting from the professional point of view. We made an effort to include a variety of styles of Web pages; from highly formal, graphical efforts to sites which focus on databases and other text based information. These pages also represent some of the more subtle points of using Web sites, such as running keyword searches or downloading documents.

A variety of pages are included, from government agencies to professional organizations to experimental sites which are being developed by individuals on an ad hoc basis. The information available on these sites ranges from raw data such as chemical properties to interpreted information such as policies and handbooks to local information such as meeting dates and telephone numbers for contacting individuals. In addition, a number of international sites are included to provide a flavor of the multi-national nature of the web.

The page images included do not attempt to show the entire home page, but rather provide some of the more interesting elements at the site. Note that these images were current in March of 1999; changes can be expected sooner than later.

Chapter 7 of this book provides a more extensive list from which to start your exploration. Remember that these pages often include pointers to related information.

AgSafe: Agricultural Safety Training Materials

http://www.cdc.gov/niosh/nasd/docs/ashome.html

AgSafe is a nonprofit coalition of groups and organizations dedicated to the reduction of injuries, illness, and fatalities in California agriculture. The Web site includes safety training materials that were developed in cooperation with the Safety Center, Inc. and the Farm Employers Labor Service. The materials are public domain for the use of and distribution by any group or organization involved in agricultural work. There are a total of twelve safety modules consisting of an English script and a Spanish script.

Additional information about the AgSafe project is available, as well as information about the preparation and distribution of the safety modules and suggestions for their use.

Safely Working around Agricultural Machinery
Protective Clothing   (Ropa Protectora)
Safely Handling Pesticides   (Manejando Pesticidas con Seguridad)
Basic First Aid   (Primeros Auxilios Basicos)
Working Safely With and Around Electricity
Working Safely in the Farm/Ranch Shop
Back Injury Prevention   (Prevención de Lesion de Espalda)
Defensive Driving   (Conduciendo a la Defensiva)
Preventing Slips and Trips   (Evitando Resbalones y Tropezones)
Proper Use of a Respirator   (Uso Apropiado de un Respirador)
Safely Working With and around
    Farm/Ranch Animals
Safely Working around Tractors

Agency for Toxic Substances and Disease Registry

http://www.atsdr.cdc.gov/atsdrhome.html

The Agency for Toxic Substances and Disease Registry prevents exposure to and adverse human health effects from hazardous substances from waste sites, unplanned releases, and other sources of pollution present in the environment. Their Web site has a variety of information about hazardous chemicals in the environment. This information includes:

- **ToxFAQ**s, fact sheets on the hazards associated with a variety of chemicals;

- A list of the Top 20 **Hazardous Substances**, based on Superfund experience; and

- A Science Corner, which includes ATSDR Special Report(s), Health and Environment Resources, Health and Environment Resources, Science Corner History and other information.

Home Page Topics

National Alerts

Announcements

- About ATSDR
- ATSDR Newsletter
- ToxFAQs
- Science Corner
- Public Health Assessments
- ATSDR Glossary

Health Advisories

Jobs

- Kids, Parents, Teachers
- ATSDR Ombudsman
- HazDat Database
- Top 20 Hazardous Substances
- Minimal Risk Levels (MRLs)

American Board of Industrial Hygiene

http://www.abih.org

The American Board of Industrial Hygiene is the body which manages the process of certifying individuals in the practice of industrial hygiene. This Web page provides a variety of information about the testing function, including qualifications for taking the test, a list of test dates and the list of formulas provided for people taking the test. In addition, there is the ability to search the roster of CIHs and a list of contacts at the office.

Specific information available includes:

- Application materials
- Bylaws
- ABIH Bulletin
- Exam Info
- CM Info
- Code of Ethics
- CIH Roster Search
- OHST/CHST Page
- UTF Info
- Other links
- Staff
- Directors

American College of Occupational and Environmental Medicine

http://www.acoem.org

The world's largest organization of occupational and environmental physicians, ACOEM is dedicated to promoting and protecting the health of workers through preventive services, clinical care, research, and educational programs. A number of categories of information are available on their Web site, including:

General Information - This section describes ACOEM itself: background, divisions, bylaws, officers and a staff directory.

Membership Information - Application forms and general membership information are provided.

Education/Conferences - Read about the State of the Art Conference, the American Occupational Health Conference and various ACOEM meetings and seminars. Detailed descriptions of their occupational medicine course and self-assessment examination are provided.

Position Statements and Guidelines - These are ACOEM documents, written for its member physicians, outlining its position on subjects such as alternative medicine and hazards of VDT exposure.

Journal of Occupational and Environmental Medicine - Abstracts of recent articles and subscription information are provided.

Other Publications - Various ACOEM and ACOEM-approved publications are described. In some cases, the publications are presented in their entirety.

Employment Referral Service - This is a bulletin listing current positions available for physicians in occupational and environmental medicine, and situations wanted by other physicians.

Association of Societies for Occupational Safety and Health (ASOSH)

http://www.asosh.org

Headquartered in South Africa, ASOSH is a group of professional societies with the mission of providing leadership in OHS for Southern Africa. It does this through facilitating the development of a holistic approach to the creation of a healthy and safe workplace. Its projects include:

- Acting as coordinator and spokesbody for OHS Involved Societies

- Influencing national policy and legislation on matters concerning OHS

- Promoting the exchange of technical information amongst OHS shareholders

- Fostering interdisciplinary understanding and promotion of Member Societies

The purpose of the Web site is to promote ASOSH, its members and OHS, to provide a comprehensive OHS Web site for Southern Africa, to provide easy and rapid access to the huge wealth of OHS&E information around the world on the Internet, and to act as a resource to Internet information for the NCOH, which houses the National OHS Information Centre (ILO-CIS Centre) for South Africa.

At the time of writing, the site had been publicly available for less than a month, but already showed promise as a unique resource—unique in the sense that, up until now, there has been very little Southern African OSH information available on the web.

The site includes details of OSH educational opportunities in Southern Africa; information about professional societies; requirements for professional status; links to industry associations, government agencies, etc.; topic-specific documents; and legislation and standards.

American Conference of Government Industrial Hygienists

http://www.acgih.org

The ACGIH promotes excellence in environmental and occupational health and provides access to high quality technical information through this Web site. Among the information here is organizational contacts, the publications catalog, and membership information.

Also available at the site are articles from the latest issue of the *ACGIH Today!* newsletter. For example, in January 1999 the issue included these articles:

1. FOHS Update: Foundation Laid, Plans Made for Lee Lecture
2. 1999 ACGIH Award Winners Chosen
3. Pierce Named Editor in Chief of ACGIH's Applied Occupational and Environmental Hygiene Journal
4. Applied Available on Web
5. ACGIH Technical Committees Plan 1999 Activities
6. Members Pass ABIH Exam
7. Honorary Member, Board Member Dies at 93
8. Q&A Today!
9. Watch Your Mailbox. . .
10. Zoom In
11. Beryllium Subject of Proposed DOE Rule
12. NEW on the Bookshelf: New Publication Offers Tips for Job Stress Prevention
13. NIH Announces New Streamlined Grant Procedures
14. Reproductive Health Center Established
15. BCPE to Hold Exams at AIHCE
16. Final Ruling on Confined Space

Welcome!	Applied	Members	Leadership
Publications	Free	Today!	Events
Home	Links	OH Talk	Classifieds

Biosafety Resource Page

http://www.absa.org/resources/resource.htm

One of the features of many Web pages is a list of links to other Web sites with related information. These can often be duplicative of one another, as one site leads to another only to lead back to the first. It is also a challenge to keep these lists up to date, as the sites often change or new sites are added as new organizations come on-line.

A good example of a well maintained links list is found at the American Biological Safety Association Web page, shown below. The links are organized into logical categories and it is easy to browse the list to determine what is available. The list is also frequently updated.

American Biological Safety Association

Home

Biosafety Resources on the Internet

- **Risk Group Classifications** for Infectious Agents (**ABSA**)
- **Material Safety Data Sheets** for Infectious Agents (**Office of Biosafety, LCDC**)
- **42 CFR 72.6**: Additional Requirements for Facilities Transferring or Receiving Select Agents (**CDC, Office of Health and Safety**)
- **Biosafety Guidelines** (BMBL, NIH Guidelines, BSC's)

Revised: 11/03/98
© 1998 American Biological Safety Association

Boston University School of Public Health

http://www-busph.bu.edu/Gallery/Gallery-Lobby.nclk

The Boston University School of Public Health has established a Web page for the exhibition of photographs related to environmental health and safety issues. The photographs are of workers in hazardous workplaces, of people living in areas affected by hazardous chemicals, and other public health issues.

In March 1999, the exhibits available at this site included the following:

GALLERY I BREATH TAKEN: THE LANDSCAPE OF ASBESTOS

GALLERY II STOLEN DREAMS: PORTRAITS OF WORKING CHILDREN

GALLERY III TOXIC TOUR OF TEXAS

GALLERY IV PESTICIDE POISONING OF AMERICA

GALLERY V COAL MINING SERIES

GALLERY VI MEMORIES COME TO U.S. IN THE RAIN AND THE WIND: ORAL HISTORIES AND PHOTOGRAPHS OF NAVAJO URANIUM MINERS AND THEIR FAMILIES

ENVIRONMENTAL HEALTH DEPARTMENT

The mission of the Environmental Health Department Gallery is to provide a human dimension to the various problems we face with respect to both environmental and occupational health hazards. Too many people today are confronting or suffering the demands and impositions of modern industrial society without sufficient notice. The victims of both occupational and environmental diseases are still strangely missing from our sight, too often, lost amid the legal and social controversies concerning liability for past exposures and future perils. To the degree we see them at all, it is usually as objects rather than as subjects, statistics to be recorded, cases to be diagnosed, or plaintiffs to be deposed. The photographs viewed in this Gallery will bring individuals and their struggles to center stage.

Bill Ravanesi MA MPH

Canadian Centre for Occupational Health and Safety

http://www.ccohs.ca

CCOHS' site provides access to a variety of information. Some information is available free and some is accessible on a subscription basis.

Free health and safety information includes OSH Answers, a collection of hundreds of OSH-related questions and answers compiled by CCOHS's Inquiries Service, information and hazard alerts from various government agencies, a collection of documents from Human Resources Development Canada, etc. A list of upcoming conferences and seminars, plus a popular directory of health and safety Internet resources are available.

CCINFOWeb is CCOHS's Web database service. Databases accessible on a subscription basis include all Canadian OSH and environmental legislation, a large MSDS database, bibliographic databases that, together, provide access to summaries of about half a million publications, etc. Searching all of these databases is free. Viewing the database records themselves requires a subscription.

Français

CCOHS

The Canadian Centre for Occupational Health and Safety (CCOHS) promotes a safe and healthy working environment by providing information and advice about occupational health and safety.

Home
Search
What's on this site?
Products & services
CCINFOweb
OSH Answers
Education & training
About CCOHS
Internet directory
Other H&S info.

Français

Today's Features

- **North American OSH Week**
  It's Everybody's Business! Why not participate in this continent-wide event?

- **NIOSH Pocket Guide**
  to Chemical Hazards - A concise, highly organized source of industrial hygiene information on hundreds of hazardous chemicals

- **In company of rodents**
  What you should know about Hantavirus, a rare but serious health hazard

- **Work in a hotel laundry?**
  Protect yourself from repetitive motion injury

- **CanOSH**
  Collection of OSH info from Canadian federal, provincial & territorial governments

- **Warehouse hazards**
  From forklifts to dangerous chemicals, CCOHS' Warehouse Workers Safety Guide helps you keep on top of the hazards in your workplace

Center for Safety in the Arts

http://artswire.org:70/1/csa

The CSA provides information on hazards in the visual arts, performing arts, children and school arts programs, museums, and general health and safety information and laws relevant to the arts. It is primarily intended for artists, performers and others working in the arts, and most of the files are written for people without a health and safety background. It is also a resource for health and safety professional who want information on art hazards. This is an electronic information center which used to have a physical counterpart in New York before being closed due to funding cuts.

Available information includes a catalog of books and videos available for sale and over 100 online documents describing various arts-related and general safety hazards. Most of these documents are written by Michael McCann and other former CSA staff members. Other categories of information include general health and safety precautions, and laws and regulations. The full text of back issues of CSA's newsletter, *Art Hazards News,* is also available.

Center for Safety in the Arts

Welcome to the **Center for Safety in the Arts gopher/web site!**

This site provides information on hazards and precautions in the arts. More information about CSA's gopher/web site can be found here.

For information on the **Center for Safety in the Arts**, click here

---

- Center for Safety in the Arts Books, Videos, and Other Publications
- Art Hazards Menu
- General Hazards Menu
- Precautions Menu
- Laws and Regulations Menu
- Resources Menu
- Other Health and Safety Gophers
- Other Health and Safety WWW Sites
- What's New on the Center for Safety in the Arts Gopher/WWW Site
- Search All Center for Safety in the Arts Menus

Last updated October 14, 1998.

We welcome comments on our gopher/web site. Please leave your comment here

---

This gopher/web site is brought to you by the Center for Safety in the Arts. It is designed and maintained by: Michael McCann.

Centers for Disease Control and Prevention (CDC)

http://www.cdc.gov

This U.S. government site contains a wealth of health and disease information. There are numerous documents and data sets, most of which are related to public health, biosafety, and infectious diseases. The major topic areas are travellers' health; health information; publications, software, and products; data and statistics; training and employment; and funding.

Some of the available information includes the following:

Travelers' Health provides specific health advisories for international travelers.

Health Information - This provides documents on a number of different diseases, including AIDS, tuberculosis, hepatitis, histoplasmosis, hantavirus, legionella, lyme disease and others.

Publications, Software, and Products - A number of full text publications and slide sets are presented. Highlights include slide sets for Epidemiology and Prevention of Viral Hepatitis A-E and HIV/AIDS, and Guidelines for Isolation Procedures in Hospitals. Downloadable software for public health applications is also available.

Current and back issues of two of CDC's serial publications, *Morbidity and Mortality Weekly Report* (MMWR) and *Emerging Infectious Diseases* are presented in their entirety. These are also available by email. Check Chapter 7 for details.

ChemFinder

http://chemfinder.com/

ChemFinder, provided by CambridgeSoft, is a chemical meta search engine. This means that it searches a number of collections of chemical information and gives you a combined result. Since it is specifically designed to search for chemical information, it allows you to search in ways that are not common to most search engines, and shows you chemical-specific information. As an example, you can search by any combination of chemical name of synonym, formula, molecular weight, melting point, boiling point, CAS registry number and structure. The search result provides you with the physical properties of your chosen chemical, then presents links to numerous other chemical information collections that have information about it.

The list of chemicals at this index is the largest single list of chemical information collections on the Web. The following types of chemical information are accessible: biochemistry, health, medications, miscellaneous, MSDS, pesticides/herbicides, physical properties, regulations, structures and usage.

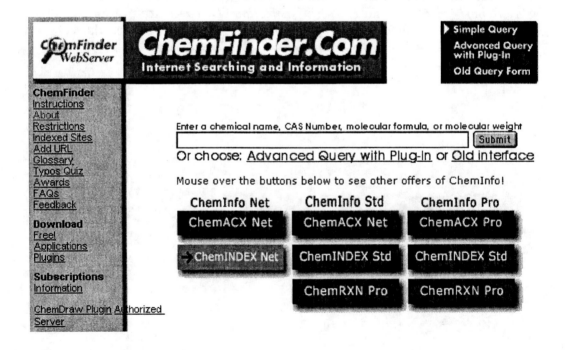

Consumer Product Safety Commission (CPSC)

http://www.cpsc.gov

CPSC is an independent U.S. federal regulatory agency. It has jurisdiction over some 15,000 types of consumer products. CPSC works in a number of different ways to reduce the risk of injuries and deaths by consumer products: develops voluntary standards with industry; issues and enforces mandatory standards or banning of consumer products; obtains the recall of products; conducts research on potential product hazards; and informs and educates consumers.

The agency's Web site contains a great deal of useful information, under some of the following categories.

Consumer - This is general information about product safety and recalls.

Library - The full text of CPSC publications is available, as well as data relevant to consumer product safety, CPSC success stories, etc.

Public Calendar - This is the agenda for the agency's meetings and activities for the next few weeks.

Recalls/News - All of the agency's recall notices dating back to 1990 are available from here.

Talk to Us - This is a fill-in form that allows you to report unsafe products to the agency.

What's Happening - Read the latest press releases and documents from CPSC.

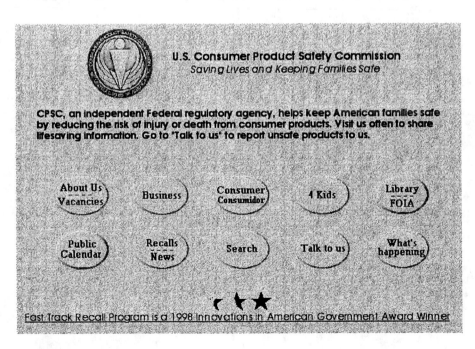

Cornell Ergonomics Web

http://ergo.human.cornell.edu/

This Web page is provided by Cornell University's Department of Design and Environmental Analysis, which offers a Human Factors and Ergonomics Program. The focus of the program is on ways to improve comfort, performance, and health in the workplace through the ergonomic design of products and environments. Issues covered by information on this Web site include the following:

- Keyboard research
- Back injuries
- Indoor air quality
- Ergonomic Software

- Cumulative trauma disorders
- Office lighting
- Carpet emissions
- Ergonomics Survey and Assessment Tools

In addition, syllabi and class work from appropriate courses in the academic program are available to be reviewed. The screen image below shows a list of some of the slide shows that are available at the site.

CUErgo Cornell University Ergonomics Web

Cornell University's <u>Department of Design & Environmental Analysis</u> offers a Human Factors and Ergonomics Program that focuses on ways to improve comfort, performance and health through the ergonomic design of products and environments. Inclusion at this site does not constitute endorsement of a product by Professor Alan Hedge or Cornell University. All materials created for this site are the copyright of Professor Alan Hedge at Cornell University.

Help with our 1999 research
Take a few minutes to complete one of our online surveys on software, tasks and products. Thanks.

What's available at this site? (content last updated 3/7/99)

- DEA 350 Homeworks
- Spring 1999 course information
- Information about all Ergonomics Courses at Cornell
- Ergonomics Guidelines - 10 tips for arranging your computer workstation
- Ergonomics information: slideshows, tutorials and a list of publications

- Information about Ergonomics Research at Cornell
- Cornell's Ergonomics Research Laboratory
- Cornell videos, reports and books
- Let me consult the CUErgo Statistics Helper to help me choose a statistical test

Department of Energy Technical Information Service

http://www.tis.eh.doe.gov/

TIS is a comprehensive collection of information services designed to empower the ES&H professional. TIS is sponsored by the Office of Environment, Safety and Health (EH).

Among the documents available are the following:

- Handbook for Occupational Health and Safety During Hazardous Waste Activities
- Perspectives on Worker Protection During DOE Hazardous Waste Activities
- Working Safely during DOE Hazardous Waste Activities
- OSH Technical Reference Manual
- Accident Analysis Documents (MORT)
- Technical Research and Applications Center - TRAC Documents
- Occupational Injury and Property Damage Summary Reports
- Implementation Guidance - Occupational Radiation Protection
- Radiological Control Technical Position Documents
- Health and Safety Plan Guidelines

TECHNICAL INFORMATION SERVICES

Connecting the World of
Environment, Safety and Health

The Occupational Disease Benefits for Energy Workers Initiative Website is available.

Duke University Occupational and Environmental Medicine

http://occ-env-med.mc.duke.edu/oem

This is the preeminent site for occupational and environmental medicine information. It is managed by Dr. Gary Greenberg of Duke's OEM program. It includes:

Information about the Occ-Env-Med-L mailing list - Read about the mailing list, find out how to subscribe and link to various archives of previous messages.

What's New on the NIOSH, EPA, OSHA, NCEH, IOM, FDA, WHO, CDC Web pages - This section "brings together links to individual pages at each of the mentioned agencies, with a description of what has been posted in recent weeks, focussing only on what would be applicable to Occupational & Environmental Medicine." It is an excellent current awareness service for physicians and others working in the field.

Occupational and Environmental Health Links - A list of links to primarily North American OEM-related Web sites is available here. They also "mirror" (maintain a current copy) of Dr. Raymond Agius' directory of mainly European sites. Various "hot topics" are described as well.

MMWR - Dr. Greenberg sorts and saves applicable articles from the Centers for Disease Control and Prevention's publication, *Morbidity and Mortality Weekly Report*. The text is available here.

This site changes on an almost daily basis. If you are interested in occupational and environmental medicine, you should access this site frequently.

Environmental Chemicals Data and Information Network

http://ecdin.etomep.net/

ECDIN - Environmental Chemicals Data and Information Network

Phatox - Pharmacological and Toxicological Data and Information Network

THE ECDIN DATABANK: Introduction

The Environmental Chemicals Data and Information Network (ECDIN) is a factual databank created under the Environmental Research Programme of the European Commission, Joint Research Centre at the Ispra (I) site, in 1974. The significant milestones of the ECDIN could be summarized as follow:

• Creation of a list of chemicals to be considered on a priority basis (criteria based on the amount of chemicals production)

• Selection of the families of data to be collected for each chemical. Definition of the design and structure of the ECDIN database.

• Hardware and software definition.

• Selection and often evaluation of the stored data performed by highly qualified experts in every data field.

• Extension of data collection to all the substances listed by the EINECS* inventory.

• INTERNET and CD-ROM dissemination.

In order to support the European Commission Policy not only in the chemicals but also in the pharmaceuticals regulatory sector, ECDIN has recently been updated with a large part of information on medicinal products, this project has been named: PHArmacological and TOXicological (PHATOX)

Environmental Protection Agency

http://www.epa.gov

The U.S. EPA's Web site contains a wide variety of information related to the environment and public health. Some of the many categories of available information include technical documents, research funding, and more; assistance for small businesses and entire industries; projects and programs; news and events; laws and regulations; databases and software; and publications. Examples of specific information available include these:

Laws and regulations - This page provides access to the full text of EPA Federal Register documents, Federal Register documents from other agencies, the Unified Agenda, Code of Federal Regulations (CFR), CFR Title 40, the United States Code, and the National Environmental Policy Act of 1969 (EPA).

Databases and software - Access a variety of data sets and download software packages. Some of the databases include the Emergency Response Notification System (ERNS), Toxic Release Inventory System (TRIS), Hazardous Waste Data, and the Pesticide Product Information System (PPIS). Available software includes ECOTOX Threshold Software and Software for Environmental Awareness.

Other information includes a comprehensive series of documents on Indoor Air Quality and access to the IRIS database of human health effects from environmental chemicals.

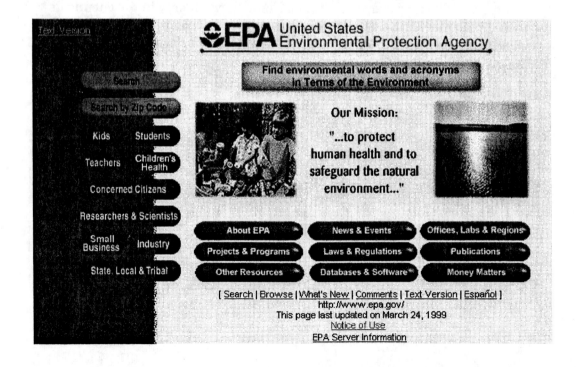

ErgoWeb

http://www.ergoweb.com

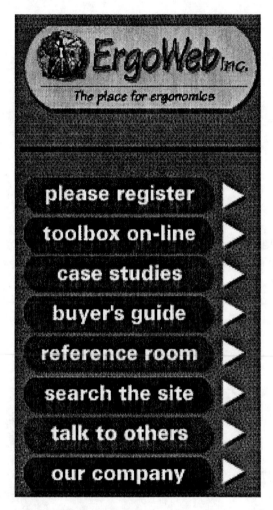

The home page says:

"ErgoWeb is the place for ergonomics on the World Wide Web. We continue to offer volumes of useful ergonomics information for free to a large world wide audience. We also offer subscription access to a sophisticated set of ergonomic job evaluation, analysis, design and redesign software through our Web site. If you're searching for a one stop spot for ergonomics information, products, case studies, instructional materials, standards and guidelines, communication opportunities, ergonomics related news, and more—all presented in an ergonomic format—you've found it!" The information here includes the following:

ErgoWeb's Job Evaluator Toolbox Software - This is an integrated suite of ergonomic evaluation, training, and design software.

ErgoWeb Buyer's Guide - This is a database of ergonomics products and services.

Reference Room - This section contains current ergonomics reference documents involving standards and U.S. regulatory compliance. Also included are a training section (Ergonomic Concepts), book/journal listings with some abstracts, over 3000 bibliographic references, and links to professional organizations in ergonomics.

Case Studies - These are real examples of cases in which ergonomics hazards have been addressed. Each case study consists of a task description, risk factors, administrative or design solutions, comments, and vendor information. Before and after illustrations accompany each study where available.

European Agency for Safety and Health at Work

http://www.eu-osha.es/

This site is intended to be a "one-stop shop" for European occupational health and safety information. The site contains European information in the following areas: legislation and standards; research; practice; strategies and programs; statistics; information; and news. At the time of publication, many of these categories were under development.

Legislation - The European Commission has prepared a number of Directives to protect the health and safety of workers at work. This page provides information on and links to Web sites covering these Directives.

Statistics - This page provides access to European Union statistics related to health and safety at work.

Information - Sources include databases, library catalogs, and publications.

Other areas under development include research, practice, strategies and programmes, and news and events. There are also links to specific workplace health and safety agencies in the EU member countries.

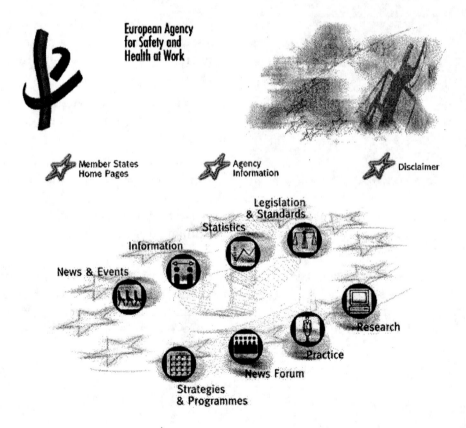

FedWorld

http://www.fedworld.gov

This site is a gateway to a wide variety of information from the U.S. federal government. The modes of access, the variety of documents available, and the technological expertise at FedWorld are expanding with changing technologies. FedWorld offers a comprehensive central access point for searching, locating, ordering, and acquiring government and business information.

In 1992, FedWorld was established by The National Technical Information Service (NTIS), to serve as the online locator service for a comprehensive inventory of information disseminated by the federal government. This new service assisted agencies and the public in electronically locating federal government information, both information housed within the NTIS repository and information made accessible through an electronic gateway of more than 100 government bulletin boards. What started out as a small dial-up access system has grown tremendously over the last four years in size, technology, and content.

Today, the FedWorld Information Network consists of an integrated network offering the public broad access to information including job announcements, databases, and Web sites.

A program of the United States Department of Commerce

About   Search   Contact   Databases

Web Site Revised 1999 March 21
File, Jobs and Web Databases Updated Daily

Browse the FedWorld Information Network

Pick From List:

| List FedWorld Databases | Go! |

Search Web Pages on the FedWorld Information Network

Enter some keywords:

[   ] Search

Search for U.S. Government Reports

Enter some keywords:

[   ] Search

Explore U.S. Government Web Sites

| About FedWorld | About NTIS | FedWorld Services For U.S. Government Agencies |
| Points of Contact at FedWorld | Privacy | Security |

Health Canada, Laboratory Centre for Disease Control

http://www.hc-sc.gc.ca/hpb/lcdc/

The Laboratory Centre for Disease Control (LCDC) is Canada's national centre for the identification, investigation, prevention, and control of human disease. LCDC is one of five directorates within the Health Protection Branch (HPB) of Health Canada, and a key component of the federal government's mandate for public health protection.

The Centre's core activities are national health surveillance, disease prevention and control. These involve the monitoring and investigation of infectious and non-infectious diseases and injuries, the study of their associated risk factors, and the evaluation of related prevention and control programs.

Some of the key documents include the following:

Laboratory safety; Biosafety guidelines; Material Safety Data Sheets for Infectious Substances; various documents related to health care workers' exposure to bloodborne pathogens; documents related HIV/AIDS and Hepatitis C

Health Canada   Santé Canada   Health Protection Branch - Laboratory Centre for Disease Control

[Search] [Bureaux] [New] [Events] [Guidelines] [Links] [Programs] [Public Health] [Publications] [Subject]

Français

Disease Prevention and Control Guidelines

- **BIOSAFETY**

- **HIV/AIDS**

- **INFECTION CONTROL**

- **LABORATORY**

- **PERINATAL HEALTH**

- **STDs**

- **TRAVEL HEALTH/QUARANTINE**

- **VACCINES AND VACCINE-PREVENTABLE DISEASES**

Howard Hughes Medical Institute Laboratory Safety Page

http://www.hhmi.org/science/labsafe

The Howard Hughes Medical Institute's site has several important resources for people working in laboratories. They host an on-line collection of Laboratory Chemical Safety Summaries written by the National Academy of Sciences Committee on Prudent Practices for Handling, Storage, and Disposal of Chemicals in Laboratories. There is also a catalog of their free laboratory safety videos, which can be ordered online. These videos are focused on biomedical research issues, but provide useful information for all laboratory people.

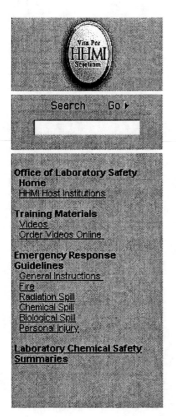

HOWARD HUGHES MEDICAL INSTITUTE

HHMI Laboratory Safety Program

OFFICE OF LABORATORY SAFETY

About the HHMI Office of Laboratory Safety

Welcome to the Office of Laboratory Safety of the Howard Hughes Medical Institute. The HHMI laboratory safety program is an integral part of the Institute's scientific research mission. The Institute strives to set an exemplary standard for safe and sensible laboratory practice.

The Office of Laboratory Safety works in partnership with the environmental health and safety programs at each HHMI host institution. Its goals are to:

- Create and maintain a safe and healthful research environment.
- Teach safe work practices.
- Promote safety awareness.
- Ensure a laboratory safety program that complements science.

The HHMI laboratory safety program provides a range of instructional and resource materials to help researchers know how to incorporate the fundamental good rules of safety into their daily work routine.

International Labor Organization

http://www.ilo.org/

Available documents include the following:

- Constitution of the International Labour Organization
- International Labour Conventions
- International Labour Recommendations
- Reports of the Committee of Experts on Application of Recommendations
- Reports of the Conference Committee on the Application of Standards
- Reports of the Committee on Freedom of Association
- Reports of Committees set up to handle the ILO's procedures
- Ratifications of International Labour Conventions (by country)
- Standing Orders of the International Labour Conference
- Interpretations of Conventions and Recommendations
- International Labour Standards, Handbook of Procedures
- Digest of Decisions (Freedom of Association)
- Multinational Enterprises, Tripartite Declaration of Principles

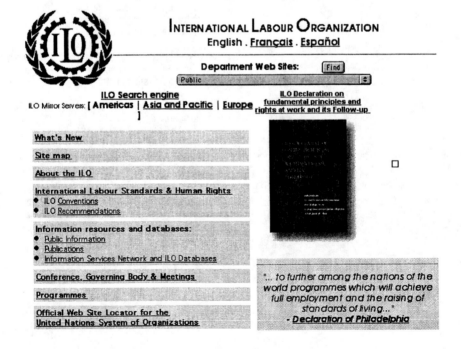

Industrial Hygiene on the WWW

http://www.cs.wku.edu/~russellg/ih/ih.html

This site offers a variety of Javascripts to implement many numeric formulas of value to industrial hygienists. Examples of formulas developed include the NIOSH lifting equation, conversion of gas concentrations between ppm and mg/m^3, diffusion and dispersion equations, and a safety management ranking system.

Java is a programming language designed to be used by Web browsers (primarily Netscape or Microsoft Explorer) to allow the programs to be used on a wide variety of computer platforms. It can be used both for straight forward numerical calculations, as it is here, or to control the presentation of information on the browser's screen. Many of the animated images seen on the Web are implemented by Java programs.

The Only Place to Be for Industrial Hygiene...
www.industrialhygiene.com

The Revised NIOSH Lifting Equation Online
By Grady Russell

International Agency for Research on Cancer (IARC)

http://www.iarc.fr

IARC's mission is to coordinate and conduct research on the causes of human cancer, and to develop scientific strategies for cancer control. The Agency is involved in both epidemiological and laboratory research, and disseminates scientific information through meetings, publications, courses and fellowships.

The site provides summaries of, and ordering information for its publications, including IARC Monographs on the Evaluation of Carcinogenic Risks to Humans, plus information related to IARC itself. It also provides access to two databases: the p53 mutation database and the "Cancer Mondial" epidemiological database.

About IARC

Staff Directory
Employment Opportunities
Directions to IARC

Upcoming meetings
Fellowship Program
Training Courses

| Home | About IARC | Press Releases | Meetings | Publications | Databases |
| Search | Contact Us | Research Units | Fellowships | Training Courses | Bookstore |

International Centre for Genetic Engineering and Biotechnology, Biosafety Web Pages

http://www.icgeb.trieste.it/biosafety/

"There is a need for further development of internationally agreed principles on risk assessment and management of all aspects of biotechnology, which should build upon those developed at the national level. Only when adequate and transparent safety and border-control procedures are in place will the community at large be able to derive maximum benefit from, and be in a much better position to accept the potential benefits and risks of, biotechnology." (Agenda 21, 16.29)

This site, based in Trieste, Italy, provides access to a variety of biosafety-related resources, including the ICGEB bibliographic database on biosafety studies, an index of selected scientific articles published on biosafety and risk assessment since 1990. Official documents on biosafety produced by international agencies. Scientific findings, articles, proceedings and workshops International biosafety regulations from Europe, the U.S.A., and other countries.

Internet Grateful Med

http://igm.nlm.nih.gov/

Internet Grateful Med (IGM), maintained by the U.S. National Library of Medicine, provides access to a series of health-related bibliographic databases. Databases include MEDLINE (including PREMEDLINE), AIDSLINE, AIDSDRUGS, AIDSTRIALS, DIRLINE, HealthSTAR, HSRPROJ, HISTLINE, OLDMEDLINE, SDILINE, SPACELINE, BIOETHICSLINE, POPLINE, TOXLINE and ChemID.

MEDLINE® contains over 9.2 million citations to the biomedical literature, referenced in more than 3800 journals. It is updated weekly. About 31,000 new citations are added each month.

TOXLINE® contains citations to journal articles, monographs, technical reports, theses, letters, and meeting abstracts, papers and reports. It is updated monthly, with approximately 9300 new citations per month. There are over 2.4 million citations in the database. Subjects include toxicology, pharmacology, and biochemical and physiological effects of drugs and other chemicals.

ChemID® (Chemical Identification) is a chemical dictionary file for over 339,000 compounds of biomedical and regulatory interest. Records include CAS Registry Numbers and other identifying numbers, molecular formulae, generic names, trivial names, other synonyms, etc.

Select Database to Search

┇MEDLINE

┇AIDSLINE

┇AIDSDRUGS

┇AIDSTRIALS

┇BIOETHICSLINE

┇ChemID

┇DIRLINE

┇HealthSTAR

┇HISTLINE

Internet Grateful Med V2.6.2

Internet Grateful Med (IGM) searches MEDLINE® using the retrieval engine of NLM's PubMed system. The default MEDLINE search is 1966 to the present and includes PREMEDLINE. This version of IGM takes advantage of PubMed's ability to display Related Articles and link to the full text of participating online journals. Specific information on new features is available.

Select a database to search from the list on the left. A brief description of each database is available at its associated information icon ┇.

Internet Grateful Med User's Guide (last updated November 18, 1998)

Internet Grateful Med Frequently Asked Questions

Loansome Doc for hardcopy document delivery

NIH Clinical Alerts

Manitoba Workplace Safety and Health Division

http://www.gov.mb.ca/labour/safety/index.html

The site provides access to information about the various branches of this provincial government Division: Workplace Safety and Health; Occupational Health; Mines Inspection; and Mechanical and Engineering.

Other information includes the full text of the Manitoba Workplace Safety and Health Act and regulations, accident statistics, the full text of past and current issues of the Division's *WorkSafe!* newsletter, and numerous full text publications.

Various types of publications are presented: bulletins, work and safety guidelines, fire safety publications, agricultural programs, and the Division's research papers.

Pour obtenir la version francaise, clique ici.

Our Mission
Create an environment that will cause employers, workers and the public to integrate safety and heal into their work as a basic right and responsibility.

The mandate of the **Workplace Safety and Health Division** is to administer legislation within the **Department of Labour** related to workplace safety and health and public safety.

MedScape

MedScape is a multi-specialty commercial Web service for clinicians and consumers. It contains an incredible amount of medically related information, including:

- thousands of full text peer reviewed clinical medicine articles;

- free, unrestricted access to the MEDLINE, AIDSLINE and TOXLINE databases from the U.S. National Library of Medicine;

- online Continuing Medical Education programs, links and self-assessment tools.

The reference section includes links to tools and documents such as a medical dictionary, practice guidelines and recommendations from the U.S. Centers for Disease Control and Prevention, and a drug search database.

Registration is required for access to some of the information, but it is free.

Specialty Sites

- Cardiology
- Diabetes & Endocrinology
- Gastroenterology
- HIV/AIDS
- Infectious Diseases
- Internal Medicine
- Managed Care
- Medical Practice
- Molecular Medicine
- Oncology
- Orthopedics
- Pediatrics
- Pharmacotherapy
- Primary Care
- Psychiatry
- Respiratory Care
- Surgery
- Urology
- Women's Health
- Multispecialty

‖ Home ‖ Site Map ‖ Marketplace ‖ My Medscape ‖ CME Center ‖ Feedback ‖ Help Desk

Medscape® The Online Resource for Better Patient Care.®

National Institute of Environmental Health Sciences

http://www.niehs.nih.gov

The National Institute of Environmental Health Sciences conducts basic research on environment-related diseases. Their Web pages outline the Institute's history and research highlights, as well as providing complete contact and visiting information.

News and Publications available from this site include press releases about scientific discoveries such as the deadly effects of asbestos exposure, the developmental impairment of children exposed to lead, and the health effects of urban pollution. There is also access to the NIEHS Clearinghouse where you can obtain free information on EMF, and a Kids' Page with games, puzzles, brain-teasers, and easy-to-understand information and descriptions about environmental health science research.

Grants and Contracts are awarded from within NIEHS's extramural science program, which supports a network of university-based environmental health science centers and provides research and training grants, as well as contracts for research and development.

Scientific Programs at NIEHS utilize onsite resources to perform, for example, the key work that led to the identification of the breast cancer/ovarian cancer gene BRCA1 and the prostate cancer suppressor gene. NIEHS is the headquarters of the National Toxicology Program, the Superfund Basic Research Program, and a host of other programs focused on major areas of environmental health research.

Employment and Training Opportunities including postdoctoral training positions, Environmental Clinical Research Training Program for Nurses and Physicians, and regular vacancy announcements.

National Institute for Working Life (Sweden)

http://www.niwl.se

From the Web site:

The development of working life conditions is rapid, multifaceted and difficult to interpret. Accepted wisdom is no longer sufficient. New knowledge must be built up for us to successfully deal with the problems and to see and make the most of opportunities.

The National Institute for Working Life is a young organization; it was formed in 1995, when the government gave it the mission of "pursuing and fostering research and learning, as well as conducting development projects concerning work, the working environment and relations within the labor market."

This means that the National Institute for Working Life has as its goal the gradual generation of a knowledge bank about what is happening in the field of work as well as about the health problems of today and the future. The Institute has some 400 employees, including scientists and others with extensive experience and knowledge of the successful Swedish developments regarding occupational health and other areas related to working life at the individual, organizational, and labor market level.

Collaboration in the international research community is growing in importance. Thus, the National Institute for Working Life works with institutes, universities and scientists from around the world. For example, the Institute offers educational opportunities to participants from developing countries and Eastern Europe.

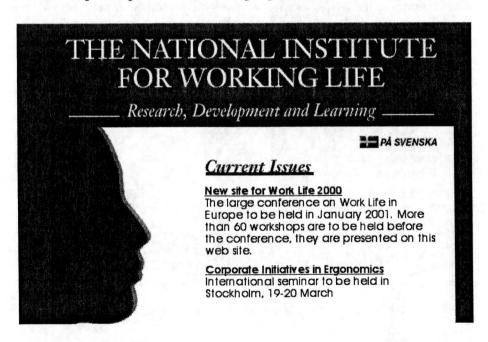

National Institute for Occupational Safety and Health (U.S.)

http://www.cdc.gov/niosh/

NIOSH's Web site contains information about the agency and its services, plus a number of useful data collections. Some of the available collections include these:

- **Certified Equipment List** - NIOSH provides a testing approval and certification program of safe personal protective devices and reliable industrial hazard measuring instruments. NIOSH develops improved performance regulations, tests and certifies (or approves) devices, and purchases approved and certified products on the open market to verify quality of manufacture. This is a list of equipment that NIOSH has certified.

- **Criteria Documents** - NIOSH recommends criteria or measures for protecting workers from serious health or safety hazards, based on a comprehensive critical review of the scientific and technical information available at the time the document was developed.

- **Documentation for Immediately Dangerous To Life or Health Concentrations (IDLHs)** - This publication documents the information that has been used by the NIOSH to determine immediately dangerous to life or health concentrations. In this document, IDLHs are listed with the basis and references for the current values as well as with the original IDLHs and their documentation.

- **NIOSH Manual of Analytical Methods** - NMAM is a collection of methods for sampling and analysis of contaminants in workplace air, and in the blood and urine of workers who are occupationally exposed.

- **Occupational Health Guidelines for Chemical Hazards** - This document summarizes information on permissible exposure limits, chemical and physical properties, and health hazards. It provides recommendations for medical surveillance, respiratory protection, and personal protection and sanitation practices for specific chemicals that have federal occupational safety and health regulations.

- **NIOSH publications** - Many of NIOSH's numbered publications, written between 1979 and 1999, are now available here in their entirety.

North American Emergency Response Guidebook (1996)

http://www.tc.gc.ca/canutec/english/guide/menug_e.htm

The 1996 North American Emergency Response Guidebook (NAERG96) was developed jointly by Transport Canada (TC), the U.S. Department of Transportation (DOT) and the Secretariat of Communications and Transportation of Mexico (SCT) for use by fire fighters, police, and other emergency services personnel who may be the first to arrive at the scene of a transportation incident involving dangerous goods. It is primarily a guide to aid first responders in quickly identifying the specific or generic hazards of the material(s) involved in the incident, and protecting themselves and the general public during the initial response phase of the incident.

Information includes hazards associated with specific chemicals, a table of placards and initial response guides to use on scene, and a table of initial isolation and protective action distances.

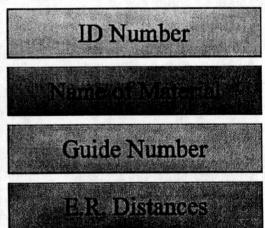

Oklahoma State University Online Training Modules

http://www.pp.okstate.edu/ehs/modules/home.htm

Oklahoma State University's health and safety people have put together several training modules which provide generic safety information to their employees. These are good examples of how the Web can be used to deliver information to a variety of audiences. This page also includes links to government training modules on the Web and Knowledgewire, a commercial Web-based training service.

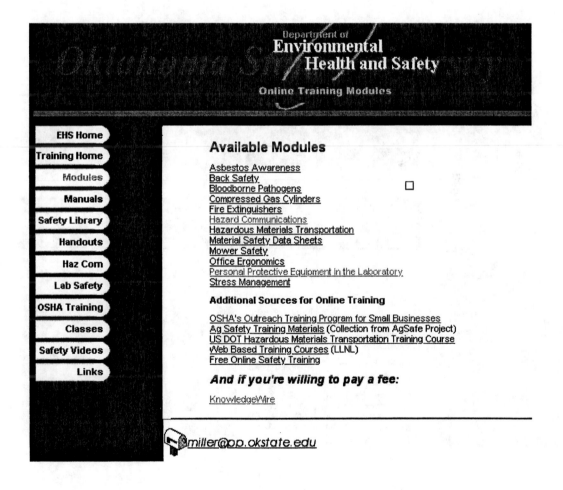

OncoLink

OncoLink is a multimedia oncology information resource. Its mission is dissemination of information relevant to the field of oncology; education of health care personnel; education of patients, families, and other interested parties; and rapid collection of information pertinent to the specialty. Information is available on a number of different types of cancer; various cancer-related medical specialties; psychosocial support; cancer-causes, screening, and prevention information; clinical trials; and much more.

Available documents include patient information pamphlets, information for health care providers, National Institute of Health publications, citations from the CANCERLIT database, and information from a variety of other sources.

celebrating
**5 years
helping
people
with cancer**

March 27, 1999

What's New

Cancer News

Editor's Choice

OncoLink TV

Book Reviews

Search OncoLink

Sponsors

University of Pennsylvania Cancer Center

AMERICAN VIATICAL CORPORATION

PROVIDING CONFIDENTIAL AND CARING FINANCIAL SOLUTIONS TO FAMILIES FACING CANCER

Copyright © 1994-1999, The Trustees of The University of Pennsylvania

March marks OncoLink's fifth anniversary, and in honor of the occasion our editors have written "No Cancer Whatsoever" - An Open Letter to Our Readers.

From left: Jason Lee, MD; Lili Duda, VMD; Debi Weil, PhD; Daniel Fleisher; Ivor Benjamin, MD; Vicky Maxon; Tom Dilling, MSIII; Maggie Hampshire, RN, BSN, OCN; Joel Goldwein, MD; George Coukos, MD, PhD; Annemarie Cardonick, RN, BSN, OCN; James Metz, MD; Katrina Klaghorn, RD; John Han-Chih Chang, MD.

OSHA

http://www.osha.gov/

In the early 1990s, OSHA began a project of collecting all of its official documents into an electronic database that would be distributed on CD-ROM disks (call OSHA Computerized Information System, OCIS). This disk is available to the general public for a nominal fee.

As the Internet began to develop in the mid-90s, the transition of this data to a Web site was a natural evolution. This gave OSHA a big head start in providing helpful technical and regulatory information over the Internet. OCIS provides access to the OSHA regulations, various summaries of these regulations, the OSHA Field Inspection Reference Manual, testimonies made by OSHA officials to various public hearings and the prefaces to the OSHA regulations.

One advantage of the Internet version of OCIS is that it is updated more regularly than the CD-ROM. In addition, search engines are available on the Web site which can be used to generate reports from the databases available at the site.

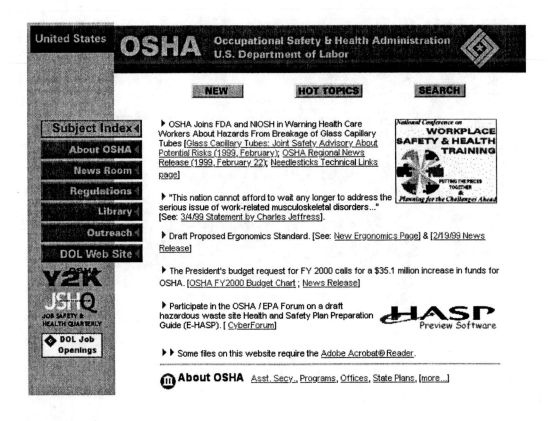

Radiation and Health Physics Home Page

http://www.sph.umich.edu/group/eih/UMSCHPS/

The site, maintained at the University of Michigan, contains information and links related to radiation safety. It has been compiled for three distinct groups: the general public, students, and the health physics community at large. Information is available in a number of categories, including:

General information - basic terminology, frequently asked questions (FAQs) on radiation and nuclear power, basics of radiation and radioactivity, etc.

Regulatory information - U.S. regulations, codes, standards and laws, federal and state regulators, etc.

Professional information resources - conferences, educational resources, radiation-related discussion groups, nuclear databases, and labs.

Radiation and health physics research information - links to numerous radiation-related databases and data sets.

Radiation specialties - emergency management, environmental radioactivity, non-ionizing radiation, and more.

SAFETY email list archives

This Web page provides interactive access to keyword searches of the SAFETY email list archives. The searches can use boolean operators and date limitations to restrict the searches to make them as effective as possible.

This index covers the messages that came over SAFETY from January, 1995 to the present (about half of the total messages). The complete archives are available at http://siri.org/mail, with a somewhat more limited search capability.

Search the SAFETY archives

Search for:

apple or pear
(green apple) or (red apple)
■ Substring search

In messages where:

● The subject is or contains:

pie or cake

● The author's address is or contains:

granny

Since: (date/time)

Until: (date/time)

June 1995
2 May 96

Start the search!

Help!

SafetyLine

http://www.safetyline.wa.gov.au

SafetyLine is an information service providing online access to the major publications issued by the WorkSafe Western Australia Commission and WorkSafe Western Australia. The objective of SafetyLine is to provide people in the workplace with immediate access to safety and health information that can be used to help improve their working environment.

Available information includes Australian and Western Australian OSH legislation, online training programmes and downloadable training materials, information about occupational health and safety in Western Australia, and OSH solutions.

OSH solutions are documents that address specific worksafe issues and problems. Numerous documents are available in the following general subject areas: hazardous substances; manual handling; plant; OSH training; noise; specific industries; work practices; work related illnesses; and others.

SafetyLine O N L I N E   WorkSafe Western Australia

Workplace Safety on the Internet

WorkSafe Western Australia has undertaken a complete overhaul of its SafetyLine information service. The new design and structural framework aim to make finding information easier. The new SafetyLine also includes a much improved search facility.

Almost all of the information from the existing SafetyLine has been converted into the new environment. The conversion will be complete by the end of March.

The new environment is open to the public for testing purposes. Any problems can be reported via email to white@worksafe.wa.gov.au Suggestions and constructive criticism are also welcome.

News
All the latest news about occupational safety and health in Western Australia.

Laws
Acts, Regulations, Codes of Practice, Guidance Notes, Prosecutions

Solutions
Essentials, Practical Solutions, Industry Information, Incident Summaries

Information
Statistics, Library Resources, OSH Links, HomeSafe, DriveSafe

Education
Education Resources, SafetyLine Institute, WorkSafe Smart Move, ThinkSafe Club

Society for Chemical Hazard Communication

http://www.schc.org

SCHC is a nonprofit organization with a mission:

- To promote the improvement of the business of hazard communication for chemicals
- To educate SCHC members on hazard communication issues
- To provide a forum for exchange of ideas and experiences
- To enhance the awareness of members and the general public of new developments in hazard communications
- To provide guidance or technical expertise to private, nonprofit groups and to government

This is an organization of companies and people interested in the development of material safety data sheets, labels, and other systems for communicating chemical risk information. Included at the Web site are organizational news, presentation files from meetings, job openings and news, national and international news affecting chemical hazard communication issues.

 SCHC Home

Toxnet

http://toxnet.nlm.nih.gov

ToxNet, provided by the U.S. National Library of Medicine, is a collection of databases on toxicology, hazardous chemicals and related subjects.

The Toxicology Data Search provides access to NLM's HSDB (Hazardous Substances Data Bank), CCRIS (Chemical Carcinogenesis Research Information System) from the U.S. National Cancer Institute, and GENE- TOX (Genetic Toxicology/ Mutagenicity Data Bank) and IRIS (Integrated Risk Information System), both from the U.S. Environmental Protection Agency (EPA).

The Toxic Releases (TRI) Search allows you to select and search any of the EPA's TRI series of files (beginning with TRI87) containing data on the estimated quantities of chemicals released to the environment or transferred off-site for waste treatment, as well as information related to source reduction and recycling.

The Toxicology Literature Search provides access to the following bibliographic files, consisting of citations to the scientific literature: DART (Developmental and Reproductive Toxicology) and its backfile ETICBACK, and EMIC (Environmental Mutagenesis Information Center) and its backfile EMICBACK. TOXLINE, a database of over 1,000,000 citations to the scientific literature, is available as part of Internet Grateful Med. See the description of Internet Grateful Med in this section.

Welcome to TOXNET ON THE WEB

This free-of-charge search interface provides access to the TOXNET system of databases on toxicology, hazardous chemicals, and related areas.

TOXNET is sponsored by the National Library of Medicine, through the Toxicology and Environmental Health Information Program of its Specialized Information Services Division.

Comments? Questions? Problems? - E-Mail TOXNET User Support.

Toxicology Data Search - Select and search any of the following files containing factual information related to the toxicity and other hazards of chemicals: HSDB (Hazardous Substances Data Bank), CCRIS (Chemical Carcinogenesis Research Information System) from the National Cancer Institute, and GENE-TOX (Genetic Toxicology/Mutagenicity Data Bank) and IRIS (Integrated Risk Information System) both from the Environmental Protection Agency (EPA).

Typing Injury FAQ

http://www.tifaq.com

The Typing Injury FAQ (frequently asked questions) and Typing Injury Archives are sources of information for people with typing injuries, repetitive stress injuries, carpal tunnel syndrome, etc. It is targeted at computer users suffering at the hands of their equipment. You'll find pointers to resources all across the net, general information on injuries, and detailed information on numerous adaptive products. Categories of information include the following:

General information about typing injuries - general advice and pointers to both published resources and those available on the Internet.

Keyboard alternatives - Frequently asked questions are answered and products used to replace or enhance standard keyboard use are covered. Products include split, contoured, and chording keyboards, and other keyboard alternatives/ accessories.

Speech recognition - A list of speech recognition systems and related topics.

Alternative pointing devices - A list of trackballs, trackpads, "funky mice," and other ways to replace your normal mouse.

Software monitoring tools - Software that reminds you to take breaks.

Furniture information - Information on ergonomic office furniture.

FAQ
Typing Injury
The RSI Community's Online Resource

Articles	RSI-Ergo Information	Industrial Ergonomics	Organizations	Services	Resellers	Archive
Furniture	Alternative Keyboards	Speech Recognition	Pointing Devices	Accessories	Software	Kids

About

Site Map

Sponsoring

Awards

Feedback

The Typing Injury FAQ (frequently asked questions) is an educational site, provided by the CTD Resource Network, Inc., containing a wide variety of information about repetitive strain injuries (RSIs), resources for dealing with these ailments, and a broad description of assistive products to reduce injury risk and symptoms. (see About for more information.)

We thank our sponsors for their support. Please visit them as they make this whole resource possible.

Also see The RSI Network newsletter

United Kingdom Health and Safety Executive

http://www.open.gov.uk/hse/

The Health and Safety Executive ensures that risks to people's health and safety from work activities are properly controlled.

As the Web site says "the law says employers have to look after the health and safety of their employees; employees and the self-employed have to look after their own health and safety; and all have to take care of the health and safety of others, for example, members of the public who may be affected by their work activity. Our job is to see that everyone does this."

The site is well organized, with an extensive table of contents and search capability. Categories of information include the following:

- What We Do
- Contacts for Enquiries
- What's New
- Press Releases
- Hazards at Work
- Publications and Videos
- HSE Statistics
- Current Events
- Freedom of Information

Welcome to the

Health & Safety Executive

Reducing Risks, Protecting People

University of Edinburgh, Health Environment and Work

http://www.med.ed.ac.uk/hew/

This educational resource is probably the largest academic site in Occupational and Environmental Health within the European Union. Hundreds of files from the University of Edinburgh are available, linked to many more elsewhere.

This site consists of hundreds of files about Environmental and Occupational Health from the University of Edinburgh, linked to hundreds more elsewhere. This site was originally conceived as an educational resource for students of the University of Edinburgh, but many of the pages have been adapted to provide a broader appeal, while retaining their academic quality. The Web site also provides a window for potential students, and informs about research, collaboration and other aspects of Environmental and Occupational Health, at the University of Edinburgh.

Highlights of the site include a series of online tutorials on occupational health topics, and an excellent directory of primarily European health and safety Internet resources.

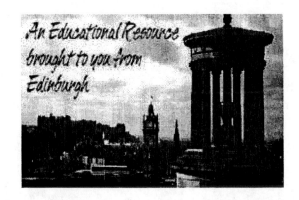

University of Minnesota Environmental Health and Safety

http://www.dehs.umn.edu

Many colleges and universities have health and safety department Web sites with valuable information available on it. Larger universities have nearly every general safety or health hazard associated with some part of their operation. Thus, policies and procedures for many different situations can be found on their sites. The University of Minnesota site is typical of a well-maintained University Web site.

About Us

Mission Statement
Industrial Hygiene and Safety Division
Radiation Protection Division
Related U of M Health and
   Safety Resources

Hazardous Waste Division
Public and Occupational Health Div.
Administrative Services

Resources

Agricultural Safety
Biosafety
Employee Right to Know
Ergonomics
Radioactive Material Services

Asbestos and Lead Management
Chemical Waste
Chemical Hygiene
Indoor Air Quality
Safety Procedures Training

Links

Asbestos & Lead
Chemical Waste
College/University EHS Depts
Environmental
Fire and Life Safety
Government-Federal
Indoor Air Quality
Occupational Health & Safety
Radiation Protection

Biosafety
Chemical Hygiene
Emergency Response
Ergonomics
Flood Clean-up Information
Government-Minnesota
MSDS
Public Health
Just for Fun

Vermont SIRI

http://siri.org, http://hazard.com

The Vermont SIRI (Safety Information Resources on the Internet) Web site is a joint project of Ralph Stuart and Dan Woodard. Its primary features are a large collection (see next entry), the archives of several of the major safety related email lists (SAFETY, occ-env-med-l, hs-Canada, ih), and a collection of public domain electronic files.

Among the files in the library are more than 400 clip art line drawings related to safety issues, a variety of articles written for the safety professional audiences, and outlines for "toolbox talks"—short discussions of safety issues related to generic workplaces.

Vermont **SIRI**
Safety Information Resources, Inc.

Bookmark
Both Servers
http://hazard.com
http://siri.uvm.edu

Ten reasons Linux is better than Windows

- Safety E-mail Archive

Bookmark all three MSDS sites for increased reliability

- SIRI MSDS-Vermont
- SIRI MSDS-Florida
- SIRI MSDS-California

- Online Library

- Safety Graphics
- Discussion Board
- Using the Internet for Safety
- The New Vermont SIRI Bookstore!

- Other Safety Websites
- OSHA and EPA Regulations
- Health and Safety Consultants List
- Contact Us

Where to Find MSDSs on the Internet

http://www.ilpi.com/msds/index.chtml

This is a one-stop shop for finding MSDSs and related software. The site includes an extensive list of free MSDS sites, MSDS software, commercial MSDS collections, and even fun MSDSs.

Where To Find
Material Safety Data Sheets
On The Internet

Table of Contents

Breaking updates (Mar-25):
- **NEW!!** Take our 7-part on-line MSDS quiz/tutorial. (Mar-26)
- We've added a question about pharmacy concerns to the MSDS FAQ.

- News Notes
- What is a MSDS?
- Two examples of a MSDS
- Further information about MSDS
- Other useful or interesting information
- Safety books and manuals
- Commercial MSDS suppliers
- MSDS software and databases

- Where can I find MSDS? (non-Internet sources)
- MSDS sites on the Internet (60+ **free** sites)
  - General sites
  - Government agencies
  - Chemical manufacturers and suppliers
  - Pesticides (herbicides, insecticides etc.)
  - Miscellaneous
- The MSDS FAQ (Mar-19)
- Humor: Create **UNSAFE** Material Data Sheets
- The on-line MSDS Quiz (**New** Mar-26)

World Health Organization

http://www.who.int

The World Health Organization came into being in 1948, when the 26th United Nations membership ratified its Constitution. The objective of WHO is the attainment by all peoples of the highest possible level of health. Health, as defined in the WHO Constitution, is a state of complete physical, mental, and social well-being and not merely the absence of disease or infirmity.

The Web site provides the following information: descriptions of WHO's major international programmes; a list of the organization's publications; the full text of the WHO Weekly Epidemiological Record; access to the WHO Statistical Information System; newsletters; international travel and health information; etc.

Specific information includes the following:

- Diseases: communicable/infectious diseases
- Diseases: tropical diseases
- Diseases: vaccine preventable diseases
- Diseases: noncommunicable diseases
- Environment and lifestyle

Young Worker Awareness,
Workplace Health and Safety Agency, Ontario, Canada

http://www.yworker.com

This site contains health and safety information for young workers, their parents, teachers, principals, employers, and others. Though the information is specific to the province of Ontario, Canada (the Young Worker Awareness schools program is only available to Ontario high schools), others may also find it useful. Information is available in the following categories:

What are the Risks? - This section describes the top five causes of injuries to young workers, the top five critical injuries to young workers, the five most common injuries to young worker's, and details of the different types of injuries.

The Law - The full text of the various acts and regulations that affect young workers in Ontario is presented.

Your Rights - Information is provided about workers' rights and responsibilities; WHMIS (Workplace Hazardous Materials Information System); Material Safety Data Sheets; workplace inspections; health and safety training; and the work refusal process.

What if I Get Hurt? - The procedures for injury reporting in Ontario are described.

Other information on the site includes a health and safety quiz, and true health and safety stories.

4 Networking on the Network

In addition to the ability to access file libraries for research purposes, the Internet provides valuable opportunities for professional networking. As does any professional network, the Internet can help you learn more about the basic technical issues of your field, get tips about approaching specific problems, be aware of new issues developing in the field, find prospective partners or consultants, and celebrate (or commiserate) with others in similar circumstances. The advantage of the Internet for networking activities is that it provides a convenient way to have ongoing discussions with geographically dispersed colleagues. These discussions can take place either in a group or individually. This section describes some of the considerations involved in using the Internet for this purpose.

General Considerations

As the Internet has grown in popularity, the time required to remain current with its content has increased as well. While the technical details of using email and Web sites have simplified significantly in the 1990s, the task of wading through all the possibly relevant information sources has become more complex. It is important that you have a clear idea of what your goals for using the Internet are; otherwise, you may find that you are devoting a lot of time to using it without much payback.

Using the Internet to network with other people with similar interests can minimize this learning curve. The benefits of networking on the Internet are that it is a low-budget and time-efficient way to be involved in your professional community. Productive professional relationships can be developed with a wide range of people without face-to-face meetings. These relationships usually start in discussion areas such as email lists or newsgroups. They often develop into private correspondence that is able to be more speculative than public discussions.

The primary costs of developing a network of professional contacts over the Internet are time, patience, and a network connection. Fortunately, a powerful Internet connection is not required, as most networking happens via email with little graphical content and small files. A good place to find more information about the process of networking and how it is affected by the Internet is *The Network Observer*, which is written by Phil Agre, a faculty member in the Communications Department at the University of California at Los Angeles. He has written several interesting articles on this topic and subscription to his email list (The Red Rock Eater list) is free. More information can be found on the Web at **http://dlis.gseis.ucla.edu/people/pagre/ tno.html**.

Styles of Participation

The way in which you participate in an Internet discussion group will vary depending on many factors. There are three primary styles of participating: daily participation, selective participation, and lurking. With experience, you may find that different styles are appropriate for different lists. The numbers below for the portion of a list membership participating at these different levels are based on experience with the SAFETY list, but are applicable to email lists in general.

Daily participation: Some people enjoy email and email discussions so much that they make a daily habit of reading all the email they can get and responding to many of the discussion topics that arise. These people are usually less than 10 percent of a list's subscribers, and often account for 30 to 50 percent of the postings to a list. While this can be annoying if you don't like someone's style, these people are important for keeping the list active.

Not only do they provide topics of discussion, but they also provide some sense of what questions are likely to be answered by the list's subscribers. It is usually easy to identify these people by following the list for a week or so.

Selective participation: Another way to participate in a list is to read the postings regularly, but avoid responding publicly unless an issue in which you have special interest arises. This is a more common practice than daily participation and about 25 percent of a list's members fall into this category. These people provide an important error-checking function for the list, in that they will usually point out occasions when incorrect information is presented as fact. They also broaden the range of questions that can be answered successfully by the list.

Lurking: The most common use of email lists is *lurking*—simply reading the postings of a list without responding, unless one has the need to answer a specific question. For most lists, the large majority of subscribers do this. Most people find active participation in more than one or two lists to be too time-consuming. The presence of lurkers is important to a list because they provide questions and comments that normally wouldn't come out of the general flow of discussion among more regular posters.

Finding a level of participation that you are comfortable with for a particular list will take some experience with the list. However, the more active you are on the list, the more likely it is that your questions to that list will generate useful responses. In addition, by actively participating in discussions, you can move a thread (a sequence of messages on the same subject) in a direction that is useful to your particular interests.

One important thing to consider is that commercial announcements are controversial on the Internet, particularly in discussion areas. Because the Internet originated as a government-supported facility, there were restrictions on commercial discussion. The spirit of these restrictions has lingered, although the government support is gone. There are now effectively no restrictions on commercial use of the Internet, but expectations of most users is that commercial announcements will be low-key and not repetitive. Usually commercial announcements from regular list contributors are more acceptable than from people who appear on the list only for that purpose. Web sites are more appropriate places for advertisements of a commercial nature.

Finding Discussion Groups

There are thousands of discussions groups operating on the Internet, either through email or in newsgroups, and sometimes with a gateway between both forms. Some are formally organized with charters (such as the SAFETY WELCOME message in Appendix 2); others are simply collections of email addresses being held together by someone's personal email software. Finding discussions that are of interest to you can be a bit of a challenge. However, there are several good places to check.

The first place to check is trade magazines and professional journals. *What's new on the Internet?* is a favorite topic for articles. These articles usually include both a list of Web sites relevant to the profession and discussion groups. It is important to note what type of forum is being discussed when reading these lists; using a newsgroup is significantly different from using an email list from the computer point of view.

Other places to look for relevant discussions are in the various "lists of lists" that exist on the Internet. One example of such a list is available at Tile.Net (**http://www.tile.net**). This site includes descriptions of a wide variety of mailing lists, organized by name, subject area, or location of the host machine. For example, a search on the word *safety* found a variety of lists, including several local safety discussions, a radiation safety list, and one which includes food safety as part of its discussion. Chapter 7 in the book provides such a list in paper format.

Remember that formal descriptions of discussions groups can often be significantly different from the actual subjects talked about within the group. You can search the actual text of many discussions within their archives. For newsgroups, this is most easily accomplished by using a Usenet search engine such as DejaNews (**http://www.dejanews.com**). By putting in keywords such as those discussed in Chapter 3, you can quickly see if a newsgroup discussion has developed around the topic of interest.

The SAFETY Mailing List

The easiest way to discuss the finer points of the use of email lists is to look at one in detail as an example. The SAFETY list is the oldest, largest, and busiest safety related list on the Internet, so information about it is the most fully developed. It will be discussed here as an example of how an email list might operate.

SAFETY started in December 1989. When the Internet started growing significantly in the early 90s, SAFETY grew with it. SAFETY now routinely has 35 to 40 messages, and hits the daily maximum of 50 messages regularly. There are currently (March 1999) over 3000 email addresses on the list. About 60 percent of them receive every posting individually, while 25 percent receive SAFETY in digest form, and 15 percent of the subscriptions are set to NOMAIL. So, every posting goes to over 2200 user-ids, and these messages may be passed on to more people, depending on their content.

The professions represented on SAFETY range from industrial hygienists and laboratory technicians to librarians and students, primarily located in the United States (about 90 percent). About 30 percent of the list is associated with U.S. academic institutions, and another third of the subscribers represent industrial and commercial sectors. The other third comes from a variety of sources, including government agencies, the military, and international sites. The qualifications of participants range from Ph.D.s and Certified Industrial Hygienists to people new to the safety field and lay people. This creates a wide range of questions and responses to the questions presented.

Asking SAFETY Questions

Getting valuable information from SAFETY can take some practice. Phrasing a question too broadly or too narrowly can result in a lot of discussion that does not answer the question you have in mind. Follow the list for a while to get a sense of the style of question that receives the most useful response. If your question is going to spark a response, responses will start coming within 24 hours. Some responses will come privately, and some will come via SAFETY. Even if you request private responses, be sure to watch SAFETY for other responses. One good way to get a specific question answered is to ask the question privately to someone who demonstrates expertise in the subject in which you're interested. It is quite likely your contacts will be more willing to discuss your situation in depth privately than publicly.

One common SAFETY question involves surveys; people need to canvass a large number of people to determine "standard practice." Short, specific surveys are likely to draw useful responses. However, it is unlikely that even the best SAFETY survey will draw enough responses to be statistically significant. Email surveys are more useful for generating a range of ideas than hard numbers.

Answering SAFETY Questions

How and when to answer questions to SAFETY is a matter of personal preference and judgment. Many people regularly read SAFETY without ever posting to the list. On the other hand, a core of people are likely to respond to just about anything that gets posted. Often, ensuing discussions bring out relevant information not directly related to the original question.

One important choice to be made when you decide to respond to the question is who the response will be directed to—the individual who asked the question or the list as a whole. If you feel that the information is so basic that the average SAFETY reader would already know it, or when you need clarification that might result in a lot of local details, it is probably best to respond directly to the individual. However, if you have some information that may be helpful and you are interested in further discussion on the subject, or when you have specific expertise on a subject and feel a need to clear up a common misconception, responding to the list is more appropriate.

If you think that a question or subject is inappropriate for the list or if you find someone is really annoying, do not complain to the list as a whole. If you are comfortable responding directly to the individual, that is the best approach. If not, send a note to the list owner to express your concerns. The list owner(s) may not be able to take immediate action, but they can use their influence to try to discourage troublesome behavior.

Using the SAFETY archives

Every posting sent to SAFETY is archived by the LISTSERV program running the list. These archives are an important resource of safety information because you can do keyword searches on the messages in the archives. These searches can be done by email or on the Web. Usually, it is most efficient to use the Web index to explore the archives and determine what search strategy is best, and then use an email job to retrieve relevant postings as a single file. The details of using the archives can be found in the SAFETY WELCOME message (Appendix 2).

Other Email Lists

Internet email lists related to health and safety topics range from technical discussions of environmental engineering problems (ENVENG-L) and medical discussions of occupational health and safety concerns (OCC-ENV-MED-L) to lists which focus on specific safety subjects (BIOSAFTY and RADSAFE). Other email lists focus on particular geographic areas, such as HS-CANADA or UNIOHS (Australia and New Zealand). A catalog of these lists can be found in Chapter 7.

5 Marketing Safety on the Internet

In addition to providing a new information resource for safety professionals, the Internet is becoming the site of routine business both in the workplace and for the general public. This provides new opportunities for safety professionals to get out their messages, whether these are aimed at changing workers' behavior to improve workplace safety, advertising the availability of their products or services, or providing public access to government information. The unique advantages of the electronic information delivery (for example, ease of updating, asynchronous delivery of information, interesting interactive graphical presentations, essentially limitless storage capability) make it ideal for providing safety information to many of the audiences that the safety professional serves.

This chapter provides an introduction to some of the conceptual issues involved in using the Internet for this purpose. It focuses on the process of developing, maintaining, and promoting a Web site; however, remember that there are other ways of using the Internet to sell safety besides Web sites. Maintaining an active email presence or pointing people to information that other people have provided on the Internet may be as effective and a more efficient way of serving a particular group.

If you decide a Web site is the way to go, it is best to consider getting professional help (computer help, that is). Just as developing an attractive piece of paper literature requires careful graphic consideration, developing an effective electronic platform for your information requires specialized skills. There are many people and commercial organizations available to help with some or all of the technical aspects of Web site planning and design. For this reason, this chapter does not concentrate on the technical computer issues involved in developing a Web site.

Planning Your Web Site

If you want your organization to have a presence on the Internet, there are a number of decisions that you need to make before you start.

Developing a Reason for a Web Site

Why does your organization want to be on the Internet? Is your main purpose to provide access to information that you would provide for free by other means? For example, many government organizations are setting up Web servers to provide access to publications that are now available on paper. Their aim is to provide improved access to their public information and, not incidentally, to reduce the cost of reproduction and shipping.

Do you want to promote your organization's products and services? If so, you should plan to provide more information to your customers than you do on paper, or else there will be little incentive to use the Web site, particularly compared to those of your competitors. As well as providing in-depth product and service information, you should be prepared to provide new information that isn't likely to be found elsewhere. If you are hoping to attract new customers via the Internet, remember that Internet users expect to get something useful for free, even from commercial sites.

As an example, if you are selling health and safety consulting services, you could provide a generic version of the full text of one or more of your reports on a health and safety subject, links to other health and safety information on the Internet, or something else that would be of interest to your customers. Exhibiting your skills online is the best way to make an impression on people.

Do you want to sell products or services electronically? Consider how you will accept credit card purchases. Secure mechanisms for credit card transactions are available on the Internet, but managing these transactions involves a significant number of customer relations decisions. However, people are becoming steadily more comfortable with electronic purchasing, and sites that do not provide that option will be less productive as sales tools.

Do you want to provide safety training and information electronically? This requires careful planning about how the electronic information will fit into the culture of your organization. Remember that the Internet is only one part of the "information ecosystem" that your audience is exposed to. Messages promoted on the Web page must be reinforced through paper documents (probably shorter as a result of the Web presence), person-to-person interactions, and upper-management commitment.

Selecting Information to Put on the Net

Once you have defined your goals, the first issue is what type of information you want to include on the Web site and in what form you want to distribute it. These choices require balancing several issues, including the types of information you are trying to deliver (*e.g.*, material safety data sheets, contact information for help in resolving safety problems, or training materials); the ease of access your audiences have to the Internet; and the level of documentation you are required to have to demonstrate regulatory compliance. Providing safety information to a group of workers provides a good example of these trade-offs.

The advantage of using a Web site is that it is possible to include a large amount of information there, most of which has already been generated for other purposes. It is relatively easy to move training handouts and overheads from paper and transparencies to a form accessible at your Web site. People in your audience can then access these materials at their convenience or use them to review the information you have already presented. At the same time, you can leverage the many other electronic resources that are available on the Internet to provide background information or more details on the issue (for example, the text of the OSHA regulation or a standard operating procedure from another institution).

The disadvantage of this approach is that a Web site must be carefully planned to make the information you have put there easy to find. For example, although safety professionals are used to working with programs on a regulatory issue basis (for example, bloodborne pathogens separately from chemical hygiene issues), this is probably a confusing way to organize information for the average worker who is looking for an answer to the question *How do I clean up this stuff I just spilled?*

Despite the challenges of providing safety information over the Internet (or an institutional Intranet), the potential advantages are attractive enough that many people are doing this. Many organizations have developed Web-based training programs that lead workers through a curriculum designed to meet regulatory requirements. The success of this training is measured by quiz results that are documented within an electronic database. Such services are also available commercially, although they are only appropriate for providing background information on a particular safety issue. Follow-up training on local specifics is always necessary.

Making Your Web Site Effective

Developing a health and safety Web site presents the same challenge as the development of other safety information: How do you gain and hold the attention of the people who should have the information you have to share?

Three criteria to keep in mind for the material you place on your Web site are these:

- It should be fun.
- It should be useful.
- It should change over time.

Fortunately, Web sites lend themselves to these criteria. As you browse the web, you will find many examples of Web sites that are some or all of these. Keep in mind that these Web sites require quite a bit of effort to develop and maintain. It is important to keep the ambition of your Web site within the scope of the resources you have to devote to it.

Designing the Web site as a whole can be a major challenge. Various guides are available on the Internet that highlight some of the issues involved. They are listed in the last section of Chapter 6. Of particular interest is the Yale Style Guide to the Internet at **http://info.med.yale.edu/caim/manual/index.html**

Technical Computer Issues

Who is going to design and maintain your Web server? You can either do this yourself or have professionals do it for you. A number of factors should be involved in your decision.

Web Page Design

Do you have people in your organization who already know how to create Web pages? Are their Web pages both functional and visually appealing? Remember that while it is relatively easy to create basic Web pages, creating Web pages that communicate your message effectively requires a specific set of skills and experience. The use of fill-in forms and image maps requires behind-the-scenes programming to process them. Consider whether you have the skill in-house to design and write these programs. As an alternative to doing it yourself, do you have money available to invest in professional Web design? As with other Internet developments, developing professional looking Web pages is becoming easier, and maintaining them is not a difficult chore. However, appropriate training and care must be taken in approaching this task. There are many examples of poorly thought out Web pages that defeat their purpose by being difficult to use.

Web Page Content

Another consideration is whether you have staff who can provide and maintain content for your Web page. One of the major strengths of the Web is the ability to update information on it, so new content should be added on an continuing basis. Many servers are completely redesigned every few months as their goals and audiences change. The availability of staff resources or a professional services budget for the ongoing maintenance of the Web server should be considered carefully.

Computer Support

Separate from Web page design issues, there are hard-core computer issues, such as the question of where the Web server will be housed. This decision can involve significant cost and manageability consequences. You can either run a WWW server on a computer at your own site, or you can rent space on an Internet service provider's computer. If your organization does not have a dedicated connection to the Internet, renting space is a good short-term and possibly long-term solution. The main advantages are as follows:

- You do not have to dedicate resources to maintaining an Internet connection, providing and maintaining the computer, and backing up the data.

- The security of your organization's internal computer system does not need to be addressed as part of the Web creation process.

- If you decide to have the Web server designed professionally, you may be able to get a package deal.

The disadvantages are that you are dependent on your Internet service provider and will have to pay the service provider on an ongoing basis. Many people have reported that some ISPs can be difficult to work with, particularly for smaller accounts. Also, the ISP industry is a young one, and consolidation and continuing business shifts among providers is to be expected.

Selecting a Domain Name

Another important issue associated with running a Web server on someone else's computer system is the address people use to reach your site. Unless you specify otherwise, your organization's name will likely be secondary to that of the service provider. As an example, if your organization is ABC Company, and your Web server is on the computer of an Internet service provider called DEF, people will reach your site by using an address similar to http://www.def.net/~abc. The alternative is to have your own *domain name* created. This is a unique name that identifies your organization on the Internet. Most Internet service providers will help you to have a domain name set up and will let you use that name for access to your site on their computer. If your ABC Company had its own domain name, your Web server would be accessible using an address such as http://www.abc.com.

One of the main advantages to using your own domain name is that you are more easily recognizable on the Internet. If an experienced Internet user is looking to see if you have a site, he/she will try http://www.abc.com and find you. No one will be able to find you by chance if your address is http://www.def.net/~abc. The other advantage is that the address of your Web server is independent of the Internet service provider and is therefore portable. If you change Internet service providers or decide to run the Web server on your own computer, the address does not have to change. Your users will continue to be able to access your site at http://www.abc.com.

If your organization has a dedicated Internet connection, you will probably choose to run a WWW server at your own site. In this case, the issues of a domain name, the integrity of your Internet connection, security, and computer maintenance will already have been dealt with. The main advantage to running your own WWW server is that you have complete control of it. If you already have a computer available to act as a WWW server, you can make your Web server available quite inexpensively.

Promoting Your Web Site

Creating your Web site is only half the battle. The other half, of course, is letting people know that it exists. The same challenge that faces people doing research on the Internet (so many options to choose from!) exists here. There are numerous ways of letting people know about your Web server, both on and off the Internet.

On the Internet

The Internet itself is the place to start promoting your Web site. You know that people using the Internet already have access to it and are comfortable with using electronic information. There are many different ways of pursuing this strategy. Some are based on understanding Internet information tools; others are based on person-to-person contact.

Alert the general Internet directories and search engines - There are many directories that list Web servers and search engines that provide access to their content. Directories such as Yahoo and Magellan and search engines such as AltaVista and Lycos are accessed several million times per day. These sites are supported by advertising revenue based on the number of hits they receive. This means that usability often runs second to commercial considerations. It is important to understand how these sites are managed in order to make best use of this strategy.

Given that, having your site registered with as many of these directories and search engines as possible provides your best opportunity for wide exposure to the Internet population. One option for listing your site with a variety of directories and search engines is to start with Submit-It at **http://www.submit-it.com**. As the name implies, this is a central site from which to submit information about your Web server to the major directories and search engines. This is a free service that helps you submit information about your site to several search engines and directories as well as a variety of commercial services for making information about your site available to some 300 other sites.

Links to and from related sites - Many Web servers provide links to related sites. If someone is browsing a site with a similar focus to yours, he/she will likely be interested in your site as well. As an example, let us assume that you have laboratory safety information available on your Web server. If someone is browsing another site containing laboratory safety information, they would find your site very easily if there were a link to it. The way to get a link added is to contact the Webmaster (that is, the individual in charge) of another Web server. Offer to put a link to his/her site on your Web server and ask him/her to do the same for yours.

Mailing lists and newsgroups - One of the best ways of reaching potential users of your Web server is through subject-oriented mailing lists and Usenet Newsgroups. If you have health and safety-related information on your Web server, it is considered entirely appropriate to announce its availability on health and safety mailing lists and newsgroups. In many cases, you must subscribe to a mailing list before you are allowed to send a message to its other subscribers. Anyone can send a message to a Usenet Newsgroup.

Chapter 8 provides information about appropriate mailing lists and newsgroups. When announcing the creation of your Web server, be sure to include both the full Uniform Resource Locator (URL) and a general, noncommercial description of what users will find there. As an example, an organization might announce its Web server with a message similar to the following:

We are pleased to announce the availability of our new World Wide Web server at "http://www.xxx.com." Available information includes the full text of selected publications, health and safety documents from other organizations, information about upcoming conferences and meetings, links to other health and safety Internet resources, and information about our services.

It is also common to announce major Web site additions or revisions on mailing lists and newsgroups.

Email signatures - Create or edit the signature file used with your email software. Add your Web site's URL. Anyone who reads your email will then know about the existence of your Web page and where to find it.

Improving Web search results - Search engines index literally millions of Web sites and hundreds of millions of pages. If someone searches for *hazardous materials*, what will make your site stand out among the thousands of sites that have hazardous materials information?

There are a number of strategies for increasing the likelihood that pages from your site will be considered relevant to health and safety-related searches performed on various search engines. Since different search engines use a variety of approaches to indexing the content of sites and establishing their relevance, there is no one-size-fits-all approach that can be taken. To quote from "Submit It!" (http://www.submit-it.com):

> Do not make it your goal to appear in the top ten list of every search engine. This would not only be a very difficult task, but would most likely end in disappointment. You are virtually guaranteed to have varying success rates with different search engines due to the collection of variables that play a role in every search result. These variables include size of database, method used for determining relevancy, policy on spamming, use of <META> tags and more.

That said, these are some of the common approaches that you can use when creating your Web pages. In most cases, a particular approach will affect both how a page is indexed with a search engine and how it will appear when its summary is displayed in search results.

Use <META> tags carefully - A *meta tag* is text that is embedded into the HTML (Web page source) coding of a Web page, but does not appear when that page is viewed in a Web browser. Browsers do not display them, but some search engines use them when indexing pages and displaying search results. There are two meta tags that can be used: "keywords" and "description." In a document describing what makes chemicals poisonous, keywords and description tags would look like this:

<meta name="keywords" content="toxic effects, toxicology, chemical exposure, toxic, toxicity, route of entry, chronic toxicity, acute toxicity, chemical hazard, toxic effects, toxicology, chemical exposure">

<meta name= "description" content="What are the routes of entry into the body? Why does the amount or dose entering the body matter? What are the differences between toxicity and hazard?">

The keywords tag allows you to specify a number of words and phrases to be indexed by search engines. They allow you to put in synonyms and references to relevant subjects that are not mentioned in the text of the particular document. Some search engines will index these and consider them when determining how relevant a Web page is to a particular search. The description tag is used by some search engines when displaying search results. It should therefore be a short summary of the content of the page.

Use the Web page title effectively - Use descriptive words in the title of the page. The title is text that appears in the title bar of the browser window—*i.e.*, the colored bar that forms the top border of the window. As an example, if your Web site were for ABC Consulting, the title on your home page should be something like "ABC Consulting, OSHA compliance and HazMat training specialists" rather than just "ABC Consulting" or "ABC Home."

There are two reasons for using descriptive titles. The first is relevance ranking. Most search engines work on the assumption that the content of the title is highly relevant to the content of the page itself. Therefore, if someone performs a search for *OSHA compliance*, your home page as described above would be considered highly relevant to the search.

The second reason is recognition of relevant search results. Some search engines display the first few lines of text from a Web page when showing search results, but many do not. All search engines display the title of a page when showing search results. If someone were viewing the results of a search on one of the engines that displays only page titles, seeing "ABC Consulting, OSHA compliance and HazMat training specialists" would give someone unfamiliar with the organization an idea of what they might find there, whereas "ABC Home" would not.

The first few lines of text on a page should summarize the content of the page. Most search engines assign higher relevance to text at the top of the page than to text in the middle of, or at the bottom of a page. Therefore, the first few lines of text may determine how relevant a page is to a certain subject. Also, some search engines show the first few lines of text from a page when displaying search results. If this text is descriptive, searchers will know what to expect when they visit the page.

Off the Internet

In addition to Internet promotion, think about how your Web site fits into other forms of organizational communication. It is becoming very common for organizations of all types to have WWW servers. Some of the ways that you can inform your clients about your site are these:

- Announce its creation in press releases and/or newsletters.

- Include the Web site address in any print advertising that you produce or buy.

- Add the Web address to print documents produced by your organization, to letterhead, and to business cards.

6 Emerging Issues on the Internet

As we have emphasized in this book, the Internet is still a rapidly developing medium. Hardware capabilities increase every year and software programmers use those capabilities almost as quickly as they are available. The "ratchet effect" of user's expectations means that each of today's innovations will be tomorrow's routine tools. Given this situation, it is difficult to make guesses as to what the Internet will look like in five or even three years.

However, there are some issues which we have seen emerge in the later 1990s that are important considerations for health and safety professional to keep in mind when thinking about the potential of the Internet. These considerations involve practical concerns of computer use and the long-term potential of the electronic communication to reshape our culture and our jobs. Discussion of some issues (privacy, commercial trends, web page design) are beyond the scope of this book but are available on the SAFETY list in the "Internet Culture" postings that occasionally appear. Others of these issues are summarized in this chapter. They are divided into the Good News and the Bad News department. We'll start with the Bad News...

The Bad News

Spam, Spam, Spam

Spam is an emotional issue for many users of the Internet. Simply put, spam is electronic junk mail, email notes that you receive unsolicited and which are usually of little interest. Spam promotes a wide variety of products and services, some of which many people find offensive. Spam is seldom used by reputable sales organizations. However, spam has attracted enough attention that laws have been proposed in the U.S. Congress to ban it. (It is an interesting idea that any one country could have effective control over information travelling over the Internet.)

Managing your email to avoid spam is possible. Carefully selecting which website to leave your email address on, avoiding posting to email lists or newsgroups, and complaining to the service providers of the addresses which send you spam are all legitimate approaches. However, these can all be time consuming and are usually easily defeated by a determined spammer.

Our experience has been that the simplest and most efficient approach to managing spam is to delete each day those messages that arrive touting the easiest way ever to get rich or the most effective way of attracting the opposite sex. One way to approach this is to use email filters. Many email software packages allow you to use "mail rules" or "filtering" capabilities. This is the ability to have your email software preview incoming messages, and perform certain actions based on their content. As an example, you can eliminate many of the "get rich quick" messages by setting up a filter that automatically deletes messages that have "dollars!" in either the subject or the message body. This filter can be controlled so that messages from people who normally discuss these matters are not deleted.

On the other hand, as email list managers, we are quite interested in seeing to it that the lists are not used in inappropriate ways. One way of protecting an email list from spam is to check each message before it is allowed to go to the list. This works in some cases but is a significant commitment for the list owner to undertake. And the membership lists of email lists are usually publicly available from the mailing list management software, so this method is not foolproof. Most mailing list management software packages provide some antispamming protection within the software itself. These are often effective in preventing the most flagrant misuse of the list. However, their strategies are like locking your house—they cannot guarantee to prevent an intrusion but they do make it inconvenient enough so that the miscreant will find another way to achieve his goal.

In summary, spam is likely to remain manageable if deleting unwanted email is not too large a chore. If you find that it is an annoying part of the Internet for your particular situation, then you will find more aggressive management strategies at various websites. **http://spam.abuse.net/spam/** is a good place to start. One email list, SPAM-L, discusses this issue in detail.

Email Attachments

Email originated as simple ASCII text files that were moved from one computer to another. The content of the messages was constrained to the words included in the message. However, this constraint removed significant content from many interactions, primarily the emotional tone of the message. In addition, many binary computer files that contained specific data to be shared could not be transmitted by ASCII codes.

One response to this limitation was the inclusion of codes within the email which could change the font, color, size, and formatting of the text in the message. Another development allowed users to send the binary data produced by a spreadsheet, word processing, or graphics program along with the email note as an attachment. However, because these codes and attachments do not follow ASCII specifications, they are not interpretable by as wide a number of programs as standard email.

This lack of standardization can create significant problems for people who receive email attachments. If they do not know what program was used to encode the attachment, they are not likely to be able to read the attachment. And even if they know what program to use and have it available, many attachments become corrupted in email transit. Therefore, a significant number of attachments sent with email are unusable.

This is particularly a problem for email lists. Because email lists have a large audience with a wide variety of computer setups and abilities, it is unlikely that a significant portion of the list membership will be able to benefit from the attachment. Therefore, most email list managers ask that attachments not be sent to the list. Some Mailing List Management software, such as LetterRip, automatically excludes attachments.

If you have a situation which calls for sending an attachment to someone, contact them first to be sure that they know that the attachment is coming and what to do with it when they get it. Many people have a policy of deleting all attachments they receive because attachments potentially contain computer viruses which can damage data.

An alternative to sending attachments to people is to use a File Transfer Protocol (FTP) program to move the file from one computer to another. This may require using a third machine as the FTP server, but most Internet Service Providers and some websites make such services available for free. (*See*, for example, **http://iMacFloppy.com/**.)

Software Footprint and Updates

The *footprint* of software is the amount of hard disk space and RAM (Random Access Memory, a measure of the amount of data your computer can handle at one time) that a particular piece of software requires to run. As hardware has become more powerful and cheaper, software has become bloated with new features that cause it to have a very large footprint (in traditional desktop computer terms). This creates a cycle which requires you to buy new hardware to take advantage of new software, which soon requires new hardware again. (Guess who's the winner in this scenario.) Several companies have attempted to come to grips with this problem by offering "lite" versions of their machines and software which have less power at less cost.

This creates management questions about how often computer setups should be upgraded. During the early 1990s, computer functionality increased with nearly every innovation, and upgrading was usually easy to justify. However, during the later part of the 1990s many of the upgrades have had marginal improvements and feature enhancements that are not of value to many or most users. Therefore, it is important to decide what your individual computer needs are before worrying about what options are available in the marketplace.

One rule of thumb is that the first three versions of new software are worth obtaining. The first version is usually "proof of concept" and probably contains significant usability problems. The second version resolves many of those problems. The third version incorporates user feedback and completes the *feature set*—the list of things that the software can do—for that type of software. After Version 3, most of the changes are minor bug fixes and rearranging of features in ways that confuse experienced users as much as they help new users.

The Good News

Connectivity Speeds and Options

As described in Chapter 1, the Internet Protocol is designed to be used by a wide variety of electronic devices. We are beginning to see the full development of this flexibility at the turn of the century. Options for collecting email, browsing the web, and supplying information to the public are multiplying weekly. Some brief examples are described here.

Palmtop Connections

Palmtop computers are pocket-sized devices which are designed to be able to collect notes, track calendars, and send and receive data from other devices. Many of these are now being used to connect to Internet functions such as email and the WorldWideWeb.

For example, Pocketmail (http://www.pocketmail.com) is a device which uses acoustic signals over ordinary phone sets to collect and send email. By using an 800 number in the United States, you can use your email while travelling without incurring long distance charges or finding a phone jack to hook a modem into. The keyboard is not suitable for writing long reports, but is functional for typing notes and email responses. The size limitation on email received on the pocketmail device is 4000 bytes, which covers about 80 percent of the traffic on the Internet. The cost for this device is US$150 and the email access is US$100/year, which is significantly cheaper than standard portable computer approaches to field work.

Other palmtop computers are being developed which provide cellular connections to email and the World Wide Web. Such systems depend on a cellular telephone infrastructure which is not fully deployed in North America. However, some systems are satellite based (such as Global Positioning Systems) and can be used anywhere.

Web-Based Email Addresses

Another option for staying in touch while on the road, or for separating your electronic correspondence into various segments, is a free web-based email address. A number of the major content providers on the Internet offer such services—Microsoft, Netscape, AltaVista, and Yahoo!, to name a few. As an example, HotMail (**http://www.hotmail.com**) allows you to create a free email address for yourself by registering on the HotMail web site. Once the email address has been created, you can send and receive email through the website. Your email address will be **user-id@hotmail.com**.

The big advantage to a web-based email account is that you can access it from any computer connected to the Internet—your own computer, a computer in one of your organization's locations, or a computer in a public library, at a conference, or a friend's house. The free email accounts can be used to keep your personal email separate from your professional email system; your professional email can be rerouted to your web-based account temporarily while you are going to be travelling; and your email address can stay the same if you change jobs, locations, or Internet service providers.

Naturally, there are a couple of disadvantages. The first is that using a web interface for accessing your email can be slower than traditional email. How much slower will depend on several factors: the time of day; the speed of your Internet connection; and, if you are accessing your mail from outside of North America, the Internet infrastructure in your country. As a general rule of thumb, accessing web-based email services is faster before noon, North American Eastern Standard Time (5:00 PM GMT), because many of the larger servers are located in the western United States.

Another disadvantage to these systems stems from the fact that these services are sustained by paid advertising. You have to view a number of online advertisements while using the service, and there is an advertisement for the email service itself tacked onto each message you send. In some situations, this is not a significant problem, but in other cases it can be rather annoying. Web-based email is probably not a practical approach for most people, but it can be a valuable tool in specific circumstances.

Major content providers are offering other free services to attract people to their site. For example, you can put your calendar online for access similar to that described above or participate in online auctions. All of these services are based on the presumption that daily use of the website is something that people are willing to commit to in order to use the services. This provides a reliable audience for the content providers to sell to advertisers.

Connection Speeds

As Internet use increases in popularity, the market in connection services has become competitive and connections speeds are increasing and cost is decreasing. Cable modems and satellite systems make speeds available to retail customers which were once thought of as being only for commercial purposes. This speed increase will enable new uses for the Internet, potentially in terms of live broadcasts over the Internet of many different audio/visual signals.

At the institutional level, networking speeds are increasing as well. One example is the Internet 2 system being developed by about 150 American educational institutions. This system will make available to members connection speeds and capacities which resemble those which usually exist on special wiring within single buildings. While access to this capability will be limited to the members of the group managing the system, it is a good example of how the Internet will develop in different directions depending on users' needs without losing touch with the rest of the world.

Accessing PDF files with Adobe Acrobat Reader

Many organizations are now making the full text of their OSH documents and publications available on their websites. Often these documents are made available as PDF files. What are PDF files and why are organizations using this format instead of creating ordinary web pages?

PDF files work best for organizations such as a government agency that has a series of printed publications that it would like to make available on its website. If these publications have been written over a number of years or by different departments, they probably exist in a variety of formats. Some may exist as word processor files, some may be in desktop publishing programs, and some may only be available on paper. Converting them into HTML (HyperText Markup Language, the standard language for web pages) could involve a significant amount of effort, depending on the format(s) of the documents.

A viable alternative is for the organization to convert the documents into Portable Document Format (PDF) files. Using Acrobat software from Adobe Inc., documents in a variety of formats—word processor or desktop publisher files or scanned images of paper documents—can be converted into PDF files. No reformatting or retyping is required. Once the documents have been saved as PDF files, they can be made available on the organization's website without any further effort.

How to View PDF Files

Organizations buy the Adobe Acrobat software which converts their documents into PDF files. The Adobe Acrobat Reader, the software used to view PDF files, is available free to anyone who wants it. The Acrobat Reader can be downloaded from a number of websites, including Adobe's own at **http://www.adobe.com**. When you install the free Acrobat Reader on your computer, it can be used as a "plug-in" to Microsoft Internet Explorer or Netscape Navigator. This means that when you come across a PDF file on a web site, your web browser will know automatically that it needs to use the Acrobat Reader to view it. Click on a link to the PDF file and the Acrobat Reader will be loaded automatically by the browser. The document will appear on your computer screen looking very much like the original document it was converted from. If you have a laser printer, you will be able to print a near original quality copy of the document.

What's the Catch?

PDF files are mostly a "win-win" approach to making information available on the web. An organization can make its material available on the web with very little effort, and anyone visiting the organization's web site can view and print the material in near original quality, using free software.

However, there are a couple of disadvantages to PDF files. Because they are essentially image files, PDF files are larger than web pages with the same amount of text. PDF files with text only can be 250KB or larger. PDF files must be downloaded in their entirety to your computer before you can view them. Regular web pages, on the other hand, display as they are loading. Secondly, unlike regular web pages, the content of PDF files is not regular text. Therefore, it is not usually indexed by web search engines such as AltaVista, nor will its content be searchable on an organization's own server. You will not be able to find on the Internet a document saved as a PDF file by using a search based on words and phrases in the document itself.

A double-edged aspect is that, because the file is graphically based, it is much more difficult to change what it says. Many organizations consider this a distinct advantage. However, many users prefer to modify documents collected from the web for their own purposes. It is possible to cut or copy text from Adobe files, but retaining the look and feel of the document requires extensive work.

The bottom line is that many government agencies and other organizations are making their documents available as PDF files. You will be at a distinct disadvantage if you do not install the Acrobat Reader on your computer and become comfortable using it. The Acrobat Reader is available for computers running various versions of Windows, for Apple Macintoshes, and for computers running other operating systems. It is available on the Adobe web site at **http://www.adobe.com/acrobat/**.

Free Databases and Document Collections

There is a general feeling that everything is free for the taking on the Internet. This is not entirely true. There are many subscription-based services on various websites. However, many databases and document collections that were formerly available only from commercial online search services or on CD-ROM are now accessible on the web. Most of these documents are from government or quasi-governmental agencies. Most of them are still available commercially, usually on CD-ROM, for people who do not want to rely on an Internet connection. This section describes a few of the "web bargains" that are of interest to health and safety professionals.

Chemical Carcinogenesis Research Information System (CCRIS)

http://toxnet.nlm.nih.gov

CCRIS is a Toxicology Data File on the US National Library of Medicine's (NLM) TOXNET® system. It is a scientifically evaluated and fully referenced data bank, developed and maintained by the National Cancer Institute (NCI). It contains some 8000 chemical records with carcinogenicity, mutagenicity, tumor promotion, and tumor inhibition test results. Data are derived from studies cited in primary journals, current awareness tools, NCI reports, and other special sources. Test results have been reviewed by experts in carcinogenesis and mutagenesis. It is available from the US National Library of Medicine's Toxicology Data Network (TOXNET).

Domestic Substances List (DSL) and the Non-Domestic Substances List (NDSL)

http://www2.ec.gc.ca/cceb1/cas_e.html

The Domestic Substances List (DSL) and the Non-Domestic Substances List (NDSL) were created in accordance with the Canadian Environmental Protection Act (CEPA) by Environment Canada. The DSL defines "existing" substances for the purposes of implementing CEPA and is the sole basis for determining whether a substance is *existing* or *new* to Canada. The NDSL specifies substances, other than those on the DSL, that were in world commerce, but not in Canada, and is based on the US Environmental Protection Agency's (EPA's) 1985 inventory compiled for the Toxic Substances Control Act (TSCA). Substances that are not on the DSL may require notification and assessment before they can be manufactured or imported into Canada. Substances on the NDSL have different notification requirements. The database is useful for chemical manufacturers, suppliers, and importers who may potentially have to submit information under CEPA, government regulators and enforcers, and regulatory compliance specialists. It is available on commercial CD-ROM collections.

ECDIN - Environmental Chemicals Data and Information Network

http://ecdin.etomep.net/

This database provides information on over 120,000 chemical compounds including all those listed in the European Inventory Substances (EINECS), which is the legal basis for the distinction between *new* and *old* substances in the European Community.

Hazardous Substances Data Bank® (HSDB)

http://toxnet.nlm.nih.gov

The Hazardous Substances Data Bank® (HSDB) contains comprehensive data profiles on the toxicity and biomedical effects of over 4,500 potentially toxic chemical substances. It is created and updated by specialists at the US National Library of Medicine (NLM). Compiled from an extensive range of authoritative sources, HSDB is widely recognized as a reliable and practical source of health and safety information. Data are derived from a core set of standard texts and monographs, government documents, technical reports, and the primary journal literature. Much of the data is scientifically peer reviewed. Each chemical record contains up to 150 data fields arranged into 10 broad subject categories. Complete references to the data sources are provided. HSDB is built, maintained, reviewed, and updated on the US National Library of Medicine's Toxicology Data Network (TOXNET).

IARC Monographs

http://193.51.164.11/monoeval/allmonos.html

IARC Monographs Programme on the Evaluation of Carcinogenic Risks to Humans. The IARC Monographs series publishes authoritative independent assessments by international experts of the carcinogenic risks posed to humans by a variety of agents, mixtures, and exposures. Since its inception in 1972, the series has reviewed more than 800 agents, and IARC Monographs have become well-known for their thoroughness, accuracy, and integrity. To aid in the selection of future topics, the programme also monitors long-term carcinogenicity testing underway in various laboratories throughout the world and publishes the results as a Directory of Agents Being Tested for Carcinogenicity.

Integrated Risk Information System (IRIS)

http://www.epa.gov/iris/

IRIS contains data concerning human health effects that may result from exposure to various chemicals in the environment. It provides summaries of health risks and EPA regulatory information on over 450 specific chemicals. It contains EPA consensus opinion on potential chronic human health effects based on chemical hazard identification and dose-response assessment.

International Chemical Safety Cards (ICSC)

http://www.cdc.gov/niosh/ipcs/icstart.html#english

An ICSC summarizes essential health and safety information on chemicals for their use at the "shop floor" level by workers and employers in factories, agriculture, construction, and other work places. The ICSCs project is an undertaking of the International Programme on Chemical Safety (IPCS). The project is being developed in the context of the cooperation between the IPCS and the Commission of the European Communities. The IPCS is a joint activity of three cooperating International Organizations: namely, the United Nations Environment Programme (UNEP), the International Labour Office (ILO), and the World Health Organization (WHO). The main objective of the IPCS is to carry out and disseminate evaluations of the hazards posed by chemicals to human health and the environment.

MEDLINE (MEDlars onLINE)

http://igm.nlm.nih.gov/

MEDLINE is the National Library of Medicine's (NLM) premier bibliographic database covering the fields of medicine, nursing, dentistry, veterinary medicine, the health care system, and the preclinical sciences. It contains bibliographic citations (*e.g.*, author, title, and journal reference) and author abstracts from over 3,900 biomedical journals published in the United States and 70 foreign countries during the previous four years. In total, it contains over 9 million records dating back to 1966. MEDLINE is available from the US National Library of Medicine's website. According to NLM, the average cost for searching MEDLINE using non-Web based access methods is $32/hour.

New Jersey Right-to-Know Hazardous Substance Fact Sheets

http://www.state.nj.us/health/eoh/rtkweb/rtkhsfs.htm

Prepared for 1,055 individual hazardous chemicals, this site contains acute and chronic health hazards, identification, workplace exposure limits, medical tests, workplace controls and practices, personal protective equipment, handling and storage, questions and answers, definitions, and emergency response information for fires, spills, and first aid. 235 fact sheets have been translated into Spanish. A minimum of 100 fact sheets are revised every year.

NIOSH Manual of Analytical Methods

http://www.cdc.gov/niosh/nmam/nmammenu.html

NMAM is the result of part of the research activities of NIOSH relating to the determination of workplace contaminants. NMAM is a collection of methods for sampling and analysis of contaminants in workplace air and in the blood and urine of workers who are occupationally exposed. These methods have been developed specifically to have adequate sensitivity to detect the lowest concentrations as regulated by OSHA and recommended by NIOSH, and sufficient range to measure concentrations exceeding safe levels of exposure. The methods have been developed or adapted by NIOSH or its contractors and have been evaluated according to established experimental protocol and evaluation criteria. NMAM also includes chapters on quality assurance, strategies for sampling airborne substances, method development, and discussions of some portable direct-reading instrumentation. NMAM is still available commercially from several vendors for $600 or more.

NIOSH Pocket Guide to Chemical Hazards

http://www.cdc.gov/niosh/npg/npgd0000.html

The NIOSH Pocket Guide to Chemical Hazards database is a practical source of information for recognizing and controlling workplace chemical hazards. It lists workplace exposure limits, recommends respirator selections and protective measures, identifies signs and symptoms of exposure, and describes first aid treatment for hundreds of critical workplace chemicals. It also provides information on measurement methods, chemical and physical properties, incompatibilities, and reactivities.

1996 North American Emergency Response Guidebook

http://www.tc.gc.ca/canutec/english/guide/menug_e.htm

The 1996 North American Emergency Response Guidebook (NAERG96) was developed jointly by Transport Canada (TC), the U.S. Department of Transportation (DOT), and the Secretariat of Communications and Transportation of Mexico (SCT) for use by fire fighters, police, and other emergency services personnel who may be the first to arrive at the scene of a transportation incident involving dangerous goods. It is primarily a guide to aid first responders in quickly identifying the specific or generic hazards of the material(s) involved in the incident and protecting themselves and the general public during the initial response phase of the incident.

OSHA Analytical Methods

http://www.osha-slc.gov/html/dbsearch.html

Analytical methods for organic and inorganic chemicals. The organic and inorganic methods are available commercially for $900 or more.

TOXLINE®

http://igm.nlm.nih.gov/

TOXLINE is the US National Library of Medicine's extensive collection of online bibliographic information covering the biochemical, pharmacological, physiological, and toxicological effects of drugs and other chemicals. TOXLINE and its backfile TOXLINE65 together contain more than 2.5 million bibliographic citations, almost all with abstracts and/or indexing terms and CAS Registry Numbers. TOXLINE is available from the US National Library of Medicine's website.

7 A Web Safety and Health Directory

While the Internet is changing daily, with information resources being added, moved, or lost on a regular basis, a snapshot of the Internet at a particular point in time is helpful in showing its value. The Web sites and discussion groups listed below are a sampling of those that were active in March 1999. Most of them can expected to be stable over time. Check in at the Internet pointer sites to see updates, and be prepared for pleasant surprises!

Web sites are categorized according to their content. For each site the Uniform Resource Locator (URL, the address of the site on the World Wide Web) is given, followed by a brief description, usually based on the Table of Contents found at that URL. Remember that almost all of these sites include pointers to other sites on the Internet with related information. So, even if a particular site sounds like it may be of marginal interest, it may be worth checking to see what other sites it points to.

This list does not attempt to be comprehensive. Also, since safety subjects overlap freely, many sites are listed in more than one category. The idea is to give you places to start your Web surfing. If you are looking for a specific piece of safety information, follow the outline in Chapter 3 to find it as efficiently as possible.

Section A

General Health and Safety Information

1. General Health and Safety Information and Statistics

Conferences

1997 Australian Vice-Chancellors' Committee (AVCC) OH&S Conference

http://www.adelaide.edu.au/HR/OH&S/avccohs/conference.htm

Read about and download the proceedings from this conference.

Institution of Occupational Safety and Health (IOSH)

http://www.iosh.co.uk/

The site contains information on all the major areas of activity that this U.K.-based Institution is involved in membership and career information documents, U.K. conferences, etc.

Safety98

http://www.safety98.org

This is a health and safety "cyberconference," held November 2 to 13, 1998. "The objective of Safety98 is to raise the profiles of engineers and scientists in health and safety and, in so doing, reach as wide an audience as possible."

Consulting Services

CAE Consultants

http://www.caeconsultants.com/

This consulting company prepares custom documents such as standard operating procedures, guidelines, process documents, etc.

Center for Hazard Information

http://www.hazardinfo.com/

This professional safety engineering firm provides consulting services, plus a monthly *Hazard Information Newletter*. Read about the company's services, and how to subscribe to the newsletter.

Center for Safety and Environmental Management

http://www.nauticom.net/www/csem/

The site contains descriptions of the company's environmental safety training, onsite staffing, consulting and field services.

Ceske pracovni zdravi a bezpecnost

http://www.markl.cz

Czech occupational health and safety. The site contains a description of this consulting company's products, a directory of OSH organizations in the Czech republic, Czech OSH legislation, etc. All information is available in Czech and some in English.

Commission de la santé et de la sécurité du travail du Québec (CSST)

http://www.csst.qc.ca/

The site describes the role and services of the Commission, the laws it administers, etc. Other information includes CSST news releases, summaries of their magazine *Prévention au travail* and a publications list. Most information is in French and some is also in English.

Compliance Management Group—OSHA Safety Plans and Labor Law Posterbook

http://www.oshaman.com/

This company sells written policy and procedures documents, complete safety plans, training materials, training videos, CDs, labor law posters, etc. View details of their products.

Comprehensive Loss Management, Inc.

http://www.clmi-training.com

This company produces safety and ergonomic training programs to help companies achieve OSHA compliance.

EHS Diagnostics Inc.

http://www.ehsdiagnostics.com

EHS Diagnostics Inc. (EDI) is a Canadian environment, health and safety consulting firm, based in Calgary, Alberta, Canada, which provides both consulting services and health and safety software development. Read about the company's services.

EHS Management Technologies Private Limited, India

http://www.corporatepark.com/ehsindia/

Read about this consulting company's environmental, health and safety, and management services.

Environmental Safety Services

http://www.envirosafety.qpg.com/

Read about the company's environmental and health and safety consulting services.

Environmental Safety Services

http://www.ps.uga.edu/ess/

Read about the company's consulting services.

Front-Line Safety

http://www.frontlinesafety.com/

This Nova Scotia-based company provides health and safety training programs, safety equipment rentals, consulting/safety audits, etc. They also provide an online WHMIS course.

Hascom

http://www.hascom.com

Find out about this U.K.-based health and safety consultancy. Information is provided about the company itself, its services, links to selected sites and an online newsletter. The newsletter is particularly useful to those involved in health and safety in the U.K.

Hazard Alert Training Supplies Canada Inc. (HATSCAN)

http://www.hatscan.com/

HATSCAN is an occupational health, safety, and environment training and consulting firm, providing consulting and training in WHMIS, TDG, OH&S, due diligence, WCB, environmental compliance, hazardous waste, and more. Find out about the company's courses, publications, and computer assisted training.

HAZCON Pty. Ltd.

http://www.dcscomp.com.au/hazcon/

Information about this Victoria, Australia-based environmental and occupational health and safety consulting company and its services.

HSE Group, Inc.

http://www.hsegroup.com/

The HSE Groups is an international consortium of health, safety, and environmental consultants. Read about available services, Web-based training, the HSE book store, alerts, etc. "Fun" MSDSs for chocolate and water are also available here.

JBF Associates Inc.

http://www.jbfa.com

JBFA is "an engineering firm that specializes in process hazard analysis, reliability and availability engineering, system safety, environmental engineering, process safety management, risk management planning and communication, quantitative risk analysis, and incident investigations." (Now known as **Risk and Reliability Division, ABS Group Inc.**)

Medical Device Link—Safety Consultants

http://www.devicelink.com/consult/Safety.html

This directory provides descriptions of companies that provide safety services.

National Environmental Safety Compliance, Inc.

http://www.albany.net/~nesc/

NESC specializes "in Health, Safety and Environmental Issues through training, consulting and site inspections." As well as providing information about the company's services, they provide some information files and links.

Nationwide Safety Seminars

http://www.safety4u.com/

This firm provides electrical safety training programs, expert witness services, safety consulting, etc.

Occupational Safety Consultants

http://www.mindspring.com/~stephens/oschome.html

Information about the company's services is provided.

OH&S Consulting Services

http://www.hronline.com/dr/

Dilys Robertson is a Toronto-based occupational health and safety consultant and writer. Read about her consulting services, which include training design and delivery and research work with joint health and safety committees.

OHS Associates, Inc.

http://www.ohsa.com/

Read about the company's services: occupational health and safety consulting services and training, NFPA ratings, HMIS ratings, litigation support services, etc.

Paint-Safe(R)

http://www.ibpat.net/paintsafe/

This Web page describes Paint-Safe(R), a company that appears to do health and safety research.

Pathfinder Associates

http://www.webcom.com/pathfndr/welcome.html

The company's Web pages provide current information on OSHA regulations, citations, etc. There is also company product and service information, various health and safety documents, etc.

Preventive Action Safety Services Ltd.

http://www.autobahn.mb.ca/~passltd/

Read about this Winnipeg Canada based health and safety company and its services.

Pro-Safety Inc.

http://members.aol.com/safetypros/prosafe.htm

Read about this company's safety services and programs, and its various safety software packages.

Professional Development Unit, uksafety.net

http://www.uksafety.net/

This U.K. safety and technical training consultancy's site provides a database of U.K. consultants; a bulletin board that includes "hot gossip," frequently-asked-questions, news of upcoming events, and a "trading post"; and a catalogue of the company's training programmes.

Professional Safety Consultants Inc.

http://www.pscsafe.com

The site provides information about the staff and services if this occupational safety consulting company.

Quality Safety Services Inc.

http://www.qssi-tx.com/

This Texas-based consulting company provides OSHA, EPA, DOT, and TNRCC safety and environmental training programs, plant safety inspections and accident prevention program surveys. Read about its services.

RACCO Safety

http://www.raccosafety.com.br/

The site describes this Brazilian company's safety, industrial hygiene, and occupational health products and services. Most information is in Portuguese, with a general description in English.

Resource Environmental Associates

http://www.rea4ehs.com/

Information is provided about this Toronto, Canada-based environment, health and safety consulting company, its services, publications, etc.

Safety Advisory Services

http://www.safetyadvisory.com/

Read about the company's loss control and substance abuse programs.

Safety Check Consulting

http://www.lights.com/safetycheck/

Information about this Saskatchewan, Canada-based safety consulting company's services: safety audits, training programs, emergency preparedness, accident/incident investigation, etc.

Safety Online

http://www.safetyonline.net

This virtual mall provides information about products and services in a variety of health and safety areas. Excerpts from, and subscription for *Industrial Safety and Hygiene News* are provided. View the full text of their *Safety Currents* newsletter.

Safety Systems Inc.

http://www.safetysystemsinc.com

The site describes this environmental health and safety consulting company's services.

Safety Training and Consulting Services

http://www.stacs.com

The site provides information about this Canadian consulting company. Its services include training, consulting, and recruiting.

Safety Training and Tracing, Inc.

http://www.standt.com/

The company's OSH consulting services and software "Saf-T-Man Safety Manager" are described.

Windsor Occupational Health Information Service (WOHIS)

http://www.mnsi.net/~wohis/

WOHIS is an occupational health and safety information/referral service located in Windsor, Canada. Its mandate is "to promote community awareness of occupational health and safety issues in order to prevent illness and injury." Read about its services, view publications, etc.

Databases

Asbestos Institute

http://www.asbestos-institute.ca

Access the Montreal-based Institute's Biomedical Data Bank, view reports and press releases, and get information on their printed publications.

Asociacion para la Prevencion de Accidentes

http://www.safetyonline.net/apa/home.htm

Information is provided about this Spanish association, its services (training, databases, publications, etc.), subscriptions to APA's journal *Prevencion*, a series of safety posters, and conference information. All information is in Spanish.

CCINFOWeb

http://ccinfoweb.ccohs.ca

This is the home page for the commercial OHS database service on the Web, provided by the Canadian Centre for Occupational Health and Safety (CCOHS). Databases include: MSDS, CHEMINFO, RTECS, NIOSHTIC, and HSELine, plus the full text of all Cdn. environmental and OSH legislation. Other databases will be added on an ongoing basis.

Health and Safety Promotion in the European Union

http://www.hsa.ie/hspro/index1.html

HSPro - EU will set up a remotely accessible telematics system providing a "one-stop shop" for occupational health and safety information. Categories of information include news and events, guidance, OSH information, OSH databases, bibliographies, library catalogs, publications, etc. More detail is available in the Top 50 Sites section.

National Technical Information Service (NTIS)

http://www.ntis.gov/health/health.htm

The site provides information about this U.S. Government agency's health and safety publications, databases, and other information.

RILOSH database

http://www.library.ryerson.ca/molndx

Search this database of over 140,000 records related to labour relations and occupational health and safety. Created by the Ontario Ministry of Labour, it is now maintained by Ryerson Polytechnic University, Toronto (formerly MOLINDEX).

SilverPlatter

http://www.silverplatter.com

The site provides information about databases commercially available from SilverPlatter, both on CD-ROM and via the Internet.

Directories

Ceske pracovni zdravi a bezpecnost

http://www.markl.cz

Czech occupational health and safety. The site contains a description of this consulting company's products, a directory of OSH organizations in the Czech republic, Czech OSH legislation, etc. All information is available in Czech, and some in English.

Directory of Safety Directors on the Net

http://www3.seton.com/safetydirectors.html

Maintained by Seton Identification Products, this growing directory provides contact information about Safety Directors at various companies academic institutions and other organizations.

International Labour Organization

http://www.ilo.org/

The site provides information about the ILO, its various international programs, publications, etc. There are also links to ILO departmental home pages, plus an ILO directory.

Medical Device Link—Safety Consultants

http://www.devicelink.com/consult/Safety.html

This directory provides descriptions of companies that provide safety services.

Oregon Occupational Safety and Health Division (OR-OSHA)

http://www.cbs.state.or.us/external/osha/index.html

The site contains a directory of OR-OSHA's services, news releases, an employer's toolkit, information about codes, publications, and videos for sale, etc. The *Oregon Health and Safety Resource*, OR-OSHA's newsletter, is also accessible.

Pro-Am Industrial Safety and Hygiene

http://www.pro-am.com

This is one of the best developed "safety malls" on the Web. Find products, services, links to OSHA standards, and a good directory of Internet resources.

Robertson's Safety and Health Information and Resources

http://www.public.usit.net/jerrywr/

This is a directory of Internet safety and health resources.

Safety, Health and Environmental Consultants

http://siri.uvm.edu/consultants/CONSULT.HTM

This is a directory of North American consultants. Descriptions of their services and links to their Web sites are provided.

Seton Identification Products

http://www.seton.com

This extensive, well-organized site provides detailed information about Seton's safety products, MSDSs for its chemical products, a health and safety resources directory, etc.

SilverPlatter Health and Safety World

http://www.silverplatter.com/oshinfo.htm

These pages, edited by Sheila Pantry, provide links to health and safety resources on the Internet, an editorial, a diary of upcoming events, plus a large directory of links to health and safety sites.

Univ. of Minnesota, Department of Environmental Health and Safety

http://www.dehs.umn.edu/

The site provides access to numerous policy documents, guidelines, manuals, etc., plus a directory of Internet resources.

Working Conditions and Work Practices: Occupational Safety and Health

http://ext.ilr.cornell.edu/osh/default.html

Cornell University's School of Industrial and Labor Relations presents an archive of OSH-related educational materials, NY state resources for OSH educators and practitioners, a directory of OSH professional and advocacy organizations, and an online forum for OSH educators and practitioners.

Educational Resources / Training

Affirmed Medical Inc.

http://www.affirmed.com

The company provides first aid and safety products. Find out about these, plus their U.S. regulatory compliance products and training videos.

American Safety and Health Training

http://www.safety-training.com

Information about this nonprofit organization's multimedia, computer-based training packages and medical monitoring services is provided.

Arbetslivsinstitutet

http://www.niwl.se/niwl.htm

The National Institute of Working Life (Arbetslivsinstitutet) is Sweden's Research and Development center for occupational health and safety, working life, and the labor market. Access is provided to information about the Insitute, its library catalog, and the full text of its publications.

BLMC Computer Based Training Programs

http://www.blmc.com/

Information is provided about this Canadian company's computer-based health and safety training programs.

BNA Communications Safety Training Programs

http://www.bna.com/bnac/sfprod/sftymain.html

Descriptions of the company's many video-based training programs are available.

Canadian Society of Safety Engineering (CSSE)

http://www.csse.org

Information is provided about the society, its programs, publications, courses, etc.

Christie Communications

http://www.mrg.ab.ca/christie/

Safety training courses and a list of safety-related Internet resources.

Cornell OSH Archive of Educational Materials

http://ext.ilr.cornell.edu/osh/edu/occindex/srchindex.html

Cornell University's School of Industrial and Labor Relations, OSH Extension, is piloting an online archive of public domain and freeware occupational safety and health educational materials and policies. Find training materials on a variety of OSH subjects.

Diving Safety Office, East Carolina University

http://ecuvax.cis.ecu.edu/academics/schdept/diving/diving.htm

"The purpose of this office is to ensure safe and effective diving practices and to provide support for the research, educational, and recreational diving programs on campus."

Don Brown Productions Inc.

http://www.digitaldbp.com/

This company produces health and safety training videos. The site includes its catalog of over 800 titles.

Griffith University, Faculty of Health and Behavioural Sciences

http://www.ua.gu.edu.au/hbk/hbs_ind.htm

Information about courses provided by this Queensland, Australia University is available. The School of Occupational Health and Safety is within this faculty.

HAZCON Pty. Ltd.

http://www.dcscomp.com.au/hazcon/

Information about this Victoria, Australia-based Environmental and Occupational Health and Safety Consulting company and its services.

HSE Group, Inc.

http://www.hsegroup.com/

The HSE Groups is an international consortium of health, safety, and environmental consultants. Read about available services, Web-based training, the HSE book store, alerts, etc. "Fun" MSDSs for chocolate and water are also available here.

Industrial Accident Prevention Association (IAPA)

http://www.iapa.on.ca

Information is provided about this Ontario, Canada safety association, its products, services, and training programs.

Lenco Enterprises

http://www.lara.on.ca/~lenco/

The site describes this Hamilton, Canada-based consulting company. Lenco gives seminars and training on electrical safety, lockout/tagout, etc.

National Environmental Safety Compliance, Inc.

http://www.albany.net/~nesc/

NESC specializes "in Health, Safety and Environmental Issues through training, consulting and site inspections." As well as providing information about the company's services, they provide some information files and links.

National Environmental Training Association

http://ehs-training.org

Read about the National Environmental Training Association, an international, educational, professional association of specialists who design, deliver, and manage environmental, health and safety training.

Occupational Training, Inc.

http://www.otrain.com

Read about this Hawaii-based OSH training company and its courses.

Oklahoma State University, Environmental Health and Safety, Online Training Modules

http://www.pp.okstate.edu/ehs/MODULES/HOME.HTM

The site provides online training for a growing number of EHS topics, including asbestos awareness, bloodborne pathogens, compressed gas cylinders, fire extinguishers, hazard communications, office ergonomics, personal protective equipment in laboratories, etc. More detail about this site is available in the Top 50 Sites section.

Quality Safety Services Inc.

http://www.qssi-tx.com/

This Texas-based consulting company provides OSHA, EPA, DOT, and TNRCC safety and environmental training programs, safety plant inspections and accident prevention program surveys. Read about its services.

Rocky Mountain Center for Occupational and Environmental Health

http://rocky.utah.edu/

Located on the campus of the University of Utah, the Rocky Mountain Center is one of 14 NIOSH Educational Research Centers. The site provides information about the center's various educational programs.

Safety Check Consulting

http://www.lights.com/safetycheck/

Information about this Saskatchewan, Canada-based safety consulting company's services: safety audits, training programs, emergency preparedness, accident/incident investigation, etc.

Safety Systems Inc.

http://www.safetysystemsinc.com

The site describes this environmental health and safety consulting company's services.

Safety Training and Consulting Services

http://www.stacs.com

The site provides information about this Canadian consulting company. Its services include training, consulting, and recruiting.

Safety Training and Tracing, Inc.

http://www.standt.com/

The company's OSH consulting services and "Saf-T-Man Safety Manager" software are described.

SafetyCare Group of Companies

http://www.safetycare.com.au

SafetyCare is a worldwide supplier of occupational health and safety video programs, courses and manuals. View their online video catalog and their list of offices in various countries.

SafetyLine Institute

http://www.safetyline.wa.gov.au/institute/

"The SafetyLine Institute is an online education and training facility established by WorkSafe Western Australia to provide high quality education and training materials in occupational safety and health." Available courses include OSH Management, OSH Information Sources, Safe Systems of Work, OSH Promotion, etc.

Talk Tools

http://www.talktools.com/

"The Talk Tools program was designed by safety and communication experts to help supervisors conduct more professional and effective safety meetings." Read about the program, which includes storyboards, interactive safety talks, etc.

University of New South Wales, Department of Safety Science

http://argus.appsci.unsw.edu.au/

Read about the department's courses, research, etc.

University of Wisconsin-Milwaukee, Environmental Health, Safety, and Risk Management

http://www.uwm.edu/Dept/EHSRM/

Information about the university's health and safety policies and guidelines is provided, along with links to other organizations.

Journals and Newsletters

African Newsletter on Occupational Health and Safety

http://www.occuphealth.fi/eng/info/anl/

The full text of current and previous issues is available at the site. The newsletter is published by the Finnish Institute of Occupational Health, in association with the ILO - CIS International Occupational Health and Safety Information Centre.

American Society of Safety Engineers, Puget Sound Chapter

http://www.exit-18.com/PSC/

Access to this ASSE chapter's newsletter and links to other related sites.

American Society of Safety Engineers, Greater San Jose Chapter

http://www.best.com/~assegsjc/

Access the chapter's newsletter, information on classes, links to other ASSE sites, and selected links to other sites.

American Society of Safety Engineers, South Texas Chapter

http://www.connecti.com/~safety/

Information is provided about this ASSE chapter. View its newsletter and selected documents produced by the chapter.

American Society of Safety Engineers, Western Pennsylvania Chapter

http://www.wpa-asse.com/

The site provides membership information, descriptions of available jobs, links to other sites, the chapter's newsletter, etc.

Arbill Industries

http://www.arbill.com

Information is provided about the company's health and safety products. You can view its newsletter, order products, etc.

Asian-Pacific Newsletter on Occupational Health and Safety

gopher://gopher.nectec.or.th/11/bureaux/ASIA-OSH/newsletter

The full text of the current and all past issues of the newsletter is available online. The newsletter is produced by the Asian-Pacific Regional Network on Occupational Safety and Health, a project of the International Labour Organization.

Center for Hazard Information

http://www.hazardinfo.com/

This professional safety engineering firm provides consulting services, plus a monthly *Hazard Information Newletter*. Read about the company's services and how to subscribe to the newsletter.

Estonian Newsletter on Occupational Health and Safety

http://www3.occuphealth.fi/eng/info/estonia/

The full text of this newsletter is available online. It is published 3 to 4 times a year by the Estonian Centre of Occupational Health, Estonian Institute of Experimental and Clinical Medicine and the Finnish Institute of Occupational Health. It is published in English, Estonian, and Russian.

FacilitiesNet

http://www.facilitiesnet.com

FacilitiesNet is "an interactive information service for facility professionals." Members can access two online magazines, reports, an online bookstore, product information, etc.

Hascom

http://www.hascom.com

Find out about this U.K.-based health and safety consultancy. Information is provided about the company itself, its services, links to selected sites, and an online newsletter. The newsletter is particularly useful to those involved in health and safety in the U.K.

Oregon Occupational Safety and Health Division (OR-OSHA)

http://www.cbs.state.or.us/external/osha/index.html

The site contains a directory of OR-OSHA's services, news releases, an employer's toolkit, information about codes, publications, and videos for sale, etc. The *Oregon Health and Safety Resource*, OR-OSHA's newsletter, is also accessible.

OSHNet, The Internet Magazine for Occupational Safety and Health

http://www.oshnet.com

Access the full text of articles an events list and links to other health and safety sites.

Safety Online

http://www.safetyonline.net

This virtual mall provides information about products and services in a variety of health and safety areas. Excerpts from, and subscription for *Industrial Safety and Hygiene News* are provided. View the full text of their *Safety Currents* newsletter.

Safety+Health

http://www.nsc.org/pubs/sh.htm

Subscription information is provided for this U.S. National Safety Council publication. There is also information about upcoming events and news, plus a series of "Healthgrams" and "Safety Clips," intended for use in employee publications and safety bulletins.

Uni-Hoist

http://www.uni-hoist.com

Find out about the company's confined space products. The site also includes a photo gallery, links to related articles and sites, OSHA regulations, a newsletter, etc.

Policy / Procedures Documents

CAE Consultants

http://www.caeconsultants.com/

This consulting company prepares custom documents such as standard operating procedures, guidelines, process documents, etc.

ERDEC Safety Office

http://www.apgea.army.mil/RDA/erdec/risk/safety/

The Edgewood Research Development and Engineering Center (ERDEC) provides health and safety information and services to its clientele, primarily the U.S. Armed Forces. Read about its services.

Louisiana State Univ., Office of Occupational and Environmental Safety

http://130.39.86.58/page1.htm

Documents describing the office's services are available. Subject areas include safety and environmental program news, emergency procedures and information, safety and environmental program documents and information, etc.

McGill University, Environmental Safety Office

http://www.mcgill.ca/eso/

This site provides information about staff and services provided by the office, links to McGill's Occupational Health Programme, and various McGill policy documents, including its Laboratory Safety Manual.

McMaster Univ., Faculty of Engineering and Science, Safety Handbook

http://www.eng.mcmaster.ca/safety/

This is the full text of the *Safety Handbook* from this Hamilton, Canada university.

NASA/LeRC Office of Safety and Mission Assurance

http://www-osma.lerc.nasa.gov

Information about the office itself is provided. Links to many related NASA resources, such as the *Spaceflight Safety Handbook.*

Office of the Assistant Deputy Under Secretary of Defense for Safety and Occupational Health — ADU.S.D(SH)

http://www.acq.osd.mil:80/ens/sh

Information is provided about safety in the U.S. military: programs and policy documents, statistics, etc.

Outdoor Workers Exposed to UV Radiation and Seasonal Heat

http://www.usyd.edu.au/su/ohs/outdoor.html

This is the full text of the University of Sydney (Australia) policy document.

Portland State University, Environmental Health and Safety

http://www.pp.pdx.edu/FAC/Safety/

The site provides access to a variety of university documents, such as their chemical hygiene program, lockout-tagout, pesticide/herbicide use, etc.

Princeton University, Occupational Health and Safety

http://www.princeton.edu/~ehs/

Information is provided about OH&S at Princeton, there are links to numerous MSDS collections, and there is a Web version of their health and safety manual.

Queen's University, Environmental Health and Safety

http://www.safety.queensu.ca:443/safety/safety.htm

The site provides access to a number of policy documents and guidelines, plus links to MSDS sites, other sites of interest, etc.

Sonoma State University, EHS

http://www.sonoma.edu/EHS/

This site is maintained by the university's environmental health and safety department. It includes the full text of many of its health and safety programs.

University of Arkansas, Environmental Health and Safety Page

http://www.phpl.uark.edu/ehs/ehs.html

The site contains a number of policy and guideline documents: radiation safety, chemical information, etc.

University of Calgary, Safety Office

http://www.acs.ucalgary.ca/~ucsafety/

View policy and procedure documents, training course information, etc.

Univ. of California Irvine, Environmental Health and Safety Department

http://www.abs.uci.edu:80/depts/ehs/

View the university's EHS policies and procedures documents, emergency procedures, safety news, etc.

University of Colorado at Colorado Springs, Risk Management and Environmental Safety

http://www.uccs.edu/~pusafety/risk1a.htm

Available documents describe health and safety policies and procedures at UCCS.

University of Memphis, Environmental Health and Safety

http://www.people.memphis.edu/~ehas/

Information is provided about the department's services. The full text of its policy documents is available. There are also links to U.S. regulatory sites, MSDSs, etc.

Univ. of Minnesota, Department of Environmental Health and Safety

http://www.dehs.umn.edu/

The site provides access to numerous policy documents, guidelines, manuals, etc., plus a directory of Internet resources.

University of Sydney, Occupational Health and Safety

http://www.usyd.edu.au/su/planning/ohs/index.html

View the University's health and safety policy and staff guide, information about the Risk Management Office, accident investigation, etc. A series of policy documents are also presented: asbestos safety, children on campus, confined spaces, manual handling, disposal of sharps, etc.

University of Utah, Environmental Health and Safety

http://www.ehs.utah.edu/

The site provides information about the department, the full text of numerous policy and procedure documents, waste minimization documentation, pollution prevention information, links to other sites, etc.

University of Waterloo, Safety Office

http://www.adm.uwaterloo.ca/infohs/homepage.html

The site provides access to information about the safety office, plus documents such as its *Health and Safety Program Manual*, *Radiation Manual*, and *Laser Safety Manual*.

University of Wisconsin System Administration, Office of Safety and Loss Prevention

http://www.uwsa.edu/oslp/oslp.htm

The page contains links to environmental health and safety, workers compensation, and risk management. Manuals and policy documents are presented.

University of Wisconsin-Milwaukee, Environmental Health Safety and Risk Management

http://www.uwm.edu/Dept/EHSRM/

Information about the University's health and safety policies and guidelines is provided, along with links to other organizations.

Products

31 Mile Equipment Co.

http://www.31mile.com

Read about the company's safety products and services.

Aearo Company

http://www.aearo.com

This site provides information about the company's safety products. Aearo was formerly Cabot Safety.

Affirmed Medical Inc.

http://www.affirmed.com

The company provides first aid and safety products. Find out about these, plus their U.S. regulatory compliance products and training videos.

Arbill Industries

http://www.arbill.com

Information is provided about the company's health and safety products. You can view its newsletter, order products, etc.

BNA Communications Safety Training Programs

http://www.bna.com/bnac/sfprod/sftymain.html

Descriptions of the company's many video-based training programs are available.

Ceske pracovni zdravi a bezpecnost

http://www.markl.cz

Czech occupational health and safety. The site contains a description of this consulting company's products, a directory of OSH organizations in the Czech republic, Czech OSH legislation, etc. All information is available in Czech, and some in English.

ChinaHawk Enterprises Ltd.

http://home.hkstar.com/~chinahwk/home.html

Information about this Beijing-based company's disposable safety products for the health care, industrial and food services industries. Offices are in Beijing and Hong Kong.

Compliance Management Group—OSHA Safety Plans and Labor Law Posterbook

http://www.oshaman.com/

This company sells written policy and procedures documents, complete safety plans, training materials, training videos, CD's, labor law posters, etc. View details of their products.

Conney Safety Products

http://www.conney.com

The site provides a portion of the company's catalog of "Safety, Industrial, Maintenance, and First Aid Supplies for the Workplace and Home Environments."

Crowcon Gas Detection

http://www.crowcon.com/

View information about gas detection technologies, plus the (U.K.) company's list of international distributors.

FacilitiesNet

http://www.facilitiesnet.com

FacilitiesNet is "an interactive information service for facility professionals." Members can access two online magazines, reports, an online bookstore, product information, etc.

Government Institutes Division, ABS Group Inc.

http://www.govinst.com/

This company provides over 200 books on topics in the environmental, health, safety, telecommunications, and Internet fields. They also provide training courses, videotapes, computer based training, etc.

GPS Gas Protection Systems Inc.

http://www.GasProtection.com/

The site provides information about the company's automatic gas shutoff systems.

Lake Marking Products

http://www.lake-marking.com

See the company's catalog of signs, equipment tags, etc.

Maxi-Signal Products Co.

http://www.computer.net/~owlseye/maxi.html

This site provides access to Maxi-Signal's product catalog of heavy-duty, visible and audible signals, sirens and horns, lighting, and safety equipment.

Mettam Safety Supply

http://www.mettam.com

Read about the company and its products. There are also links to selected U.S. resources.

National Safety Equipment Outlet

http://www.safety-source.com/

Product information from this mail-order company is provided.

Pathfinder Associates

http://www.webcom.com/pathfndr/welcome.html

The company's Web pages provide current information on OSHA regulations, citations, etc. There are also company product and service information, various health and safety documents, etc.

Pro Grip Safety Sole

http://www.uleth.ca/~gadd/index.htm

Information about Pro-Grip nonslip safety soles.

Pro-Am Industrial Safety and Hygiene

http://www.pro-am.com

This is one of the best developed "safety malls" on the Web. Find products, services, links to OSHA standards, and a good directory of Internet resources.

Pro-Am Safety Product Store

http://www.pro-am.com

This is a "safety cybermall" with a large number of safety products available for sale.

Production Video L. M. Inc.

http://www.generation.NET/~pvlm/

This commercial site provides detailed information about Production Video's occupational health and safety training videos.

Protecta Fall Arrest Systems

http://www.protecta.com

Detailed information is provided about the company's fall protection, confined space entry and retrieval equipment, rescue devices, and heavy-load fall arrestors.

Quick-Aid, Inc.

http://www.quick-aid.com

Read about products and services from this medical and safety supply company.

RACCO Safety

http://www.raccosafety.com.br/

The site describes this Brazilian company's safety, industrial hygiene, and occupational health products and services. Most information is in Portuguese, with a general description in English.

Safeco Inc.

http://www.safecoinc.com/

Information is provided about this distributor of health and safety products, its distributors, product catalog, etc.

Safety Connection

http://www.safetydeck.com

Information about the company's products, plus selected links to other sites, is provided.

Safety Solutions Inc.

http://safetysolutions.com/

Find out about this full range safety products company, view product descriptions, ordering information, etc.

Safety Superstore

http://www.safetysuperstore.com

The site contains this company's catalog of safety products. This Canadian company will ship anywhere in the world.

SafetyCare Group of Companies

http://www.safetycare.com.au

SafetyCare is a worldwide supplier of occupational health and safety video programs, courses, and manuals. View their online video catalog and their list of offices in various countries.

Seton

http://www.seton.com

View Seton's identification and safety products catalog.

Seton Identification Products

http://www.seton.com

This extensive, well-organized site provides detailed information about Seton's safety products, MSDSs for its chemical products, a health and safety resources directory, etc.

Simply Safety Store

http://www.island.net/~dnagle/

The store offers training, information and products for safety in the workplace and at home. Read about the company's offerings.

SKC Online

http://www.skcinc.com/

View air sampling equipment information from the manufacturer. The site also contains frequently asked questions about the company's air sampling instrumentation, fill-in forms for requesting more information, etc.

Uni-Hoist

http://www.uni-hoist.com

Find out about the company's confined space products. The site also includes a photo gallery, links to related articles and sites, OSHA regulations, a newsletter, etc.

Publications

American Federation of State, County, and Municipal Employees, Occupational Safety and Health

http://www.afscme.org/afscme/health/content.htm

This page provides access to a series of health and safety documents, alerts, newsletters, etc. More detail about this site is available in the Top 50 Sites section.

American Society of Safety Engineers, Greater San Jose Chapter

http://www.best.com/~assegsjc/

Access the chapter's newsletter, information on classes, links to other ASSE sites, and selected links to other sites.

Arbetslivsinstitutet

http://www.niwl.se/niwl.htm

The National Institute of Working Life (Arbetslivsinstitutet) is Sweden's Research and Development center for occupational health and safety, working life and the labor market. Access is provided to information about the insitute, its library catalog, and the full text of its publications.

Asbestos Institute

http://www.asbestos-institute.ca

Access the Montreal-based Institute's Biomedical Data Bank, view reports and press releases, and get information about their printed publications.

Asian-Pacific Network on Occupational Safety and Health Information

http://www.ilo.org/public/english/270asie/asiaosh/

Presented by the ILO/FINNIDA Asian-Pacific Regional Programme on Occupational Safety and Health, this site was established with a view to sharing OSH information in and about the Asian-Pacific region.

Association paritaire pour la santé et la sécurité du travail, secteur affaires municipales

http://www.apsam.com/

This site contains a description of this Quebec health and safety agency, its mission, and its services. Bulletins, data sheets, and guidelines are also available. All information is in French.

Canadian Society of Safety Engineering (CSSE)

http://www.csse.org

Information is provided about the society, its programs, publications, and courses.

Commission de la santé et de la sécurité du travail du Québec (CSST)

http://www.csst.qc.ca/

The site describes the role and services of the commission, the laws it administers, etc. Other information includes CSST news releases, summaries of their magazine *Prévention au travail*, and a publications list. Most information is in French, but some is also in English.

Cornell OSH Archive of Educational Materials

http://ext.ilr.cornell.edu/osh/edu/occindex/srchindex.html

Cornell University's School of Industrial and Labor Relations, OSH Extension, is piloting an online archive of public domain and freeware occupational safety and health educational materials and policies. Find training materials on a varietyof OSH subjects.

Elsevier Science

http://www.elsevier.nl

View information about the company's publications. Subjects include chemistry and chemical engineering, clinical medicine, and environmental science and technology.

FacilitiesNet

http://www.facilitiesnet.com

FacilitiesNet is "an interactive information service for facility professionals." Members can access two online magazines, reports, an online bookstore, and product information.

Finnish Institute of Occupational Health

http://www.occuphealth.fi

Information is provided about the institute, its purpose, its work, etc. Lists of institute publications and journals are provided. There is also a list of OH&S conferences.

Government Institutes Division, ABS Group Inc.

http://www.govinst.com/

This company provides over 200 books on topics in the environmental, health, safety, telecommunications, and Internet fields. They also provide training courses, videotapes, and computer based training.

Hazardous Materials and Safety: Guide to Reference Sources

http://www.lib.lsu.edu/sci/chem/guides/srs103.html

This is a bibliography of HazMat and Safety publications, compiled by Louisiana State University libraries.

Health and Safety Authority

http://www.hsa.ie/osh/welcome.htm

Information is provided about the Health and Safety Authority (Ireland), a list of publications, a list of acts and regulations, information about the authority's programs, etc.

Health and Safety Promotion in the European Union

http://www.hsa.ie/hspro/index1.html

HSPro - EU will set up a remotely accessible telematics system providing a "one-stop shop" for occupational health and safety information. Categories of information include news and events, guidance, OSH information, OSH databases, bibliographies, library catalogs, and publications. More detail is available in the Top 50 Sites section.

Hong Kong Occupational Safety and Health Association

http://www.hk.super.net/~hkosha/

Information is provided about the association, Hong Kong safety and health legislation and standards, the association's publications, and links to other sites.

Industrial Accident Prevention Association (IAPA)

http://www.iapa.on.ca

Information is provided about this Ontario, Canada safety association, its products, services, and training programs.

Injury Prevention Research Unit, University of Otago, New Zealand

http://www.otago.ac.nz/Web_menus/Dept_Homepages/IPRU/

The site provides information about the unit, research it undertakes, and publications produced.

J.J.Keller & Associates

http://www.jjkeller.com/

View the company's safety product catalog, download sample software, and view descriptions of its printed publications.

Manitoba Workplace Safety and Health Division

http://www.gov.mb.ca/labour/safety/index.html

Information about the various branches of this provincial government Division (Workplace Safety and Health, Mining Inspection and Occupational Health), plus access to the full text of a number of its WorkSafe publications. More detail about this site is available in the Top 50 Sites section.

McGill University, Environmental Safety Office

http://www.mcgill.ca/eso/

This site provides information about staff and services provided by the office, links to McGill's Occupational Health Programme, and various McGill policy documents, including its Laboratory Safety Manual.

Michael Blotzer's Home Page

http://ourworld.compuserve.com/homepages/MBlotzer/

This site provides links to various health and safety resources, including information about and excerpts from Mike's book, *Internet User's Guide for Safety and Health Professionals.*

NASA/LeRC Office of Safety and Mission Assurance

http://www-osma.lerc.nasa.gov

Information about the office itself is provided. Links to many related NASA resources, such as the *Spaceflight Safety Handbook.*

National Institute for Occcupational Safety and Health (NIOSH)

http://www.cdc.gov/niosh/homepage.html

This home page has been established to provide information about NIOSH and related activities. Content includes NIOSH publications, respirator information, meetings and symposia, etc. More detail about this site is available in the Top 50 Sites section.

National Occupational Health and Safety Commission (WorkSafe Australia)

http://www.worksafe.gov.au

Read about this national OH&S agency and view statistical information, Australian regulatory requirements, fact sheets, alerts, etc.

National Safety Council

http://www.nsc.org/

Read about the council, its activities and research, publications, etc.

National Technical Information Service (NTIS)

http://www.ntis.gov/health/health.htm

The site provides information about this U.S. government agency's health and safety publications, databases, and other information.

New Zealand's Health and Safety Net

http://www.OSH.dol.govt.nz/

This is the home page of the Occupational Safety and Health Service of the Department of Labour, New Zealand. The site provides information about the organization, work hazards, health and safety law, training programs, etc.

OH&S Consulting Services

http://www.hronline.com/dr/

Dilys Robertson is a Toronto-based occupational health and safety consultant and writer. Read about her consulting services, which include training design and delivery and research work with joint health and safety committees.

OHS Canada

http://www.ohscanada.com/

The site provides information about *OHS Canada* magazine and selected content from the current and previous issues.

OHS&E—The Journal of Occupational Health, Safety and Environment

http://www.ohse.co.uk

To quote the editors of this U.K. magazine: "Besides containing details of the magazine and our previous and forthcoming features, the site has news, features, legislation information, and an Open Forum in which occupational health,safety and environment matters can be aired and responded to by anyone in the world."

Ontario Ministry of Labour

http://www.gov.on.ca/LAB/main.htm

The site contains information about the ministry, provincial government OSH discussion papers, a list of ministry publications, etc.

Oregon Occupational Safety and Health Division (OR-OSHA)

http://www.cbs.state.or.us/external/osha/index.html

The site contains a directory of OR-OSHA's services, news releases, an employer's toolkit, and information about codes, publications, and videos for sale. The *Oregon Health and Safety Resource*, OR-OSHA's newsletter, is also accessible.

OSH Answers

http://www.ccohs.ca/oshanswers/

This is a collection of frequently asked health and safety questions, presented by the Canadian Centre for Occupational Health and Safety (CCOHS).

Rådet för arbetslivsforskning

http://www.ralf.se

The Swedish Council for Work Life Research is a government agency which plans and funds Swedish research on subjects related to the work environment and working life. The site provides contact information, descriptions of its programmes, a newsletter, and its publications. Information is in Swedish and English.

Safeguard Publications

http://www.OSH.dol.govt.nz/order/safeguard/index.html

View a description of, and ordering information for *Safeguard* magazine, *Safeguard Update*, and *Safeguard Buyers Guide*, a trio of New Zealand health and safety publications from Colour Workshop.

Safety Link

http://www.safetylink.com

Information about *International Product Safety News*, a series of safety articles and essays, Frequently-asked-questions documents (FAQs), and links to various other sites.

Safety Network

http://www.safetynews.com/index.html

This Australian site describes itself as "the online hub for people interested in Occupational Health and Safety issues." It provides access to the online edition of *Australian Safety News*, the National Safety Council of Australia and other safety resources.

Safety Smart! Magazine

http://safetysmart.com

The site provides information about this magazine (selected articles available) and the company's other safety publications, posters, clip art, etc.

Safety+Health

http://www.nsc.org/pubs/sh.htm

Subscription information is provided for this U.S. National Safety Council publication. There is also information about upcoming events and news, plus a series of "Healthgrams" and "Safety Clips," intended for use in employee publications and safety bulletins.

SAREC Bookstore

http://www.sarec.ca/books/

SAREC catalogues occupational health and safety books, and provides online ordering capabilities in association with Amazon.com.

Swedish Occupational Safety and Health Administration

http://www.arbsky.se/arbskeng.htm

View Swedish work environment legislation, selected documents produced by the organization, and information about the organization itself. Content is in Swedish and English.

UAW Health and Safety

http://www.uaw.org/publications/h&s/index.html

The full text of current and past issues of the United Auto Workers' "On the Job Health and Safety" are available.

University of Edinburgh, Health and Safety Services

http://www.admin.ed.ac.uk/safety/

This site contains a mission statement, contact information about the various health and safety services at the university and "health and safety news flashes."

Vermont SIRI

http://siri.org

The site contains the archives of the SAFETY & Occ-Env-Med-L mailing lists, a large MSDS database, safety clip art, various safety-related documents, regulatory information, etc. More detail about this site is available in the Top 50 Sites section.

Windsor Occupational Health Information Service (WOHIS)

http://www.mnsi.net/~wohis/

WOHIS is an occupational health and safety information/referral service located in Windsor Canada. Its mandate is "to promote community awareness of occupational health and safety issues in order to prevent illness and injury." Read about its services, view publications, etc.

Workers Health and Safety Centre

http://www.whsc.on.ca/

The centre is a worker-driven health and safety delivery organization in Ontario, Canada. The site provides information about the centre and its services, including the full text of publications and links to labor and health and safety resources.

Worksite News

http://www.networkx.com/worksitenews/

Selections from, and subscription information about this Western Canadian occupational health, safety, and environment news magazine.

Programs

Arbetslivsinstitutet

http://www.niwl.se/niwl.htm

The National Institute of Working Life (Arbetslivsinstitutet) is Sweden's Research and Development center for occupational health and safety, working life, and the labor market. Access is provided to information about the insitute, its library catalog, and the full text of its publications.

Asian-Pacific Network on Occupational Safety and Health Information

http://www.ilo.org/public/english/270asie/asiaosh/

Presented by the ILO/FINNIDA Asian-Pacific Regional Programme on Occupational Safety and Health, this site was established with a view to sharing OSH information in and about the Asian-Pacific region.

Canadian Society of Safety Engineering (CSSE)

http://www.csse.org

Information is provided about the society, its programs, publications, and courses.

Griffith University, Faculty of Health and Behavioural Sciences

http://www.ua.gu.edu.au/hbk/hbs_ind.htm

Information about courses provided by this Queensland, Australia university is available. The School of Occupational Health and Safety is within this faculty.

Hill Air Force Base, Utah, Environment Safety and Health

http://137.241.179.15/esoh/index.htm

Read about their ESH programs and find links to related sites.

International Labour Organization

http://www.ilo.org/

The site provides information about the ILO, its various international programs, and publications. There are also links to ILO departmental home pages and an ILO directory.

Office of the Assistant Deputy Under Secretary of Defense for Safety and Occupational Health — ADUSD(SH)

http://www.acq.osd.mil:80/ens/sh

Information is provided about safety in the U.S. military: programs, policy documents, and statistics.

Oregon Occupational Safety and Health Division (OR-OSHA)

http://www.cbs.state.or.us/external/osha/index.html

The site contains a directory of OR-OSHA's services, news releases, an employer's tool kit, information about codes, publications and videos for sale, etc. The *Oregon Health and Safety Resource*, OR-OSHA's newsletter, is also accessible.

Pro-Safety Inc.

http://members.aol.com/safetypros/prosafe.htm

Read about this company's safety services and programs and its various safety software packages.

Rocky Mountain Center for Occupational and Environmental Health

http://rocky.utah.edu/

Located on the campus of the University of Utah, the Rocky Mountain Center is one of 14 NIOSH Educational Research Centers. The site provides information about the center's various educational programs.

Rådet för arbetslivsforskning

http://www.ralf.se

The Swedish Council for Work Life Research is a government agency which plans and funds Swedish research on subjects related to the work environment and working life. The site provides contact information, descriptions of its programmes, a newsletter and its publications.Information is in Swedish and English.

Safety Advisory Services

http://www.safetyadvisory.com/

Read about the company's loss control and substance abuse programs.

University of Colorado at Colorado Springs, Risk Mgmt. and Environmental Safety

http://www.uccs.edu/~pusafety/risk1a.htm

Available documents describe health and safety policies and procedures at UCCS.

Software

3D Safety CD-ROM

http://www.3dsafety.com/

View information about, and previews from this CD-ROM of safety and health-related graphics files.

American Safety and Health Training

http://www.safety-training.com

Information about this nonprofit organization's multimedia computer-based training packages and medical monitoring services is provided.

BLMC Computer Based Training Programs

http://www.blmc.com/

Information is provided about this Canadian company's computer-based health and safety training programs.

Dawnbreaker Consultants

http://204.112.119.2:80/dawnbreaker/

Information about and downloadable evaluation copies of their software are available. There are several programs of interest, including WorkSafe Tools (Occupational Safety and Health Software) and EnviroSafe (Environmental Risk Assessment Software).

EHS Diagnostics Inc.

http://www.ehsdiagnostics.com

EHS Diagnostics Inc. (EDI) is a Canadian environment, health and safety consulting firm, based in Calgary, Alberta, Canada, which provides both consulting services and health and safety software development. Read about the company's services.

Environment and Safety Data Exchange (ESDX)

http://www.esdx.org/esdhome.html

Find out about ESDX, an industry association organized to participate in setting standards for more cost-effective management information technology in environmental, health and safety (EHS) activities.

EnviroWin Software Inc.

http://www.envirowin.com

The site contains this company's EHS software catalog: chemical inventory, emergency response, OSHA compliance, and risk assessment.

J.J.Keller & Associates

http://www.jjkeller.com/

View the company's safety product catalog, download sample software, and view descriptions of its printed publications.

KnowledgeWare Communications Corp.

http://www.kccsoft.com/

This N. Vancouver, BC company provides several software packages: Simply Safety! Safety Management, WHMIS CBT Suite, and Trax & Fax MSDS Software. View information and download demonstrations of these software packages.

Pro-Safety Inc.

http://members.aol.com/safetypros/prosafe.htm

Read about this company's safety services and programs and its various safety software packages.

Safety Assistant—Software for Workplace Health and Safety

http://www.safetyassistant.com

View descriptions and sample screens from this Australian company's various OSH software packages.

Safety Training and Tracing, Inc.

http://www.standt.com/

The company's OSH consulting services and software "Saf-T-Man Safety Manager" are described.

SafetyScape Corp.

http://www.safetyscape.com/

The company produces Safety Navigator (TM) and Safety Trakker (TM) occupational health and safety management software. View information about these packages, and download a demo.

WorkTime

http://www.timedomain.com/

Information is provided about WorkTime, an employee and resource scheduling program for Windows and Macintosh, created by Time Domain, Inc.

---

2. Environmental Protection

Conferences

Indoor Air 99

http://www.ia99.org/

The site provides detailed information about the Indoor Air 99 conference in Edinburgh Scotland, August 8-13, 1999.

Consulting Services

Advanced Waste Management Systems Inc.

http://www.awm.net/

AWMS is an environmental consulting firm offering services which include HAZWOPER Worker Right-to-Know Emergency Planning and Community Right-to-Know Act. This site describes their services and qualifications.

Aerotech Laboratories

http://www.aerotechlabs.com

Aerotech Laboratories is an independent analytical laboratory, consulting firm and medical device manufacturer. Read about its services and access its Indoor Air Quality (IAQ) Compendium.

American Institute of Chemical Engineers (AIChE), Consultant and Expert Witness Electronic Directory

http://www.resume-link.com/consultants/

This searchable database references chemical engineering consultants and expert witnesses from various backgrounds and expertise including: environmental engineering, process design, and hazardous waste management.

Center for Safety and Environmental Management

http://www.nauticom.net/www/csem/

The site contains descriptions of the company's environmental safety training, onsite staffing, consulting and field services.

DuPont Safety and Environmental Management Services (SEMS)

http://www.dupont.com/safety/

This site describes DuPont SEMS' services. They provide safety seminars, training materials, and consulting services.

EHS Diagnostics Inc.

http://www.ehsdiagnostics.com

EHS Diagnostics Inc. (EDI) is a Canadian environment, health and safety consulting firm, based in Calgary, Alberta, Canada, which provides both consulting services and health and safety software development. Read about the company's services.

EHS Management Technologies Private Limited, India

http://www.corporatepark.com/ehsindia/

Read about this consulting company's environmental, health and safety, and management services.

EIC Environmental Health and Safety

http://members.aol.com/eicehs/EIC.html

As well as providing information about the services of this Washington state consulting company, the site provides access to a Weekly Hazardous Waste Inspection Checklist, articles, and an extensive set of links.

EnviroInfo Research Service

http://members.aol.com/elsevierhr/index.htm

Read about Elsevier's Enviroinfo Research Service, which "specializes in finding information for industry and environmental professionals." Environmental compliance information is their specialty.

Environmental Safety Services

http://www.envirosafety.qpg.com/

Read about the company's environmental and health and safety consulting services.

Environmental Safety Services

http://www.ps.uga.edu/ess/

Read about the company's consulting services.

HAZCON Pty. Ltd.

http://www.dcscomp.com.au/hazcon/

Information about this Victoria, Australia-based environmental and occupational health and safety consulting company and its services.

HRP Consultants

http://www.hrpconsultants.com/

Read about this Ontario, Canada-based consulting company that specializes in "assisting organizations to add value to their Environmental, Safety and ISO functions by developing programs and systems that will enhance organization success."

National Environmental Safety Compliance, Inc.

http://www.albany.net/~nesc/

NESC specializes "in Health, Safety and Environmental Issues through training, consulting and site inspections." As well as providing information about the company's services, they provide some information files and links.

National HVAC

http://www.nationalhvac.com/

The content of the site describes this Canadian company's services, including: indoor environment consulting; building ventilation assessments; field investigations, air testing and remediation management services.

Paskal Environmental Services

http://www.erols.com/paskal/index.html

Read about the company's environmental and OSH training services.

Resource Environmental Associates

http://www.rea4ehs.com/

Information is provided about this Toronto, Canada-based environment, health and safety consulting company, its services, and publications.

Restoration Environmental Contractors

http://Home.InfoRamp.Net/~restcon/

This Canadian company provides services such as asbestos abatement, lead abatement, fire restoration, mold and fungus removal, and plant cleanups. The Web site provides documents outlining step by step actions to be taken for most of these types of cleanup/restoration.

Safety Systems Inc.

http://www.safetysystemsinc.com

The site describes this environmental health and safety consulting company's services.

Waste Management, Inc.

http://www.wastemanagement.com

Read about this Illinois-based waste management company.

Water Technology International

http://www.wti.cciw.ca

WTI is an environmental technology and services company. Read about their products and services (hazardous waste, site remediation, etc.). Of particular interest is their 40-hour health and safety course, which satisfies general Canadian (federal and provincial) and U.S. (29 CFR1910.120, "HAZWOPER") health and safety training requirements for contaminated site workers.

Databases

Canadian Centre for Occupational Health and Safety (CCOHS)

http://www.ccohs.ca

CCOHS's site provides access to Canadian environmental and OSH legislation and major OSH databases (*e.g.*, NIOSHTIC, RTECS, HSELine) on a subscription basis. Other content includes information and hazard alerts and a large directory of health and safety Internet resources. More detail about this site is available in the Top 50 Sites section.

Envirofacts Warehouse, Master Chemical Integrator

http://www.epa.gov/enviro/html/emci/emci_query.html

This page allows you to obtain the acronyms, chemical identification numbers, and chemical names reported by the U.S. Environmental Protection Agency's Envirofacts program system databases (AFS, PCS, RCRIS and TRIS). There are links to fact sheets, containing information regarding health risks, exposure limits, handling regulations, and chemical properties.

Environmental Protection Agency (EPA)

http://www.epa.gov

This well-developed U.S. government site contains a huge amount of environmental information, including environmental regulations, a variety of database systems, and numerous indoor air quality documents. More detail about this site is available in the Top 50 Sites section.

Integrated Risk Information System (IRIS)

http://www.epa.gov/ngispgm3/iris/

"The Integrated Risk Information System (IRIS), prepared and maintained by the U.S. Environmental Protection Agency (U.S. EPA), is an electronic data base containing information on human health effects that may result from exposure to various chemicals in the environment.

RTK NET

http://rtk.net

The Right-to-Know Network provides free access to "numerous databases, text files, and conferences on the environment, housing, and sustainable development."

UNEP Chemicals (International Registry of Potentially Toxic Chemicals—IRPTC)

http://irptc.unep.ch/irptc/

UNEP Chemicals is the center for all chemicals-related activities of the United Nations Environment Programme. Highlights include an inventory of information sources on chemicals, the UNEP Chemicals database, and information about UNEP programs.

Directories

American Institute of Chemical Engineers (AIChE), Consultant and Expert Witness Electronic Directory

http://www.resume-link.com/consultants/

This searchable database references chemical engineering consultants and expert witnesses from various backgrounds and expertise including environmental engineering, process design, and hazardous waste management.

Canadian Centre for Occupational Health and Safety (CCOHS)

http://www.ccohs.ca

CCOHS's site provides access to Canadian environmental and OSH legislation and major OSH databases (*e.g.*, NIOSHTIC, RTECS, HSELine) on a subscription basis. Other content includes information and hazard alerts and a large directory of health and safety Internet resources. More detail about this site is available in the Top 50 Sites section.

Hardin Meta Directory—Public, Occupational, and Environmental Health

http://www.lib.uiowa.edu/hardin/md/publ.html

This is a "list of lists," providing links to a variety of health-related directories.

National Registry of Environmental Professionals

http://www.nrep.org/

Information is provided about this professional association. Read about the association itself, search a list of registered professionals, and find out about training programs and employment opportunities.

University of Edinburgh—Health, Environment, and Work

http://www.med.ed.ac.uk/hew/

The site provides a long list of "tutorials" (educational resources) on a variety of occupational and environmental health topics, along with information on Edinburgh's academic programs. There is also an excellent directory of primarily European Internet resources. More detail about this site is available in the Top 50 Sites section.

Educational Resources / Training

Center for Occupational and Environmental Health, University of California

http://ehs.sph.berkeley.edu/coeh/

View information about this research center, its academic programs, continuing education courses, and research projects.

Continuing Education Online Courses—Facility and Maintenance Courses

http://www.conedu.com

Descriptions of a number of online courses, including "Hazardous Waste Regulations for Facility Managers."

DuPont Safety and Environmental Management Services (SEMS)

http://www.dupont.com/safety/

This site describes DuPont SEMS' services. They provide safety seminars, training materials, and consulting services.

Environmental Support Solutions

http://www.environ.com

View information about the company's services, including a number of indoor air quality (IAQ) courses, and a Heating, Ventilation and Air Conditioning (HVAC) mall.

Griffith University, Faculty of Environmental Sciences

http://www.ens.gu.edu.au

View information about the faculty itself and find links to other Australian and international environmental sites.

HAZCON Pty. Ltd.

http://www.dcscomp.com.au/hazcon/

Information about this Victoria, Australia-based environmental and occupational health and safety consulting company and its services.

Interactive Media Communications

http://www.safetysite.com

Information is provided about the company and its multimedia occupational and environmental health and safety training programs on CD-ROM, videodisc, and videotape.

MGMT Alliances Inc.

http://www.mgmt14k.com

MGMT Alliances develops and presents training courses on Environmental Management Systems, ISO 14000, ISO 9000, and Integrated Management Systems.

National Environmental Safety Compliance, Inc.

http://www.albany.net/~nesc/

NESC specializes "in Health, Safety and Environmental Issues through training, consulting and site inspections." As well as providing information about the company's services, they provide some information files and links.

National Environmental Training Association

http://ehs-training.org

Read about the National Environmental Training Association, an international, educational, professional association of specialists who design, deliver, and manage environmental, health and safety training.

Paskal Environmental Services

http://www.erols.com/paskal/index.html

Read about the company's environmental and OSH training services.

Safety Systems Inc.

http://www.safetysystemsinc.com

The site describes this environmental health and safety consulting company's services.

SASSI Training Systems

http://www.mindspring.com/~sassitng/index.html

Read about the company's environmental counseling, compliance training, and ISO 14000 training.

Seagull Environmental Training

http://www.seagulltraining.com/

This U.S.-wide company provides training on subjects such as asbestos, lead paint, hazardous materials, and environmental site assessment. Read about the company's courses.

University of Edinburgh, Health Environment and Work

http://www.med.ed.ac.uk/hew/

The site provides a long list of "tutorials" (educational resources) on a variety of occupational and environmental health topics, along with information about Edinburgh's academic programs. There is also an excellent directory of primarily European Internet resources. More detail about this site is available in the Top 50 Sites section.

Water Technology International

http://www.wti.cciw.ca

WTI is an environmental technology and services company. Read about their products and services (hazardous waste, site remediation, etc.). Of particular interest is their 40-hour health and safety course, which satisfies general Canadian (federal and provincial) and U.S. (29 CFR1910.120, "HAZWOPER") health and safety training requirements for contaminated site workers.

Journals and Newsletters

Biosafety Information Network and Advisory Service (BINAS)

http://binas.unido.org/binas/binas.html

BINAS, a service of the United Nations Industrial Development Organization, monitors global developments in regulatory issues in biotechnology. Site highlights include links to regulations in many countries, UNIDO publications, the BINAS newsletter, and OECD Environment Monographs.

Campus Safety Health and Environmental Management Association (CSHEMA)

http://www.ualberta.ca/~rrichard/cshema.html

The site includes information on the CSHEMA organization, its newsletter, and training and policy information. More detail about this site is available in the Top 50 Sites section.

EHIS Publications

http://ehpnet1.niehs.nih.gov/docs/publications.html

Access is provided to various publications of the U.S. National Institute of Environmental Health Sciences: *Environmental Health Perspectives, National Toxicology Program Technical Reports*, etc.

Scandinavian Journal of Work, Environment and Health

http://www.occuphealth.fi/eng/dept/sjweh/

The site provides information about upcoming articles in the journal and how to subscribe to the paper version. The journal appears bimonthly.

Society of Environmental Toxicology and Chemistry (SETAC)

http://www.setac.org/

Available information includes details membership in this professional society, a newsletter, activities and meetings, and publications.

Policy / Procedures Documents

Campus Safety Health and Environmental Management Association (CSHEMA)

http://www.ualberta.ca/~rrichard/cshema.html

The site includes information on the CSHEMA organization, its newsletter, training and policy information. More detail about this site is available in the Top 50 Sites section.

Lawrence Livermore National Laboratory (LLNL), Environment, Safety and Health

http://www.llnl.gov/es_and_h/

A number of documents are provided, including LLNL's health and safety manual, environmental guidelines documents, and an environmental compliance manual.

Louisiana State University, Office of Occupational and Environmental Safety

http://130.39.86.58/page1.htm

Documents describing the office's services are available. Subject areas include safety and environmental program news, emergency procedures and information, and safety and environmental program documents and information.

McGill University, Environmental Safety Office

http://www.mcgill.ca/eso/

This site provides information about staff and services provided by the office, links to McGill's Occupational Health Programme, and various McGill policy documents, including its Laboratory Safety Manual.

Portland State University, Environmental Health and Safety

http://www.pp.pdx.edu/FAC/Safety/

The site provides access to a variety of university documents, such as their chemical hygiene program, lockout-tagout, and pesticide/herbicide use.

University of California Irvine, Environmental Health and Safety Department

http://www.abs.uci.edu:80/depts/ehs/

View the university's EHS policies and procedures documents, emergency procedures, and safety news.

University of Illinois (Urbana), Division of Environmental Health and Safety

http://phantom.ehs.uiuc.edu/dehs.html

This site provides links to a number of documents and other resources from the division's chemical safety and waste management, occupational safety and health, and radiation safety sections.

University of Utah, Environmental Health and Safety

http://www.ehs.utah.edu/

The site provides information about the department, the full text of numerous policy and procedure documents, waste minimization documentation, pollution prevention information, and links to other sites.

University of Virginia, Office of Environmental Health and Safety

http://www.virginia.edu/~enhealth/

This well-developed site provides access to a number of documents and information about the office's programs. Highlights include UVa's Laboratory Survival Manual and a Laboratory Safety Checklist.

University of Wisconsin System Administration, Office of Safety and Loss Prevention

http://www.uwsa.edu/oslp/oslp.htm

The page contains links to Environmental Health and Safety, Workers Compensation and Risk Management. Manuals and policy documents are presented.

Products

3M Environmental Safety and Energy Performance Products

http://www.mmm.com/market/government/env/envindex.html

Information about 3M's products.

Environmental Support Solutions

http://www.environ.com

View information about the company's services, including a number of indoor air quality (IAQ) courses and a Heating, Ventilation and Air Conditioning (HVAC) mall.

Major Safety Service

http://www.pilot.infi.net/~majsaf/

Products and services provided by this environmental instrument and safety equipment supplier are described.

Northern Industrial Supply Company

http://www.nisco.net/

This Ontario, Canada company sells industrial heating, ventilation, and air conditioning equipment. Read about the products they provide.

OMCO Air Treatment

http://www.omco.be/

Read about this Belgian company's ventilation products. Information is in French and Dutch.

W. S. Wood Associates Ltd.

http://www.wswood.com/

This Canadian company sells, services, and inventories static elimination products for industrial and electronic/cleanroom applications, ESD consumable products, as well as mold release agents. Read about the company's products and services.

Wessoclean Environmental Hygiene

http://www.wessoclean.com/

Read about the company's products for cleaning and disinfecting "water supplying pieces of equipment and plants."

Publications

Air and Waste Management Association

http://www.awma.org

AWMA provides information about the association, its publications (*e.g.*, *A&WMA Journal*), and membership.

Biosafety Information Network and Advisory Service (BINAS)

http://binas.unido.org/binas/binas.html

BINAS, a service of the United Nations Industrial Development Organization, monitors global developments in regulatory issues in biotechnology. Site highlights include links to regulations in many countries, UNIDO publications, the BINAS newsletter, and OECD Environment Monographs.

Canadian Centre for Occupational Health and Safety (CCOHS)

http://www.ccohs.ca

CCOHS's site provides access to Canadian environmental and OSH legislation, as well as major OSH databases (*e.g.*, NIOSHTIC, RTECS, HSELine), on a subscription basis. Other content includes information and hazard alerts and a large directory of health and safety Internet resources. More detail is available in the Top 50 Sites section.

Canadian Environmental Law Association

http://www.web.net/cela/

CELA is a nonprofit, public interest organization established in 1970 to use existing laws to protect the environment and to advocate environmental law reforms. The site provides information about CELA, a list of publications, subscription information for its newsletter, and descriptions of current cases and causes.

Compliance Online

http://www.ieti.com/taylor/compliance.html

Compliance Online is an "Internet newsletter for finding your way through the maze of environmental regulations. This site contains an archive of newsletter issues back to Jan. 1996."

Corporate Health and Safety—Managing Environmental Issues in the Workplace

http://www.boxelectronics.com/

View a description and ordering information for this industrial hygiene guide.

Duke University Occupational and Environmental Medicine

http://occ-env-med.mc.duke.edu/oem

This site contains numerous documents and links to resources in occupational and environmental medicine. It is also the "home" of the occ-env-med-l mailing list. More detail about this site is available in the Top 50 Sites section.

Elsevier Science

http://www.elsevier.nl

View information about the company's publications. Subjects include chemistry and chemical engineering, clinical medicine, and environmental science and technology.

Environmental Health Information Service

http://ehis.niehs.nih.gov/

This is "a full-service site that produces, maintains, and disseminates information on the environment in the form of a searchable directory." Contents include the monthly journal, *Environmental Health Perspectives*, andthe National Toxicology Program's 8th Report on Carcinogens.

Environmental Protection Agency (EPA)

http://www.epa.gov

This well-developed U.S. government site contains a huge amount of environmental information, including environmental regulations, a variety of database systems, and numerous indoor air quality documents. More detail about this site is available in the Top 50 Sites section.

Environmental Protection Agency, Indoor Air Quality Home Page

http://www.epa.gov/iaq/

This is a very content-rich site provided by the U.S. EPA. It contains documents on common indoor air pollutants, sources of indoor air pollution, and problems in different types of buildings. More detail about this site is available in the Top 50 Sites section.

Environmental Protection Agency, Office of Pesticide Programs

http://www.epa.gov/pesticides/

Access is provided to detailed U.S. EPA pesticide information.

EPA—Technical Information Packages (TIPS)

http://www.epa.gov/oia/tips/

This page contains descriptions of eleven Technical Information Packages (TIPs). Aimed at the international community, the packages focus on key environmental and public health issues being investigated by the U.S. Environmental Protection Agency (EPA).

European Environment Agency

http://www.eea.dk/

The site provides a description of the agency, along with numerous documents and document descriptions related to the environment in Europe.

Extoxnet—The EXtention TOXicology NETwork

http://ace.orst.edu/info/extoxnet/

EXTOXNET is a cooperative effort of University of California-Davis, Oregon State University, Michigan State University, and Cornell University. Various types of pesticide toxicology and environmental chemistry information are available.

Harvard Environmental Resources Online

http://environment.harvard.edu/

The site provides information about environmental resources at Harvard University (U.S.A.), an international environmental policy resource guide, and archives for the ENVCONFS-L and ENVREFLIB-L mailing lists.

Health Canada, Health Information Network

http://www.hc-sc.gc.ca/

This Canadian federal government department presents its health alerts, public health documents, news releases, etc., as well as links to its various directorates—Laboratory Centre for Disease Control, Environmental Health Directorate, etc. More detail about this site is available in the Top 50 Sites section.

Institute for Systems, Informatics and Safety

http://www.jrc.org/isis/index.asp

This Joint Research Centre of the European Commission is involved in the multidisciplinary analysis of industrial, socio-technical and environmental systems; the innovative application of information and communication technologies, and the science and technology of safety management. Read about its research activities.

McGill University, Environmental Safety Office

http://www.mcgill.ca/eso/

This site provides information about staff and services provided by the office, links to McGill's Occupational Health Programme, and various McGill policy documents, including its Laboratory Safety Manual.

New York State Department of Health, Environmental and Occupational Health

http://www.health.state.ny.us/nysdoh/consumer/environ/homeenvi.htm

This site provides access to a series of basic documents on environmental and occupational health. The content is geared to the public.

Pesticide Information Fact Sheets

http://fshn.ifas.ufl.edu/pest/default.htm#FactSheets

This is a series of documents describing various environmental and occupational aspects of pesticides. The site is maintained by the Pesticide Information Office, Institute of Food and Agricultural Sciences, University of Florida.

RTK NET

http://rtk.net

The Right-to-Know Network provides free access to "numerous databases, text files, and conferences on the environment, housing, and sustainable development."

Society of Environmental Toxicology and Chemistry (SETAC)

http://www.setac.org/

Available information includes details about membership in this professional society, a newsletter, activities and meetings, and publications.

South Carolina Department of Health and Environmental Control

http://www.state.sc.us/dhec/

Read about this state department's functions, programs, and publications.

Volvo Group—Environment

http://www.volvo.se/environment/index.html

Volvo's Environmental policies, programs, and environmental report are provided. Volvo's "blacklist" of chemicals is included in the 1995 environmental report.

Programs

Center for Occupational and Environmental Health, University of California

http://ehs.sph.berkeley.edu/coeh/

View information about this research center, its academic programs, continuing education courses, and research projects.

Department of Energy, Environment Safety and Health, International Health Programs

http://www.eh.doe.gov/ihp/

View information about health effects of radiation and related environmental hazards. Three programs are described: those of Europe, Japan, and the Marshall Islands.

Griffith University, Faculty of Environmental Sciences

http://www.ens.gu.edu.au

View information about the faculty itself and links to other Australian and international environmental sites.

Hill Air Force Base, Utah, Environment Safety and Health

http://137.241.179.15/esoh/index.htm

Read about their ESH programs and links to related sites.

Radiological and Environmental Sciences Laboratory (RESL) at the Idaho National Engineering Laboratory

http://www.inel.gov/resl/

Read about this U.S. federal laboratory's ongoing programs.

South Carolina Department of Health and Environmental Control

http://www.state.sc.us/dhec/

Read about this state department's functions, programs, and publications.

Volvo Group—Environment

http://www.volvo.se/environment/index.html

Volvo's environmental policies, programs, and environmental report are provided. Volvo's "blacklist" of chemicals is included in the 1995 environmental report.

Software

Bara Environmental Compliance

http://www.bara.com

Bara Environmental Solutions is a Texas-based consulting firm that "creates advanced environmental safety solutions through computer tracking of hazardous materials and safety equipment." Read about the company's services.

Dawnbreaker Consultants

http://204.112.119.2:80/dawnbreaker/

Information about and downloadable evaluation copies of their software are available. There are several programs of interest, including WorkSafe Tools (Occupational Safety and Health Software) and EnviroSafe (Environmental Risk Assessment Software).

EHS Diagnostics Inc.

http://www.ehsdiagnostics.com

EHS Diagnostics Inc. (EDI) is a Canadian environment, health and safety consulting firm, based in Calgary, Alberta, Canada, which provides both consulting services and health and safety software development. Read about the company's services.

EnviroDx

http://www.auhs.edu/envirodx/

"EnviroDx is a multimedia, case-focused, computer-based learning program on environmental-related diseases. The organizing metaphor for EnviroDx is an exploratory 'virtual clinic' affiliated with a busy medical school. The program user takes the part of a practicing physician faced with a patient with an unknown disease or condition that is possibly caused by exposure to environmental factors."

Environment and Safety Data Exchange (ESDX)

http://www.esdx.org/esdhome.html

Find out about ESDX, an industry association organized to participate in setting standards for more cost-effective management information technology in environmental, health and safety (EHS) activities.

EnviroWin Software Inc.

http://www.envirowin.com

The site contains this company's EHS software catalog: chemical inventory, emergency response, OSHA compliance, and risk assessment.

Interactive Media Communications

http://www.safetysite.com

Information is provided about the company, and its multimedia occupational and environmental health and safety training programs on CD-ROM, videodisc, and videotape.

Modern Technologies Corp., Environmental Technologies Division

http://www.etd.modtechcorp.com/

The site provides information about the company's environmental consulting, environmental management software, industrial hygiene, and safety services.

Text-Trieve Environmental Health and Safety Page

http://www.halcyon.com/ttrieve/

Read about the company's electronic text-management systems that "present government regulations in easy-to-use electronically published formats." A number of U.S. regulatory subject areas are described.

---

3. Commercial Services

Consulting Services

Advanced Chemical Safety

http://www.chemical-safety.com

The site provides information about this consulting company's services, including chemical safety training and U.S. regulatory compliance documentation.

Advanced Ergonomics

http://www.advergo.com

AEI provides a variety of ergonomic services to corporations and individuals. Their products and services are described.

Advanced Waste Management Systems Inc.

http://www.awm.net/

AWMS is an environmental consulting firm offering services which include HAZWOPER Worker Right-to-Know Emergency Planning and Community Right-to-Know Act. This site describes their services and qualifications.

Aerotech Laboratories

http://www.aerotechlabs.com

Aerotech Laboratories is an independent analytical laboratory, consulting firm, and medical device manufacturer. Read about its services and access its Indoor Air Quality (IAQ) Compendium.

Alex Hartov Consulting

http://www.newcc.com/alex/

Read about his consulting services, including medical and biomedical instrumentation.

Amolins and Associates Inc.

http://www.amolins.com

Read about the consulting company's workers' compensation auditing services.

Assure Health Management

http://www.assure.ca

This company designs and delivers occupational health and disability management programs to Canadian business and industry. Read about the company and its services.

Behavioral Science Technology, Inc.

http://www.bscitech.com/

Read about the company's services related to behavior-based safety: seminars, research, etc.

CAE Consultants

http://www.caeconsultants.com/

This consulting company prepares custom documents such as standard operating procedures, guidelines, and process documents.

Center for Hazard Information

http://www.hazardinfo.com/

This professional safety engineering firm provides consulting services and a monthly *Hazard Information Newsletter*. Read about the company's services and how to subscribe to the newsletter.

Center for Safety and Environmental Management

http://www.nauticom.net/www/csem/

The site contains descriptions of the company's environmental safety training, on-site staffing, consulting, and field services.

Ceske pracovni zdravi a bezpecnost

http://www.markl.cz

Czech occupational health and safety. The site contains a description of this consulting company's products, a directory of OSH organizations in the Czech republic, and Czech OSH legislation. All information is available in Czech, and some in English.

Chemical Safety Associates

http://www.chemical-safety.com

Provides information about this consulting company's services, including site remediation, MSDS preparation, and assistance with safe handling of hazardous materials.

Compliance Control Center

http://users.aol.com/comcontrol/comply.htm

Provides information about handwashing compliance monitoring services provided by the organization, as well as information about personal hygiene in food preparation.

Compliance Management Group—OSHA Safety Plans and Labor Law Posterbook

http://www.oshaman.com/

This company sells written policy and procedures documents, complete safety plans, training materials, training videos, CD's, and labor law posters. View details of their products.

Comprehensive Loss Management, Inc.

http://www.clmi-training.com

This company produces safety and ergonomic training programs to help companies achieve OSHA compliance.

Continental Hoisting Consultants

http://www.chcelevator.com/

Safety training for elevators—entrapment and fire emergency. Read about their services.

Controlled Risk International

http://www5.electriciti.com/risk/

Read about the company's risk management services, courses, and OSHA and CAL/OSHA's most frequently cited serious violations.

DuPont Safety and Environmental Management Services (SEMS)

http://www.dupont.com/safety/

This site describes DuPont SEMS' services. They provide safety seminars, training materials, and consulting services.

Dynamic Scientific Controls, Inc.

http://www.fallsafety.com

They are consultants in occupational fall protection safety planning and commercial egress systems. The site provides a list of the company's publications, courses, and videos.

EHS Diagnostics Inc.

http://www.ehsdiagnostics.com

EHS Diagnostics Inc. (EDI) is a Canadian environment, health and safety consulting firm, based in Calgary, Alberta, Canada. It provides both consulting services and health and safety software development. Read about the company's services.

EHS Management Technologies Private Limited, India

http://www.corporatepark.com/ehsindia/

Read about this consulting company's environmental, health and safety, and management services.

EIC Environmental Health and Safety

http://members.aol.com/eicehs/EIC.html

As well as providing information about the services of this Washington state consulting company, the site provides access to a Weekly Hazardous Waste Inspection Checklist, articles, and an extensive set of links.

Emilcott-dga Inc.

http://www.emilcott-dga.com

Find out about the company's services, including industrial hygiene, safety and environmental consulting, and training.

EnviroInfo Research Service

http://members.aol.com/elsevierhr/index.htm

Read about Elsevier's Enviroinfo Research Service, which "specializes in finding information for industry and environmental professionals." Environmental compliance information is their specialty.

Environmental Safety and Health of Alaska

http://www.alaska.net/~esha/

The site contains descriptions of this company's safety training courses, industrial hygiene and marine chemist Services, plus an industrial hygiene news and current events section, and news from the Midnight Sun Section of AIHA.

Environmental Safety Services

http://www.envirosafety.qpg.com/

Read about the company's environmental and health and safety consulting services.

Environmental Safety Services

http://www.ps.uga.edu/ess/

Read about the company's consulting services.

Ergonomic Central

http://www.ergocentral.com/

This site is an electronic mall of ergonomics products and services.

Front-Line Safety

http://www.frontlinesafety.com/

This Nova Scotia-based company provides health and safety training programs, safety equipment rentals, and consulting/safety audits. They also provide an online WHMIS course.

Gravitec Systems Inc.

http://www.gravitec.com/

This U.S. company specializes in engineering, training, and consulting related to fall protection. Read about the company's services.

Hascom

http://www.hascom.com

Find out about this U.K.-based health and safety consultancy. Information is provided about the company itself, its services, links to selected sites, and an online newsletter. The newsletter is particularly useful to those involved in health and safety in the U.K..

Hazard Alert Training Supplies Canada Inc. (HATSCAN)

http://www.hatscan.com/

HATSCAN is an occupational health, safety and environment training and consulting firm, providing consulting and training in WHMIS, TDG, OH&S, due diligence, WCB, environmental compliance, hazardous waste, and more. Find out about the company's courses, publications, and computer-assisted training.

HAZCON Pty. Ltd.

http://www.dcscomp.com.au/hazcon/

Information about this Victoria, Australia-based environmental and occupational health and safety consulting company and its services.

HealthGate

http://www.healthgate.com/HealthGate/MEDLINE/search.shtml

HealthGate provides free access to Medline, along with information about their commercial health, wellness, and biomedical information services.

HG & Associates, Inc.

http://www.hgassociates.com

The site provides a description of this Florida construction industry consulting company.

HRP Consultants

http://www.hrpconsultants.com/

Read about this Ontario, Canada-based consulting company that specializes in "assisting organizations to add value to their Environmental, Safety and ISO functions by developing programs and systems that will enhance organization's success."

HSE Group, Inc.

http://www.hsegroup.com/

The HSE Groups is an international consortium of health, safety, and environmental consultants. Read about available services, Web-based training, the HSE book store, alerts, etc. "Fun" MSDSs for chocolate and water are also available here.

Integrated Environmental Management, Inc.

http://www.iem-inc.com/

IEM delivers "health physics (radiation safety), nuclear engineering, and environmental services to both government and commercial clients." As well as providing information about its services, the company's site provides a section on Radioactivity Basics, and a Tool Box.

JBF Associates Inc.

http://www.jbfa.com

JBFA is "an engineering firm that specializes in process hazard analysis, reliability and availability engineering, system safety, environmental engineering, process safety management, risk management planning and communication, quantitative risk analysis, and incident investigations." (Now known as **Risk and Reliability Division, ABS Group, Inc.**)

Life in Motion

http://www.rockies.net/~eberry/

This company provides occupational health nurses and registered nurses with programs for use in both the workplace and the community.

M.O.S.T. Consulting Services

http://www.mostsafety.com/

Medical and Occupational Safety and Training (M.O.S.T.) provides consulting services for safety programs in medical facilities. Read about the company's services.

Mathews, Dinsdale & Clark

http://www.mdclabourlaw.com/

This is a Toronto, Canada-based law practice "restricting its practice to labour and employment law." The site provides a number of interpretive documents about Ontario law and quizzes to test your knowledge.

McCuaig Russell Barristers

http://www.mccuaigrussel.on.ca

This Ottawa, Canada law firm specializes in workers' compensation cases, representing management. Read about the firm's services.

Medical Device Link—Safety Consultants

http://www.devicelink.com/consult/Safety.html

This directory provides descriptions of companies that provide safety services.

Medical Horizons Unlimited (TM)

http://www.medhorizons.com

This company is a medical educational consulting firm. Read about its educational services in a variety of medically related fields, including occupational health and safety.

MedTox Health Services

http://home.earthlink.net/~medtox/

Read about the services provided by this California-based occupational medicine services company.

MJW Corporation

http://www.mjwcorp.com/

MJW provides a variety of radiological consulting services and software solutions for health physics and other technical applications. Read about the company's services and download demos of their software.

National Access and Rescue Centre

http://www.narc.co.uk/

This U.K. organization "is a training, development, and equipment supply centre for rope access, line rescue, confined space entry, mast climbing, fall prevention, and all work involving high places or other difficult access." Read about the organization's training programmes and services.

National Environmental Safety Compliance, Inc.

http://www.albany.net/~nesc/

NESC specializes "in Health, Safety and Environmental Issues through training, consulting, and site inspections." As well as providing information about the company's services, they provide some information files and links.

National HVAC

http://www.nationalhvac.com/

The content of the site describes this Canadian company's services, including indoor environment consulting, building ventilation assessments, field investigations, air testing, and remediation management services.

Nationwide Safety Seminars

http://www.safety4u.com/

This firm provides electrical safety training programs, expert witness services, and safety consulting.

Nevada Technical Associates

http://www.ntanet.net

The company specializes in training in radiation safety, radiochemistry, and related areas. Read about upcoming courses and the company's related services.

Occupational Medicine Center of Tuscarawas County (Ohio, U.S.A)

http://web1.tusco.net/omctc/index.htm

Read about the center's occupational health care services for businesses and their employees.

Occupational Safety Consultants

http://www.mindspring.com/~stephens/oschome.html

Information about the company's services is provided.

OH&S Consulting Services

http://www.hronline.com/dr/

Dilys Robertson is a Toronto-based occupational health and safety consultant and writer. Read about her consulting services, which include training design and delivery and research work with joint health and safety committees.

OHS Associates, Inc.

http://www.ohsa.com/

Read about the company's services: occupational health and safety consulting services and training, NFPA ratings, HMIS ratings, and litigation support services.

Operation Safe Site

http://www.opsafesite.com

This is a commercial site from the above-named consulting company. The site provides information about their products and services, along with links to sites and files of interest to the construction industry.

OSHA-Data

http://www.oshadata.com

OSHA-Data's site describes the company's OSHA compliance services, including searching its database of OSHA violation and enforcement records.

OSHEX/ESA

http://www.oshex.com

OSHEX is a State of New York safety equipment provider of noise control and equipment safety solutions. Environmental Safety Associates (ESA) provides safety and noise-related consulting services, including litigation. Information about the company's services is provided.

Paint-Safe(R)

http://www.ibpat.net/paintsafe/

This Web page describes Paint-Safe(R), a company that appears to do health and safety research.

Paskal Environmental Services

http://www.erols.com/paskal/index.html

Read about the company's environmental and OSH training services.

Pathfinder Associates

http://www.webcom.com/pathfndr/welcome.html

The company's Web pages provide current information on OSHA regulations and citations. There are also company product and service information and various health and safety documents.

Plant and Machinery Assessing Services

http://www.cyberstrategies.net/pamas/

This Australian consultancy develops safety programmes for the building industry. Services include preparation of safety plans, risk assessment for personal protective equipment, certificates of competency, and training programmes.

Preventive Action Safety Services Ltd.

http://www.autobahn.mb.ca/~passltd/

Read about this Winnipeg, Canada-based health and safety company and its services.

Pro-Safety Inc.

http://members.aol.com/safetypros/prosafe.htm

Read about this company's safety services and programs and its various safety software packages.

Professional Development Unit, uksafety.net

http://www.uksafety.net/

This U.K. safety and technical training consultancy's site provides a database of U.K. consultants; a bulletin board, which includes "hot gossip," frequently asked questions, news of upcoming events and a "trading post"; and a catalogue of the company's training programmes.

Professional Safety Consultants Inc.

http://www.pscsafe.com

The site provides information about the staff and services if this occupational safety consulting company.

Quality Safety Services Inc.

http://www.qssi-tx.com/

This Texas based consulting company provides OSHA, EPA, DOT, and TNRCG safety and environmental training programs, safety plant inspections, and accident prevention program surveys. Read about its services.

RACCO Safety

http://www.raccosafety.com.br/

The site describes this Brazilian company's safety, industrial hygiene, and occupational health products and services. Most information is in Portuguese, with a general description in English.

Rebernigg Associates

http://www.sulphurcanyon.com/magiers/default.htm

Read about the company's safety engineering, fire protection, and laser safety services.

Resource Environmental Associates

http://www.rea4ehs.com/

Information is provided about this Toronto, Canada-based environment, health and safety consulting company, its services, and publications.

Restoration Environmental Contractors

http://Home.InfoRamp.Net/~restcon/

This Canadian company provides services such as asbestos abatement, lead abatement, fire restoration, mold and fungus removal, and plant cleanups. The Web site provides documents outlining step-by-step actions to be taken for most of these types of cleanup/restoration.

Safe-At-Work.com

http://www.safe-at-work.com/

This site is the host for HearSaf 2000 and James Anderson and Associates. Various documents related to hearing conservation and demonstration software are available for download.

Safe-T-Proof

http://www.safe-T-Proof.com/

The company specializes in earthquake preparedness products and services. They have offices in Los Angeles and Tokyo.

Safety Advantage

http://www.SafetyAdvantage.com

This safety and industrial hygiene consultants' home page provides information about their services and training packages.

Safety Advisory Services

http://www.safetyadvisory.com/

Read about the company's loss control and substance abuse programs.

Safety Check Consulting

http://www.lights.com/safetycheck/

Information about this Saskatchewan, Canada-based safety consulting company's services: safety audits, training programs, emergency preparedness, and accident/incident investigation.

Safety Management Services

http://www.cyberg8t.com/~jprice/

This consulting company's site provides access to its Online Safety Program, including an Injury and Illness Prevention Program, Hazardous Communications Program, MSDSs, and links to other sites.

Safety Mining and Eng.

· http://www.instanet.com/~pfc/SME/safety_1.html

Safety Mining and Eng. is a group of safety professionals who provide safety consulting and training to both mining and industry.

Safety Online

http://www.safetyonline.net

This virtual mall provides information about products and services in a variety of health and safety areas. Excerpts from, and subscription for *Industrial Safety and Hygiene News* are provided. View the full text of their *Safety Currents* newsletter.

Safety Systems Inc.

http://www.safetysystemsinc.com

The site describes this environmental health and safety consulting company's services.

Safety Training and Consulting Services

http://www.stacs.com

The site provides information about this Canadian consulting company. Its services include training, consulting and recruiting.

Safety Training and Tracing, Inc.

http://www.standt.com/

The company's OSH consulting services and software "Saf-T-Man Safety Manager" are described.

SAREC

http://www.sarec.ca

Read about the expertise and services of this Canadian occupational hygiene consulting company.

Scientech Inc.

http://www.scientech.com

Read about Scientech, a worldwide technical services company specializing in environmental and safety services to the nuclear industry.

Seton's Technical and Regulatory Assistance Center (SmartTrac)

http://www3.seton.com/smarttrac.html

To quote from the site itself, Seton's "Technical and Regulatory Assistance Center will help you get the answers you need to your important safety and regulatory questions." This appears to be a free service available to the general Web-surfing public.

Shiftwork Services

http://host.mwk.co.nz/shiftwork/

Shiftwork Services is a division of the Auckland (New Zealand) Sleep Management Centre. The site provides access to the full text of the organization's newsletter, sleep tips, and guidelines for managers.

Span Corporation Occupational Health Services

http://www.spancorp.com/

The site describes this company's occupational medicine, compliance, prevention, hazard recognition, and information services.

Sulowski Fall Protection Inc.

http://ourworld.compuserve.com/homepages/AndrewSulowski/sulowski.htm

This consulting company's site provides information about its services, fall protection courses, publications, and upcoming events.

SynchroTech

http://www.schedule-masters.com/

SyncroTech is a consulting firm specializing in shiftwork, sleep, and schedule management. The company provides customized educational programs and work scheduling systems to all levels of personnel in a wide range of operational settings. Read about their services.

Thorburn Associates

http://www.ta-inc.com/

Read about this company's acoustical consulting and audiovisual system design consulting services.

Turning Point Group Inc.

http://www.turningpointgroup.com/

This company specializes in emergency preparedness, crisis management, and business recovery. Read about their services and participate in their emergency management bulletin board.

Ultimate Lifestyle Fitness

http://www.interlynx.net/ultimate/

This Hamilton, Canada-based company designs and delivers workplace wellness programs. Read about its services.

Usernomics

http://www.usernomics.com/

Ergonomics for hardware, software, and training. Find out about the company's services, or access links related to various aspects of ergonomics.

VDT Solution

http://home.earthlink.net/~ergo1/

The site provides a variety of types of information related to computer workstations and ergonomics: preferred equipment and services, a PC and laptop ergonomic guide, ergonomic product reviews, etc.

W. H. Interscience

http://www.gj.net/~whicotom/

View information about the company's safety and industrial hygiene services: consulting, publications, and software.

Waste Management, Inc.

http://www.wastemanagement.com

Read about this Illinois-based waste management company.

Water Technology International

http://www.wti.cciw.ca

WTI is an environmental technology and services company. Read about their products and services (hazardous waste, site remediation, etc.). Of particular interest is their 40-hour health and safety course, which satisfies general Canadian (federal and provincial) and U.S. (29 CFR1910.120, "HAZWOPER") health and safety training requirements for contaminated site workers.

West Arm Consulting Services

http://www.westarm.bc.ca/

This company specializes in WWW page development for EMS. Their page has extensive links to various EMS and police sites.

Products

21st Century Eloquence

http://www.voicerecognition.com/

This Florida-based company is a reseller of voice recognition products for computers. View information about their products and services.

---

31 Mile Equipment Co.

http://www.31mile.com

Read about the company's safety products and services.

3M Environmental Safety and Energy Performance Products

http://www.mmm.com/market/government/env/envindex.html

Information about 3M's products.

3M Occupational Health and Safety

http://www.mmm.com/occsafety/

The site provides several categories of information: regulatory updates (including a version of 3M's Respiratory Protection Regulations Handbook for downloading), product information, newsletters, and information about 3M's training courses.

Accu*Aire Controls, Inc.

http://www.accuaire.com/

Information is presented about the company's laboratory fume hoods and exhaust products, selected articles, and free software.

Advanced Ergonomics

http://www.advergo.com

AEI provides a variety of ergonomic services to corporations and individuals. Their products and services are described.

Aearo Company

http://www.aearo.com

This site provides information about the company's safety products. Aearo was formerly Cabot Safety.

Affirmed Medical Inc.

http://www.affirmed.com

The company provides first aid and safety products. Find out about these and their U.S. regulatory compliance products and training videos.

Alphabet Signs

http://www.alphabetsigns.com/

This company makes signs of all types. The Web site also has a "searchable database of standardized signs A to Z." View images of all these signs.

Ansell Edmont

http://www.anselledmont.com/

Information about the company's safety gloves, the company itself, and its distributors.

Arbill Industries

http://www.arbill.com

Information is provided about the company's health and safety products. You can view its newsletter and order products.

Back Be Nimble

http://www.backbenimble.com/

The company's online catalog provides information and products for back and body care. The site also includes basic information on ergonomics, back supports, and back surgery.

BackHealth U.S.A

http://www.backhealth.com

Some information is provided about back pain causes and treatments. Most of the information is about BackHealth 2000, the company's back exercise equipment.

Best Gloves

http://www.bestglove.com

Information about the company's safety gloves is presented. There is also a free downloadable program, "Comprehensive Guide to Chemical-Resistant Best Gloves."

BNA Communications Safety Training Programs

http://www.bna.com/bnac/sfprod/sftymain.html

Descriptions of the company's many video-based training programs are available.

Bouton Eye Protection

http://www.hlbouton.com/

This is a searchable catalog of the company's safety eyewear.

Canberra Industries

http://www.canberra.com

This radiation detection and analysis instrumentation manufacturer provides information about products and services, frequently asked questions (FAQ), and application notes.

Carpal Tunnel Syndrome

http://www.netaxs.com/~iris/cts/

Access summary information about carpal tunnel syndrome, along with links to other carpal tunnel resources, ergonomic products, etc. Information is also provided about Metro Smallwares' Compfort keyboard extender.

Ceske pracovni zdravi a bezpecnost

http://www.markl.cz

Czech occupational health and safety. The site contains a description of this consulting company's products, a directory of OSH organizations in the Czech republic, and Czech OSH legislation. All information is available in Czech, and some in English.

ChinaHawk Enterprises Ltd.

http://home.hkstar.com/~chinahwk/home.html

Information about this Beijing-based company's disposable safety products for the health care, industrial, and food services industries. Offices are in Beijing and Hong Kong.

Combo Ergonomic and Contract Furnishings

http://www.combo.com

View the company's ergonomic furniture catalog and a "collection of information and solutions pertaining to office ergonomics."

Compliance Management Group—OSHA Safety Plans and Labor Law Posterbook

http://www.oshaman.com/

This company sells written policy and procedures documents, complete safety plans, training materials, training videos, CD's, and labor law posters. View details of their products.

Conney Safety Products

http://www.conney.com

The site provides a portion of the company's catalog of "Safety, Industrial, Maintenance, and First Aid Supplies for the Workplace and Home Environments."

Conveyerail Systems Ltd.

http://www.conveyerail.com/

View descriptive information about VacuMove, a vacuum lifting system.

Crouch Fire and Safety Products

http://www.jtcrouch.com/

Find out about the company and its fire protection and public safety lighting products and services.

Crowcon Gas Detection

http://www.crowcon.com/

View information about gas detection technologies, along with the (U.K.) company's list of international distributors.

DeskEx Body Awareness System

http://www.iftech.com/products/deskex/

The site describes this computer workstation wellness program, consisting of an online wellness manual, exercise reminders, break reminders, and an activity monitor.

E H Lynn Industries

http://www.ehlynn.com/

The company distributes liquid and dry bulk handling equipment, tank truck equipment, hose, valves, gaskets, etc., for the chemical, petroleum, and dry bulk markets. Read about the company and itsproducts.

E. D. Bullard Co.

http://www.bullard.com

The site contains this company's catalogs of personal protective equipment: respiratory protection, head protection, and fire and rescue. A list of distributors is also provided.

Effective Prevention

http://www.interlog.com/~effprev/

Information is provided about Effective Prevention, a skin cream that "repels bodily fluids and many biological hazards in the medical and food services profession."

ENMET Corporation

http://www.enmet.com

Read about this company's industrial gas and vapor monitoring instrumentation.

Entropy Software, OOS Software

http://www.albatross.co.nz/~miker/main.htm

The site describes this New Zealand company's OOS software, a program that tells you when to take a break from your computer to help prevent/alleviate OOS (occupational overuse syndrome, or repetitive strain injuries). Download a copy of the software from this site.

Environmental Support Solutions

http://www.environ.com

View information about the company's services, including a number of indoor air quality (IAQ) courses, a heating, ventilation, and air conditioning (HVAC) mall, etc.

Ergonomic Central

http://www.ergocentral.com/

This site is an electronic mall of ergonomics products and services.

Ergonomic Design Inc.

http://www.ergodesign.com/

View information about the company's computer-related ergonomic products: wrist rests, keyboard holders, and document holders.

Ergonomic Sciences Corporation

http://www.ergosci.com

The goal of this Web site is to provide up-to-date information about the science of workspace design and function and present information that will help you choose the most beneficial ergonomic products to address specific ergonomic issues.

FacilitiesNet

http://www.facilitiesnet.com

FacilitiesNet is "an interactive information service for facility professionals." Members can access two online magazines, reports, an online bookstore, and product information.

First Aid 1st

http://www.seadrive.com/first.aid.1st/

This Alberta-based company provides first aid supplies and services to business, sports groups, nonprofit organizations, and individuals. View product descriptions and locate distributors.

Fisher Scientific

http://www.fisher1.com

View Fisher's product catalogs and MSDSs and order products online.

Flinn Scientific

http://www.flinnsci.com

This science supply company site provides a lot of useful information, including laboratory design tips and numerous documents on chemical safety.

Government Institutes Division, ABS Group Inc.

http://www.govinst.com/

This company provides over 200 books on topics in the environmental, health, safety, telecommunications, and Internet fields. They also provide training courses, videotapes, and computer-based training.

GPS Gas Protection Systems Inc.

http://www.GasProtection.com/

The site provides information about the company's automatic gas shutoff systems.

Grace Sales, Inc.

http://www.gracesales.com/

Read about the company's safety products for the firefighting community.

HBA International

http://www.hbainternational.com/

This company is a supplier of medical equipment for hospitals and emergency medical services. Read about HBA's various product lines.

Inter Consulting Systems

http://www.wionline.com/ics/

Jerry E. Smith, former Los Angeles City Fire Captain and retired California OES Fire-rescue Assistant Chief, has written a number of books for firefighters. This site describes his publications. This is also home to the "Emergency Grapevine," a Web-based discussiongroup for emergency discussions.

Lab Safety Supply

http://www.labsafety.com

As well as providing access to the company's catalog of lab safety equipment, the site presents a number of useful documents in a variety of health and safety categories.

Lake Marking Products

http://www.lake-marking.com

See the company's catalog of signs, equipment tags, etc.

LeMitt

http://www.lemitt.com

Commercial information about this temperature-regulated therapeutic cold pack for hands and wrists. There is also a link to some repetitive strain injury information.

Mac's Fire and Safety

http://www.macsfire.com

This site provides a catalog of rescue tools, fire fighting equipment, and safety gear.

Major Safety Service

http://www.pilot.infi.net/~majsaf/

Products and services provided by this environmental instrument and safety equipment supplier are described.

Matcor Global Products

http://www.matcor.ca/

Information is provided about this company's sensing products, including a UV intensity meter.

Maxi-Signal Products Co.

http://www.computer.net/~owlseye/maxi.html

This site provides access to Maxi-Signal's product catalog of heavy-duty, visible-audible signals, sirens horns, lighting, and safety equipment.

Mettam Safety Supply

http://www.mettam.com

Read about the company and its products. There are also links to selected U.S. resources.

Micronite

http://www.micronite.com

This commercial site provides information about the company's repetitive strain injury prevention software and publications.

Mouse Escalator

http://www.4amouse.com/

Read a description of this ergonomic product. It is a "positive inclined work surface" for a computer mouse.

Mouse Escalator

http://www.ergonomic-pad.com/

Description of this ergonomic product. It is a "positive inclined work surface" for a computer mouse.

National Safety Equipment Outlet

http://www.safety-source.com/

Product information from this mail-order company is provided.

Nikos International

http://www.antinfortunistica.it

The site describes the company's industrial cleaning products and personal protective equipment. Most information is in Italian.

North American Detectors, Inc.

http://www.nadi.com/

A description of the company's carbon monoxide detector is provided. There is also a frequently-asked-questions (FAQ) document about carbon monoxide.

Northern Industrial Supply Company

http://www.nisco.net/

This Ontario, Canada company sells industrial heating, ventilation, and air conditioning equipment. Read about the products they provide.

OMCO Air Treatment

http://www.omco.be/

Read about this Belgian company's ventilation products. Information is in French and Dutch.

Operation Safe Site

http://www.opsafesite.com

This is a commercial site from the above-named consulting company. The site provides information about their products and services, as well as links to sites and files of interest to the construction industry.

OSHEX/ESA

http://www.oshex.com

OSHEX is a State of New York safety equipment provider of noise control and equipment safety solutions. Environmental Safety Associates (ESA) provides safety and noise-related consulting services, including litigation. Information about the companies' services is provided.

Pathfinder Associates

http://www.webcom.com/pathfndr/welcome.html

The company's Web pages provide current information on OSHA regulations, citations, etc. There is also company product and service information and various health and safety documents.

PearlWeave Safety Netting

http://www.pearlweave.com

View information on fall protection debris containment and OSHA compliance, and PearlWeave products for the construction and safety industries.

Pioneer Products Inc.

http://www.pioneers.com

The site contains information about products from this specialty chemical company.

Pre-ventronics

http://www.primenet.com/~prevent/

Read about products and services provided by this commercial alarm and security company.

Pro Grip Safety Sole

http://www.uleth.ca/~gadd/index.htm

Information about Pro-Grip nonslip safety soles.

Pro-Am Industrial Safety and Hygiene

http://www.pro-am.com

This is one of the best developed "safety malls" on the Web. Find products, services, links to OSHA standards, and a good directory of Internet resources.

Pro-Am Safety Product Store

http://www.pro-am.com

This is a "safety cybermall" with a large number of safety products available for sale.

Production Video L. M. Inc.

http://www.generation.NET/~pvlm/

This commercial site provides detailed information about Production Video's occupational health and safety training videos.

Proformix

http://www.proformix.com

The company provides ergonomic keyboard platforms and related products and a suite of ergonomic management software.

Protecta Fall Arrest Systems

http://www.protecta.com

Detailed information is provided about the company's fall protection, confined space entry and retrieval equipment, rescue devices, and heavy-load fall arrestors.

ProtectAide Inc.

http://www.protectaide.com

This site provides an illustrated catalog of the company's personal protective equipment products and information about its bloodborne pathogens training program on CD-ROM.

Questec Input Devices

http://www.questecmouse.com

Read about this company's computer related products: mice, mice pens, trackballs, ergonomic keyboards, and antiglare filters.

Quick-Aid, Inc.

http://www.quick-aid.com

Read about products and services from this medical and safety supply company.

RACCO Safety

http://www.raccosafety.com.br/

The site describes this Brazilian company's safety, industrial hygiene, and occupational health products and services. Most information is in Portuguese, with a general description in English.

RMI Media Productions

http://www.rmimedia.com/products/index.html

View/search the company's catalog of training videos, including a number of safety and health-related titles.

Saf-T-Gard International

http://www.saftgard.com/

The site provides access to the company's catalog of industrial safety products, its newsletter, and selected links to other sites.

Safe Computing

http://www.safecomputing.com

Access the company's extensive catalog of computer ergonomics-related products.

Safe Shop Tools

http://www.safeshop.com/

Descriptive information is provided about the company's safety products for the truck repair industry.

Safe-T-Proof

http://www.safe-T-Proof.com/

The company specializes in earthquake preparedness products and services. They have offices in Los Angeles and Tokyo.

Safe-tec Canada

http://www.safetec.com

This is an online catalog of the company's infection control and personal safety products.

Safeco Inc.

http://www.safecoinc.com/

Information is provided about this distributor of health and safety products, its distributors, and product catalog.

Safemate International

http://www.safemate.com

Read about this company's antislip products.

SafeSun Personal UV Meter

http://www.SafeSun.com/

Read about this device that "lets you know how much UV you've absorbed throughout the day." The site also provides basic UV information and technical scientific data.

Safety Boot

http://members.aol.com/safetmaker/index.htm

Read a description of Safety Boot, "a reusable base for constructing free-standing temporary guardrails that exceed OSHA standards."

Safety Connection

http://www.safetydeck.com

Information about the company's products and selected links to other sites is provided.

Safety Solutions Inc.

http://safetysolutions.com/

Find out about this full range safety products company, view product descriptions, and ordering information.

Safety Superstore

http://www.safetysuperstore.com

The site contains this company's catalog of safety products. This Canadian company will ship anywhere in the world.

SafetyCare Group of Companies

http://www.safetycare.com.au

SafetyCare is a worldwide supplier of occupational health and safety video programs, courses, and manuals. View their online video catalog and list of offices in various countries.

Seton

http://www.seton.com

View Seton's identification and safety products catalog.

Seton Identification Products

http://www.seton.com

This extensive, well-organized site provides detailed information about Seton's safety products, MSDSs for its chemical products, and a health and safety resources directory.

Shiftwork Systems, Inc.

http://members.tripod.com/~Shiftwork/

Read about the company's "Circadian Lighting Technology," plus links to a number of documents and sites containing shiftwork-related information.

Simply Safety Store

http://www.island.net/~dnagle/

The store offers training, information, and products for safety in the workplace and at home. Read about the company's offerings.

SKC Online

http://www.skcinc.com/

View air sampling equipment information from the manufacturer. The site also contains frequently asked questions about the company's air sampling instrumentation and fill-in forms for requesting more information.

Stress Assessment Profile

http://www.opd.net/stress.html

Read about this tool for stress assessment, produced by Organizational Performance Dimensions.

Surety Manufacturing & Testing Ltd.

http://www.Suretyman.com/

This Edmonton, Canada-based company manufactures and sells equipment for fall protection and confined space entry. View the company's catalogue and read about training videos.

Surviving the Workplace Jungle

http://vvv.com/m2/swj/

The site presents detailed information about this workplace violence video and workbook.

Technika Scientific and Industrial Hygiene Measurement Instruments

http://www.technika.com/

Technika is a supplier of health, safety, lab, and field meters and detectors. "Our site also has general information on various areas of safety, how metering in that particular area is done, and what specifications you should look for."

Uni-Hoist

http://www.uni-hoist.com

Find out about the company's confined space products. The site also includes a photo gallery, links to related articles and sites, OSHA regulations, and a newsletter.

Uvex—Serving Safety

http://www.uvex.com

The company's various lines of protective eyewear are highlighted.

VST Chemical Corp.

http://ecki.com/vst/

Information about this company's firefighting chemicals is available. There are also links to other fire sites.

W. S. Wood Associates Ltd.

http://www.wswood.com/

This Canadian company sells, services, and inventories static elimination products for industrial and electronic/cleanroom applications, ESD consumable products, as well as mold release agents. Read about the company's products and services.

Waldmann Lichttechnik

http://www.waldmann.de/

View information about this German company's range of workplace lighting products.

Wessoclean Environmental Hygiene

http://www.wessoclean.com/

Read about the company's products for cleaning and disinfecting "water supplying pieces of equipment and plants."

Western Hard Hats

http://www.westernhardhat.com/

Information is provided about the western-style hard hats this company distributes.

Safety Software

3D Safety CD-ROM

http://www.3dsafety.com/

View information about, and previews from, this CD-ROM of safety and health related graphics files.

Accu*Aire Controls, Inc.

http://www.accuaire.com/

Information is presented about the company's laboratory fume hoods and exhaust products, selected articles, and free software.

AcuTech Consulting Inc.

http://www.acutech-consulting.com/

AcuTech provides process risk management services to industries handling hazardous materials. The site describes the company's consulting services, training, and risk management software.

Aurora Data Systems

http://www.alaska.net/~aurora/

Read about the company's database applications: Chemical Information Management System, Equipment Maintenance System, and Training Records Management System.

Bara Environmental Compliance

http://www.bara.com

Bara Environmental Solutions is a Texas-based consulting firm that "creates advanced Environmental Safety solutions through computer tracking of hazardous materials and safety equipment." Read about the company's services.

BLMC Computer Based Training Programs

http://www.blmc.com/

Information is provided about this Canadian company's computer-based health and safety training programs.

Chemical Abstracts Service

http://info.cas.org

Read about CAS, its products, and services. Find out about the STN database service and view a copy of the CAS registry structured searching manual.

Chempute Software

http://www.chempute.com

This South African company produces and provides a variety of software packages for chemical process industries. Plant safety and quality control programs are available. Download demos and tutorials.

Dakota Decision Support Software

http://www.dakotasoft.com

Read about and view sample screens from Dakota's EH&S auditing software.

DataChem Software

http://www.datachemsoftware.com

The company's site provides information about its various software packages "that help you study and practice for your professional certification examination."

Dawnbreaker Consultants

http://204.112.119.2:80/dawnbreaker/

Information about and downloadable evaluation copies of their software are available. There are several programs of interest, including WorkSafe Tools (Occupational Safety and Health Software) and EnviroSafe (Environmental Risk Assessment Software).

Dyadem International

http://www.dyadem.com

Read about and/or download trial versions of this company's "software tools for the Risk Industry."

Eclipse Software Technologies

http://www.eclipsesoft.com

Information about and downloadable demonstrations of "MSDS Wizard" and "MSDS Scan Wizard," MSDS management software.

EHS Diagnostics Inc.

http://www.ehsdiagnostics.com

EHS Diagnostics Inc. (EDI) is a Canadian environment, health and safety consulting firm, based in Calgary, Alberta, Canada. It provides both consulting services and health and safety software development. Read about the company's services.

Entropy Software, OOS Software

http://www.albatross.co.nz/~miker/main.htm

The site describes this New Zealand company's OOS software, a program that tells you when to take a break from your computer to help prevent/alleviate OOS (occupational overuse syndrome or repetitive strain injuries). Download a copy of the software from this site.

EnviroDx

http://www.auhs.edu/envirodx/

"EnviroDx is a multimedia, case-focused, computer-based learning program on environmental-related diseases. The organizing metaphor for EnviroDx is an exploratory 'virtual clinic' affiliated with a busy medical school. The program user takes the part of a practicing physician faced with a patient with an unknown disease or condition that is possibly caused by exposure to environmental factors."

EnviroWin Software Inc.

http://www.envirowin.com

The site contains this company's EHS software catalog: chemical inventory, emergency response, OSHA compliance, and risk assessment.

Euroware Associates Software

http://www.euroware.com/index.html

Information is provided about the company's MSDS and chemical-related software packages.

Fire Safety Engineering Group

http://fseg.gre.ac.uk/

This Greenwich U.K. consultancy group describes itself as "one of the largest research groups in the world dedicated to the development and application of mathematical modelling tools suitable for the simulation of fire related phenomena." Read about the company's services.

First Report—Workers' Compensation and OSHA Recordkeeping Software

http://www.firstreport.com/

Really Useful Software provides information about this software package. They also provide "commentary and information" about workers' compensation and safety issues.

Genium Publishing

http://www.genium.com

This site provides information about Genium's safety and health software programs and MSDS collection. An online safety and health glossary and database of hazard labelling are also provided.

Hyginist

http://home.wxs.nl/~ihpc/

Produced by Scheffers Industrial Hygiene Publishing and Consultancy in the Netherlands, HYGINIST is an "industrial hygiene statistical tool. It evaluates the exposure data collected with the standard exposure assessment strategies."

Interactive Media Communications

http://www.safetysite.com

Information is provided about the company and its multimedia occupational and environmental health and safety training programs on CD-ROM, videodisc, and videotape.

InteractiveWare Inc.—HAZCOM training software

http://www.interactiveware.com

Read about and download a demo of this Windows-based hazard communication training software.

J.J.Keller & Associates

http://www.jjkeller.com/

View the company's safety product catalog, download sample software, and view descriptions of its printed publications.

KnowledgeWare Communications Corp.

http://www.kccsoft.com/

This N. Vancouver, BC company provides several software packages: Simply Safety! Safety Management, WHMIS CBT Suite, and Trax & Fax MSDS Software. View information and download demonstrations of these software packages.

Map 80 Systems Ltd.

http://www.map80.co.uk/

Read about this U.K. company's labelling software for chemical, pharmaceutical, industrial, and retail settings.

Micronite

http://www.micronite.com

This commercial site provides information about the company's repetitive strain injury prevention software and publications.

MJW Corporation

http://www.mjwcorp.com/

MJW provides a variety of radiological consulting services and software solutions for health physics and other technical applications. Read about the company's services and download demos of their software.

Modern Technologies Corp., Environmental Technologies Division

http://www.etd.modtechcorp.com/

The site provides information about the company's environmental consulting, environmental management software, industrial hygiene, and safety services.

MSDS Pro

http://www.msdspro.com

Download a demonstration of this MSDS management software by Aurora Data Systems.

MSDS2

http://www.whmis.com

Read about and "test drive" BC Hydro's material safety data sheet management system.

Occupational Health Research

http://ohr.systoc.com/index.htm

Read about the company's software, SYSTOC, a program to manage patient care. There are also a lot of links to resources for occupational health professionals: U.S. regulatory info, hazardous materials, etc.

Patient's Guide to Cumulative Trauma Disorder

http://www.sechrest.com/mmg/ctd/

Fairly detailed information is provided about cumulative trauma disorders or repetitive motion injuries. Information about Medical Media Group and their software is also provided.

Pro-Safety Inc.

http://members.aol.com/safetypros/prosafe.htm

Read about this company's safety services and programs and its safety software packages.

Proformix

http://www.proformix.com

The company provides ergonomic keyboard platforms and related products, plus a suite of ergonomic management software.

Reason—Decision Systems Inc.

http://www.rootcause.com/

View a description of the company's REASON software—software that "brings professional level Root Cause Analysis capabilities for operations problem-solving to the Quality / Safety / Operations professional."

RISK—An OSH Risk Assessment and Control System

http://www.safetyassistant.com/RISK.html

"The RISK program provides a step by step approach to risk assessment. Each hazard is assessed individually, providing you with a quantitative value and an action plan for each. Then an action plan is developed for the assessment as a whole. The final printed report becomes your 'worksheet'."

RiskSafe 98

http://www.lican.com/lrs/risksafe_home.html

This site describes and provides access to a downloadable demo of RiskSafe 98, a software package to help you "identify, prioritize, and decide where to spend your resources to manage identified risks." The software is produced and provided by Liberty Risk Services.

Rmis.com (Risk Management Insurance Safety)

http://www.rmis.com

This content-rich site provides access to over 8000 risk management-related resources: databases, publications, templates and checklists, and software. Some material is free and some is fee-based.

Safety Assistant—Software for Workplace Health and Safety

http://www.safetyassistant.com

View descriptions and sample screens from this Australian company's various OSH software packages.

Safety Officer II

http://www.safetyofficer.com/

Safety Officer II is a chemical information software system for converting data associated with chemicals into end-user specific MSDSs, custom Avery labels, and management reports. Detailed descriptions of the software are provided.

Safety Training and Tracing, Inc.

http://www.standt.com/

The company's OSH consulting services and software "Saf-T-Man Safety Manager" are described.

Safety4

http://www.safety4.com/

Safety4, provider of chemical protection wear, hosts a free online chemical protection guide, complete with hundreds of chemicals and their associated permeation rate.

SafetyScape Corp.

http://www.safetyscape.com/

The company produces Safety Navigator (TM) and Safety Trakker (TM) occupational health and safety management software. View information about these packages and download a demo.

Solutions Software Corporation

http://www.env-sol.com/

This site contains descriptions of the company's CD-ROM products, many of them health and safety related, plus links to various sites.

Star Solutions

http://www.safestar.com

Read about "Safe Star," a software package that manages employee safety information and accident reporting.

Text-Trieve Environmental Health and Safety Page

http://www.halcyon.com/ttrieve/

Read about the company's electronic text-management systems that "present government regulations in easy-to-use electronically published formats." A number of U.S. regulatory subject areas are described.

WHMIS 96

http://www.bbsi.net/50para/p_web.html-ssi

Read about this CD-ROM-based WHMIS (Canadian right-to-know legislation) training package produced by C.D. 50e PARALLELE INC.

WHMIS Software Training

http://www.niagara.com/blmc/

This site contains a brief description of BLMC's WHMIS training software, with contact information.

WorkTime

http://www.timedomain.com/

Information is provided about WorkTime, an employee and resource scheduling program for Windows and Macintosh, created by Time Domain Inc.

---

4. Academic Sites

Academic Programs in Health and Safety

Center for Ergonomics, University of Michigan

http://www.engin.umich.edu/dept/ioe/C4E/

Information about the center's education and research in ergonomics, programs, and publications.

Center for Industrial Ergonomics, University of Louisville (Kentucky)

http://www.spd.louisville.edu/~ergonomics/

Read about the center's research and educational activities, which focus on integrating people, organization and technology at work, and improving quality and productivity through ergonomics, safety, and health management.

Center for Occupational and Environmental Health, University of California

http://ehs.sph.berkeley.edu/coeh/

View information about this research center, its academic programs, continuing education courses, and research projects.

Center for Prevention of Violence and Control, University of Minnesota

http://www.umn.edu/cvpc/

The Center for Violence Prevention and Control was developed to facilitate interdisciplinary collaboration in research and graduate education that can ultimately affect the prevention and control of violence. Read about the center's research, funding opportunities, and educational programs.

Cornell OSH Certificate

http://www.ilr.cornell.edu/depts/extension/oshcert.htm

Read about Cornell University's "Occupational Safety and Health" Certificate program. Some of the courses can be taken on the Web.

Cornell University, School of Industrial and Labor Relations

http://www.ilr.cornell.edu/

The purpose of this site is to provide information about the School of Industrial and Labor Relations at Cornell University and its library, and to disseminate information on all aspects of employer-employee relations and workplace issues.

Environmental and Occupational Health Resources

http://astro.ocis.temple.edu/~rpatters

Dr. Patterson's home page contains links to Temple University's Environmental Health Master's degree program, plus various environment and occupational health-related Web sites.

Faculty of Occupational Medicine, Royal College of Physicians (U.K.)

http://www.facoccmed.ac.uk/

The faculty is "a professional and academic body empowered to develop and maintain high standards of training, competence and professional integrity in occupational medicine." Read about its various educational programmes, publications, and newsletter.

Griffith University, Faculty of Environmental Sciences

http://www.ens.gu.edu.au

View information about the faculty itself, along with links to other Australian and international environmental sites.

Griffith University, Faculty of Health and Behavioural Sciences

http://www.ua.gu.edu.au/hbk/hbs_ind.htm

Information about courses provided by this Queensland, Australia university is available. The School of Occupational Health and Safety is within this faculty.

Health Promotion Center, University of California Irvine

http://www.seweb.uci.edu/users/dstokols/hpc.html

"UCIHPC is a research and consulting unit operating within the School of Social Ecology at the University of California, Irvine. We are interested in all aspects of health promotion. Our research focuses on comprehensive, integrated approaches to health promotion, especially on workplace-based health promotion."

Hudson Valley Community College, WWW courses

http://www.hvcc.edu/academ/dlc/cornell/index.html

Presented in collaboration with Cornell University's School of Industrial and Labor Relations, online courses are offered as part of Cornell's Labor Studies Program

---

Injury Prevention Research Unit, University of Otago, New Zealand

http://www.otago.ac.nz/Web_menus/Dept_Homepages/IPRU/

The site provides information about the unit, research it undertakes, and publications produced.

Inver Hills Community College Emergency Health Services program

http://www.ehs.net/IHCC/

The site provides information about this Minnesota college's EHS program, along with links to emergency medical services information.

Massey University (New Zealand), Dept. of Human Resource Management

http://www.massey.ac.nz/~wwhrm/teachosh.htm

This site contains a description of Massey's Postgraduate Diploma in Occupational Safety and Health (Dip OSH).

McGill University, Occupational Health Sciences

http://www.mcgill.ca/occh/

The site provides information about their academic programs, activities/news/seminars, research groups/labs, and publications.

NIOSH Educational Research Centers (ERCs)

http://rocky.utah.edu/erc.html

This page, provided by the Rocky Mountain Center for Occupational and Environmental Health, provides links to all 14 U.S. NIOSH ERCs.

Oklahoma State University, Environmental Health and Safety, Online Training Modules

http://www.pp.okstate.edu/ehs/MODULES/HOME.HTM

The site provides online training on a growing number of EHS topics, including asbestos awareness, bloodborne pathogens, compressed gas cylinders, fire extinguishers, hazard communications, office ergonomics, and personal protective equipment in laboratories. More detail about this site is available in the Top 50 Sites section.

Rocky Mountain Center for Occupational and Environmental Health

http://rocky.utah.edu/

Located on the campus of the University of Utah, the Rocky Mountain Center is one of 14 NIOSH Educational Research Centers. The site provides information about the center's various educational programs.

Southern Africa Occupational Health Web

http://www.und.ac.za/und/med/comhlth/occ_hlth.html

Maintained by the Occupational Health Programme, Dept. of Community Health, University of Natal, the site describes the university's Occupational Health Programme, there is a discussion paper on women in the workplace, and there are links to various occupational health sites.

Tulane University, Center for Applied Environmental Public Health

http://caeph.tulane.edu

Read about this university's masters level programs in OSH, including a two-year Internet-based MPH degree program in OSH Management.

University of Alberta, Medical Laboratory Science Program

http://www.ualberta.ca/~medlabsc/

Read about their undergraduate and graduate programs and Internet resources available to students.

University of Birmingham (U.K.), Institute of Occupational Health

http://www.bham.ac.uk/IOH/

Read about the institute's postgraduate programs and upcoming local, national, and international conferences and meetings.

University of British Columbia, Occupational Hygiene Program

http://www.interchg.ubc.ca/occhyg/

The site describes UBC's Master's program in occupational hygiene.

University of California Davis, Department of Epidemiology and Preventive Medicine

http://www-oem.ucdavis.edu

Information is provided about the department, its programs, and research.

University of Edinburgh, Health Environment and Work

http://www.med.ed.ac.uk/hew/

The site provides a long list of "tutorials" (educational resources) on a variety of occupational and environmental health topics, as well as information on Edinburgh's academic programs. There is also an excellent directory of primarily European Internet resources. More detail about this site is available in the Top 50 Sites section.

University of New South Wales, Department of Safety Science

http://argus.appsci.unsw.edu.au/

Read about the department's courses and research.

University of Vienna (Austria), Department of Occupational Medicine

http://www.univie.ac.at/Innere-Med-4/Arbeitsmedizin/

Information is provided about the department, its research, and personnel. Most information is in German. Summary information is in English.

University of Washington, Department of Environmental Health Library

http://weber.u.washington.edu/~dehlib/

The site provides a series of links to environmental health publications, theses, and academic programs at the University of Washington and elsewhere.

Université catholique de Louvan, Unité de Toxicologie et de Medecine du Travail

http://www.md.ucl.ac.be/entites/esp/toxi/

The site contains a description of the unit, its members, educational programs, and a list of scientific publications.

Work Environment Program, University of Massachusetts Lowell

http://www.uml.edu/Dept/WE

Read about the Work Environment Program (WEP), an academic and research graduate program whose main role is to educate scientists to study and evaluate workplace factors that affect the health of workers.

University and College Health and Safety Departments

Baylor College of Medicine, Environmental Safety

http://www.bcm.tmc.edu/envirosafety/

Read biosafety, chemical safety, and radiation safety policy documents and guidelines at Baylor.

Baylor College of Medicine, Office of Environmental Safety

http://research.bcm.tmc.edu/es/enviro.html

The site provides access to various policy and guidelines documents and manuals. As examples, their Chemical Hygiene Plan, Biological Safety Exposure Control Plan, and Radiation Safety Manual are available.

California State Polytechnic University Pomona, Environmental Health and Safety

http://www.csupomona.edu/~ehs/

The site provides access to MSDSs and health and safety manuals and guidelines.

College and University EH&S Units

http://www.uky.edu/FiscalAffairs/Environmental/otherehs.html

This is a directory of hundreds of North American university and college environmental health and safety Web sites.

Denison University Safety

http://www.denison.edu/sec-safe/safety.html

The university's chemical hygiene plan, MSDS links and information, and training information are provided. Check out Madame Curie in protective clothing.

Diving Safety Office, East Carolina University

http://ecuvax.cis.ecu.edu/academics/schdept/diving/diving.htm

"The purpose of this office is to ensure safe and effective diving practices and to provide support for the research, educational, and recreational diving programs on campus."

Hong Kong University of Science and Technology, Safety and Environmental Protection Office

http://www.ab.ust.hk/sepo/

The site contains the university's health and safety manual and various guidance documents.

Indiana University Environmental Health and Safety

http://www.indiana.edu/~ehswww/

The site provides information about and from the various divisions of the department: environmental management, occupational safety and health, and research safety. View documents such as their Emergency Procedures Handbook and Safety Handbook.

Indiana University, Physical Plant

http://www.indiana.edu/~phyplant/pubs/safeman.html

The university's safety manual is available here.

Louisiana State Univ., Office of Occupational and Environmental Safety

http://130.39.86.58/page1.htm

Documents describing the office's services are available. Subject areas include safety and environmental program news, emergency procedures and information, and safety and environmental program documents and information.

McGill University, Environmental Safety Office

http://www.mcgill.ca/eso/

This site provides information about staff and services provided by the office, links to McGill's Occupational Health Programme, and various McGill policy documents, including its Laboratory Safety Manual.

McMaster Univ., Faculty of Engineering and Science, Safety Handbook
http://www.eng.mcmaster.ca/safety/

This is the full text of the Safety Handbook from this Hamilton, Canada university.

Michigan State Univ., Office of Radiation, Chemical and Biological Safety
http://www.orcbs.msu.edu

Various documents are presented, including MSU's chemical hygiene program, their chemical lab safety checklist, radiation safety manual, and *Safe Science* newsletter.

Oklahoma State University, Environmental Health and Safety
http://www.pp.okstate.edu/ehs/index.htm

Access compliance manuals, programs, safety training information, safety related information sheets, and checklists.

Portland State University, Environmental Health and Safety
http://www.pp.pdx.edu/FAC/Safety/

The site provides access to a variety of university documents, such as their chemical hygiene program, lockout-tagout, and pesticide/herbicide use.

Princeton University, Occupational Health and Safety
http://www.princeton.edu/~ehs/

Information is provided about OH&S at Princeton, there are links to numerous MSDS collections, and there is a Web version of their Health and Safety manual.

Queen's University, Environmental Health and Safety
http://www.safety.queensu.ca:443/safety/safety.htm

The site provides access to a number of policy documents and guidelines, along with links to MSDS sites and other sites of interest.

San Diego State University, Environmental Health and Safety
http://tns.sdsu.edu/~ehs/

This site provides EHS contact information and some radiation safety information.

Sonoma State University, EHS

http://www.sonoma.edu/EHS/

This site is maintained by the university's environmental health and safety department. It includes the full text of many of its health and safety programs.

Southern Illinois Univ., Center for Environmental Health and Safety

http://www.cehs.siu.edu

The site contains a description of the center, a virtual tour, and staff directory. Access is also provided to numerous documents, such as biological and radiological safety training modules and its chemical hygiene plan.

University of Adelaide, Occupational Health and Safety Unit

http://www.adelaide.edu.au/HR/OH&S/

Detailed information is provided about the unit. Read the University's guidelines and hazard alerts.

University of Arkansas, Environmental Health and Safety Page

http://www.phpl.uark.edu/ehs/ehs.html

The site contains a number of policy and guideline documents: radiation safety, chemical information, etc.

University of Calgary, Safety Office

http://www.acs.ucalgary.ca/~ucsafety/

View policy and procedure documents and training course information.

University of California Irvine, Environmental Health and Safety Department

http://www.abs.uci.edu:80/depts/ehs/

View the university's EHS policies and procedures documents, emergency procedures, and safety news.

University of California, Davis—Environmental Health and Safety

http://www.ehs.ucdavis.edu/

This site provides information on a variety of health and safety related subjects— biosafety, chemical/lab safety, environmental protection, hazardous waste, and radiation safety.

University of Canterbury (New Zealand), Health and Safety

http://www.mech.canterbury.ac.nz/dinfo/helth.htm

Read about the university's health and safety department.

University of Colorado at Colorado Springs, Risk Mgmt. and Environmental Safety

http://www.uccs.edu/~pusafety/risk1a.htm

Available documents describe health and safety policies and procedures at UCCS.

University of Edinburgh, Health and Safety Services

http://www.admin.ed.ac.uk/safety/

This site contains a mission statement, contact information about the various health and safety services at the university, and "health and safety news flashes."

Univ. of Illinois (Urbana), Division of Environmental Health and Safety

http://phantom.ehs.uiuc.edu/dehs.html

This site provides links to a number of documents and other resources from the division's chemical safety and waste management, occupational safety and health, and radiation safety sections.

University of Illinois at Urbana-Champaign, Radiation Safety Section

http://phantom.ehs.uiuc.edu/~rad/

This site contains the university's radiation safety manual, a laser safety tutorial, and assorted links and documents. Information about radioactive materials is also provided, including a radioactive decay calculator.

University of Maryland at College Park, Environmental Safety

http://www.inform.umd.edu/DES/

Information at this site includes the university's Chemical Hygiene Plan and Laboratory Safety Guide, along with search forms for various MSDS collections.

University of Memphis, Environmental Health and Safety

http://www.people.memphis.edu/~ehas/

Information is provided about the department's services. The full text of its policy documents is available. There are also links to U.S. regulatory sites, MSDSs, etc.

University of Minnesota Duluth, Environmental Health and Safety
http://www.d.umn.edu/~mlabyad/ehso.html

View information about the department, summary information about its services and policies, and related links.

Univ. of Minnesota, Department of Environmental Health and Safety
http://www.dehs.umn.edu/

The site provides access to numerous policy documents, guidelines, manuals, etc., plus a directory of Internet resources.

University of Otago (New Zealand), Health and Safety
http://wsweb.otago.ac.nz/wsweb/health.htm

Information about health and safety at the university is provided.

University of South Carolina, Health and Safety Programs Unit
http://129.252.111.69/hs/

Health and safety policies and procedures documents are available, along with links to other health and safety resources.

University of Sydney, Occupational Health and Safety
http://www.usyd.edu.au/su/planning/ohs/index.html

View the university's health and safety policy and staff guide, information about the Risk Management Office, and accident investigation. Various policy documents are also presented: asbestos safety, children on campus, confined spaces, manual handling, and disposal of sharps.

University of Technology, Sydney—Environment Health and Safety
http://www.hru.uts.edu.au/ehs/index.htm

View the university's policies and procedures for finding and fixing hazards, making your workplace safe, reacting to and reporting accidents, disposing of hazardous waste, and managing emergencies.

University of Toronto, Office of Environmental Health and Safety
http://www.utoronto.ca/safety/ehshome.htm

The site provides access to U of T's health and safety manual, emergency procedures documents, and MSDS sites.

University of Utah, Environmental Health and Safety

http://www.ehs.utah.edu/

The site provides information about the department, the full text of numerous policy and procedure documents, waste minimization documentation, pollution prevention information, and links to other sites.

University of Virginia, Office of Environmental Health and Safety

http://www.virginia.edu/~enhealth/

This well-developed site provides access to a number of documents and information about the office's programs. Highlights include UVa's Laboratory Survival Manual and a Laboratory Safety Checklist.

University of Waikato (New Zealand), Health and Safety

http://www.waikato.ac.nz/fmd/hsc/

A number of documents are available, including the university's Occupational Health and Safety Policy and information about Occupational Overuse Syndrome.

University of Waterloo Safety Office

http://www.safetyoffice.uwaterloo.ca/

The site provides access to information about the safety office, plus documents such as their Health and Safety Program Manual, Radiation Manual, and Laser Safety Manual.

Univ. of Wisconsin, System Admin., Office of Safety and Loss Prevention

http://www.uwsa.edu/oslp/oslp.htm

The page contains links to environmental health and safety, workers' compensation, and risk management. Manuals and policy documents are presented.

Univ. of Wisconsin-Milwaukee, Environmental Health Safety and Risk Management

http://www.uwm.edu/Dept/EHSRM/

Information about the University's health and safety policies and guidelines is provided, along with links to other organizations.

Utah State University, Environmental Health and Safety

http://www.ehs.usu.edu/

General information is provided about health and safety at this university.

5. Professional Organizations

ABSA Canada

http://www.nanook.com/absa_canada/

ABSA Canada is a Canadian affiliate of the American Biological Safety Association. Its objective is to establish a Canadian network of individuals interested in a wide variety of biological safety issues.

Academy of Certified Hazardous Materials Managers (ACHMM)

http://www.achmm.org

Read about the organization, its annual conference, and membership.

Air and Waste Management Association

http://www.awma.org

AWMA provides information about the association, its publications (*e.g.*, *A&WMA JOURNAL*), and membership.

Alberta Construction Safety Association

http://www.compusmart.ab.ca/acsa/

The site provides information about this industry association, its training courses, a description of resources available to members, and products.

American Association of Occupational Health Nurses

http://www.aaohn.org/

AAOHN's site provides a variety of types of information: publications, *AAOHN Journal* information, practice resources, continuing education, and legislative updates.

American Association of Poison Control Centers

http://198.79.220.3/aapcc/

This site provides a directory of centers, educational materials, and a newsletter.

American Biological Safety Association

http://www.absa.org

Information is provided about the association itself, the biosafety profession, the BIOSAFTY mailing list, and an extensive directory of biosafety resources on the Internet.

American Board of Industrial Hygiene (ABIH)

http://www.abih.org

Information is provided about ABIH, CIH certification, examinations, and ABIH publications. More detail about this site is available in the Top 50 Sites section.

American Chemical Society

http://www.acs.org

Information is provided about the society itself, membership, programs, technical divisions, and local sections.

American College of Occupational and Environmental Medicine (ACOEM)

http://www.acoem.org

Categories of available information include information about ACOEM, membership information, courses/conferences, statement papers/guidelines, publications, press releases, a self-assessment exam, employment service, and other OEM links. More detail about this site is available in the Top 50 Sites section.

American Conference of Government Industrial Hygienists (ACGIH)

http://www.acgih.org/

View information about ACGIH, a catalog of its publications, demos of the organization's electronic products, upcoming events, and an "OH Talk" Web chat section. More detail about this site is available in the Top 50 Sites section.

American Industrial Hygiene Association (AIHA)

http://www.aiha.org

This site provides access to information about AIHA, its conferences, publications, and lists of accredited laboratories. More detail is available in the Top 50 Sites section.

American Industrial Hygiene Association, Pacific Northwest Section

http://users.aol.com/nohc96/pnsaiha.html

Read about this AIHA section, plus access links to a substantial number of health and safety resources.

American Institute of Chemical Engineers (AIChE), Consultant & Expert Witness Electronic Directory

http://www.resume-link.com/consultants/

This searchable database references chemical engineering consultants and expert witnesses from various backgrounds and expertise including environmental engineering, process design, and hazardous waste management.

American Nuclear Society

http://www.ans.org

The professional society makes available division section and local chapter information. Also listed are a bulletin board of upcoming conferences meetings or workshops, nuclear engineering resources, and links to other professional societies.

American Public Health Association

http://www.apha.org

Read about APHA, news releases, information about and abstracts from its journal, etc. Links are also provided to legislative, science policy, and other public health-related sites.

American Society of Heating, Refrigerating and Air-Conditioning Engineers (ASHRAE)

http://www.ashrae.org

The site contains information about ASHRAE, its meetings and conferences, publications, research, and standards. The society's online journal is also accessible from the site.

American Society of Safety Engineers (ASSE)

http://www.asse.org

View information about the society, membership, and continuing education. Information is also provided about professional accreditation and relevant U.S. regulatory develop-ments. More detail about this site is available in the Top 50 Sites section.

American Society of Safety Engineers—Colonial Virginia Chapter

http://www.mindspring.com/~jacksull/safety/asse.htm

Find out about this ASSE chapter, its meetings, and training courses.

American Society of Safety Engineers, North Carolina Chapter

http://www.nc-asse.org

The site provides information about this ASSE chapter, membership, its newsletter, and upcoming meetings.

Asociacion para la Prevencion de Accidentes

http://www.safetyonline.net/apa/home.htm

Information is provided about this Spanish association, its services (training, databases, publications, etc.), subscriptions to APA's journal *Prevencion*, a series of safety posters, and conference information. All information is in Spanish.

Association for Canadian Registered Safety Professionals (ACRSP)

http://www.acrsp.ca/

ACRSP's site provides information about association membership, professional designation, and the association's Code of Ethics.

Association of Ontario Health Centres

http://www.aohc.org/

The Association of Ontario Health Centres (AOHC) is the nonprofit organization that represents community health centers (CHCs) and some health service organizations (HSOs) in the province of Ontario, Canada. The site contains information about AOHC, its position papers, and health promotion programs.

Association of Professional Industrial Hygienists

http://industrial.hygienist.com/

The Association of Professional Industrial Hygienists, Inc. (APIH) was established to offer credentialing to industrial hygienists who meet the education and experience requirements found in Tennessee Code Annotated, Title 4, Chapter 3. APIH adopted the Tennessee Code as its basis for credentialing because it was the first legal definition in the U.S. of an industrial hygienist.

Association of Public Safety Communications Officials (APCO)

http://www.apco.ca/

Information about this nonprofit association is provided: membership information, conferences, and training. This is the Web site of the Canadian chapter of APCO International.

Association of Societies for Occupational Safety and Health (Southern Africa)

http://www.asosh.org/

The initial focus of this site will be on OHS in South Africa and will be expanded to provide information on OHS in the Southern African Development Community (SADC) region. Find out about the member associations, their resources, training programmes, and legislation. More detail about this site is available in the Top 50 Sites section.

Australian and New Zealand Society of Occupational Medicine

http://www.anzsom.org.au/

"ANZSOM is a professional and social organisation which provides a focal point for the advancement of knowledge in those registered medical practitioners who are actively involved in or are interested in occupational medicine.

Australian Institute of Occupational Hygienists (AIOH)

http://www.curtin.edu.au/org/aioh/

This is the official home page of the AIOH. It contains information about the institute's publications, membership, office holders, and conferences.

Bioelectromagnetics Society

http://biomed.ucr.edu/bems.htm

Read about BEMS, a nonprofit organization of biological and physical scientists, physicians and engineers interested in the interactions of nonionizing radiations with biological systems.

British Occupational Hygiene Society (BOHS)

http://www.bohs.org/

This page describes the society, its mission, members, publications, conferences, and meetings. There are also links to related sites.

British Safety Council

http://www.britishsafetycouncil.co.uk

The site provides information about the council, its mission, history, and membership.

British Toxicology Society

http://www.bts.org

The site provides information about the society, its publications, and upcoming events.

Campus Safety Health and Environmental Management Association (CSHEMA)

http://www.ualberta.ca/~rrichard/cshema.html

The site includes information on the CSHEMA organization, its newsletter, and training and policy information. More detail about this site is available in the Top 50 Sites section.

Canada Safety Council (CSC)

http://www.safety-council.org

CSC is an independent, national, nonprofit membership safety organization. The Web site offers information about the organization, its training programs, and publications.

Canadian Chemical Producers' Association (CCPA)

http://www.ccpa.ca

Information about CCPA and its responsible care program is provided.

Canadian Medical Association, Occupational Medicine, Clinical Practice Guidelines

http://www.cma.ca/cpgs/occup.htm

Links are provided to a series of documents describing specific guidelines.

Canadian Registration Board of Occupational Hygienists

http://www.crboh.ca/

The CRBOH is a national, not-for-profit organization, which sets standards of professional competence for occupational hygienists and occupational hygiene technologists in Canada. The site provides information about ROH and ROHT certification, exams, and maintenance.

Canadian Society of Safety Engineering (CSSE)

http://www.csse.org

Information is provided about the society, its programs, publications, and courses.

Canadian Society of Safety Engineering, Edmonton Chapter

http://www.freenet.edmonton.ab.ca/csse/

The site provides information about this CSSE chapter, membership, and meetings.

Chartered Institute of Environmental Health (U.K.)

http://www.cieh.org.uk

The Chartered Institute of Environmental Health (CIEH) is a nongovernmental organisation dedicated to the promotion of environmental health and the dissemination of knowledge about environmental health issues for the benefit of the public. It is also responsible for the training and professional development of environmental health officers and provides educational services to the profession

---

Construction Industry Research and Information Association

http://www.ciria.org.uk/

CIRIA is a nonprofit U.K. organization that provides best practice guidance to professionals. Read about the organization, its research, and publications.

Construction Safety Association of Ontario

http://www.csao.org/

As well as providing information about the association itself, the site provides a number of basic safety awareness documents. The full text of their magazine *Construction Safety* and their newsletter is also available.

Education Safety Association of Ontario

http://www.esao.on.ca/

The Education Safety Association of Ontario Inc., is a not-for-profit Ontario "Safe Workplace Association." The site has links to documents, facts and figures, information about the organization itself, and a health and safety discussion group for the educational sector.

Electrical Utilities Safety Association of Ontario

http://www.eusa.on.ca/

Read about this safe workplace association, its consulting services, training programs, and accident prevention solutions.

Environment and Safety Data Exchange (ESDX)

http://www.esdx.org/esdhome.html

Find out about ESDX, an industry association organized to participate in setting standards for more cost-effective management information technology in environmental, health and safety (EHS) activities.

Ergonomics Society

http://www.ergonomics.org.uk/

The U.K.-based Ergonomics Society Web site contains information about the society and its members, lists of primarily U.K. ergonomics consultants, training courses, and jobs.

European Biological Safety Association

http://biosafety.ihe.be/EBSA/HomeEBSA.html

EBSA was formed in 1996 "to promote biosafety as a scientific discipline and serve the growing needs of the biosafety professionals throughout Europe."

Farm Safety Association of Ontario

http://www.fsai.on.ca

The site provides information about the association itself, plus safety information for agriculture/ horticulture and landscapers.

Health Care Health and Safety Association

http://www.hchsa.on.ca

HCHSA is a designated "Safe Workplace Association" in Ontario, Canada. The site provides information about the association, its mandate, news, etc.

Health Physics Society

http://www.hps.org/hps/

HPS is a professional organization dedicated to the development, dissemination, and application of both the scientific knowledge of, and the practical means for, radiation protection. Read its newsletter and find out about upcoming meetings.

Hong Kong Occupational Safety and Health Association

http://www.hk.super.net/~hkosha/

Information is provided about the association, Hong Kong safety and health legislation and standards, the association's publications, and links to other sites.

Human Factors and Ergonomics Society

http://hfes.org/

Read about the society itself, its news, publications, and meetings.

Human Factors Association of Canada

http://www.hfac-ace.ca/

Read about the association itself, ergonomics conferences in Canada, programs and courses in ergonomics, and view a directory of consultants.

Industrial Accident Prevention Association (IAPA)

http://www.iapa.on.ca

Information is provided about this Ontario, Canada safety association, its products, services, and training programs.

Institute of Electrical and Electronics Engineers (IEEE)

http://www.ieee.org

Available information includes frequently-asked-questions documents (FAQs), information about IEEE standards, technical activities, conferences, and society membership.

Institute of Occupational Hygienists

http://www.ed.ac.uk/~alfred/ioh.html

Read about this British institute's membership, code of ethics, and officers.

Institute of Safety in Technology and Research

http://www.bham.ac.uk/istr/

"The Institute was founded as the Institute of University Safety Officers (IUSO) by a group of U.K. university safety officers who were seeking to establish a body to represent their individual interests on a professional basis and in their specialist area of research and high technology."

Institution of Occupational Safety and Health (IOSH)

http://www.iosh.co.uk/

The site contains information on all the major areas of activity that this U.K.-based institution is involved in: membership and career information documents, U.K. conferences, etc.

International Association of Electrical Inspectors, Palm Beach Florida Chapter

http://www.iaei.com

Read about the organization's membership and upcoming meetings.

International Association of Emergency Managers

http://www.emassociation.org/

IAEM is a nonprofit educational organization dedicated to promoting the goals of saving lives and protecting property during emergencies and disasters. Read about the association, its conference, etc.

International Institute of Risk and Safety Management

http://www.britishsafetycouncil.co.uk/iirsm/Default.htm

Information is provided about the institute itself and membership options that are available.

International Radiation Protection Agency

http://www.irpa.at

Information is provided about this international organization, its congresses, membership, and newsletter, along with links to associate societies.

International Society for Occupational Ergonomics and Safety

http://biz-comm.com/isoes/

The site provides information about the society (membership, officers, by-laws, etc.) plus a member's directory, newsletters, and information about their annual conference.

International Society of Indoor Air Quality And Climate (ISIAQ)

http://www.cyberus.ca/%7Edsw/index.html

ISIAQ "is an international, independent, multi-disciplinary, scientific, nonprofit organization whose purpose is to support the creation of healthy, comfortable and productivity-encouraging indoor environments." Read about the organization, membership, publications, conferences, and symposia.

Mines and Aggregates Safety and Health Association

http://www.masha.on.ca/

MASHA is a designated "Safe Workplace Association" in Ontario, Canada. The site provides information about the association, its publication catalogue, and a calendar of events.

Minnesota State Fire Chiefs Association

http://www.msfca.org/

Information is provided about the association, membership, and services.

Municipal Health and Safety Association of Ontario

http://www.mhsao.com/

Read about the association, its mission, publications, and courses.

National Assn of Fire Equipment Distributors (NAFED)

http://www.halcyon.com/NAFED/HTML/Welcome.HTML

This site provides links to many resources related to fire prevention, including national, regional, and municipal organizations, as well as educational programs.

National Association of Healthcare Safety and Risk Practitioners

http://www.nahsrp.org.uk/

NAHSRP is an association for safety and risk professionals that work in the healthcare sector in the United Kingdom. Its aim is to promote best practice in the provision of safer health care. Read about the organization and its membership, view back issues of its newsletter, and participate in its message forum.

National Environmental Health Association

http://www.neha.org

NEHA is a (U.S.) national professional society for environmental health practitioners. Read about NEHA's annual conference and requirements for NEHA certification.

National Fire Protection Association (NFPA)

http://www.nfpa.org

The site provides information about the NFPA, its departments, publications, seminars, and educational programs.

National Registry of Environmental Professionals

http://www.nrep.org/

Information is provided about this professional association. Read about the association itself, search a list of registered professionals, and find out about training programs and employment opportunities.

National Safety Council

http://www.nsc.org/

Read about the council, its activities and research, and publications.

National Sanitation Foundation

http://www.nsf.com

The site describes NSF International and provides a database of NSF-certified products (biohazard cabinets, water purification systems, etc.) and online newsletters.

Nova Scotia Environmental Industry Association

http://www.nseia.ns.ca/

Information is provided about the association, membership, newsletter, and codes of practice. Also provides Nova Scotia environmental legislation, permit and licensing information, and related info.

Occupational Health and Safety Information Group

http://panizzi.shef.ac.uk/oshig/oshig.html

This professional association set up in 1990 "caters for the needs of health and safety information specialists." Read about membership in the association, meetings, and its newsletter.

Occupational Hygiene Association of Ontario (OHAO)

http://www.ohao.org/

Read about this provincial association, membership, upcoming events, and the association's newsletter.

Ontario Service Safety Alliance

http://www.ossa.com

"The OSSA was developed to meet the needs of the service industry by providing education and training to assist in preventing injuries and illnesses." View information about the organization, its educational programmes, and research and studies.

Radiation Research—Official Journal of the Radiation Research Society

http://www.cjp.com/radres/

View the table of contents for the current issue, "mini abstracts" of articles from previous issues, and subscription/membership information.

Radiation Research, Official Journal of the Radiation Research Society

http://www.cjp.com/radres/index1.htm

Abstracts and tables of contents from this journal are provided.

Risk Assessment and Policy Association

http://www.fplc.edu/tfield/rapa.htm

Read about RAPA, its goals, and membership. Information about and articles from RAPA's journal *Risk* are also presented.

Risk: Health, Safety and Environment

http://www.fplc.edu/tfield/profRisk.htm

Risk is the official journal of the Risk Assessment and Policy Association. View tables of contents of past and current issues, subscription information, etc.

Royal Society of Chemistry

http://www.rsc.org/

"The Royal Society of Chemistry is the Learned Society for chemistry and the Professional Body for chemists in the U.K. with 46,000 members worldwide." Read about membership, the society's information resources, library, etc.

Safety Central

http://www.safetycentral.org

A joint site for the Industrial Safety Equipment Association (ISEA) and the Safety Equipment Distributors Association (SEDA), it offers information on standards and regulations, a buyer's guide for safety equipment, a Washington Insider, and links to the members' sites.

Society of Chemical Industry

http://sci.mond.org

Information is provided about this U.K.-based learned society, its membership, meetings, and publications. There are details about their health and safety group.

Society of Environmental Toxicology and Chemistry (SETAC)

http://www.setac.org/

Available information includes details about membership in this professional society, a newsletter, activities and meetings, and publications.

Society of Occupational Medicine

http://www.ed.ac.uk/~rma/som/

This U.K. Society aims to stimulate interest, research, and education in occupational medicine. Read about its organization, activities, and publications.

Swedish Society of Radiation Physics

http://www.fysik.lu.se/~radiofys/sfrengho.htm

Read about the society itself, upcoming meetings and conferences, and links to radiation sites. Content is in Swedish and English.

Transportation Safety Association of Ontario

http://www.tsao.org

This safe workplace association's site provides access to publications, information about training programmes, and the association's services.

Vessel Operators Hazardous Materials Association (VOHMA)

http://www.vohma.com/

VOHMA is "an international organization comprised of representatives of the ocean common carriers of the world, operating under the flags of several nations, dedicated to improving the understanding and uniform application of rules and regulations governing maritime transportation of dangerous goods."

---

6. Government Agencies

Agency for Toxic Substances and Disease Registry (ATSDR)

http://atsdr1.atsdr.cdc.gov:8080/

This U.S. government agency's site contains numerous documents and data sets (*e.g.*, HazDat database and ToxFAQ documents) and links to other related organizations.

Alberta Health

http://www.health.gov.ab.ca

The site provides information about this provincial government department. It includes public health information and details about emergency services.

Alberta Labour

http://www.gov.ab.ca/lab/index1.html

This Alberta government department's site provides access to the full text of acts and regulations, accident and injury statistics, and publications.

Arbetslivsinstitutet

http://www.niwl.se/niwl.htm

The National Institute of Working Life (Arbetslivsinstitutet) is Sweden's Research and Development center for occupational health and safety, working life and the labor market. Access is provided to information about the institute, its library catalog, and the full text of its publications.

Arizona State Mine Inspector

http://www.indirect.com/www/mining/

The site contains information about this state government agency, its annual report, and upcoming conferences.

Army Industrial Hygiene Program

http://chppm-www.apgea.army.mil/Armyih/

This U.S. Army site provides access to Army industrial hygiene documents, publications and policies, along with links to a number of other documents and sites of interest to industrial hygienists.

At Work With Julie—Workers' Compensation questions and answers

http://www.abag.ca.gov/govnet/julie/julie.html

This Ann Landers-style site provides workers' compensation information in a question and answer format. The site, and therefore presumably the answers, are California-based. Julie (Carroll) is director of the Association of Bay Area Governments' Workers Compensation Admin. Program.

Australian Institute of Health and Welfare, National Injury Surveillance Unit

http://www.nisu.flinders.edu.au/welcome.html

The National Injury Surveillance Unit (NISU) of the Australian Institute of Health and Welfare provides access to epidemiological and statistical studies conducted by the agency.

Belgian Biosafety Server

http://biosafety.ihe.be

This server, located at the Institute of Hygiene and Epidemiology (IHE) in Brussels, provides detailed information about biosafety, particularly in the European Union.

Brookhaven National Lab, Safety and Environmental Protection Div.

http://sun10.sep.bnl.gov/seproot.html

Information at this site includes health and safety questions and answers, EHS policies, standards, and notices, and BNL's employee safety handbook.

Bureau of Labor Statistics

http://stats.bls.gov/blshome.html

View statistical data, publications, and research papers. Highlights include the U.S. *Occupational Outlook Handbook* and various research papers.

California Department of Industrial Relations (DIR)

http://www.dir.ca.gov/

This site contains information about workers compensation, occupational safety and health, and California labor law.

California Office of Emergency Services

http://www.oes.ca.gov

Information is provided about the Governor's Office of Emergency Services, the California Emergency Plan, current conditions (weather, roads, etc.), and the office's earthquake program.

Canadian Centre for Occupational Health and Safety (CCOHS)

http://www.ccohs.ca

CCOHS's site provides access to Canadian environmental and OSH legislation, plus major OSH databases (*e.g.*, NIOSHTIC, RTECS, HSELine) on a subscription basis. Other content includes information and hazard alerts and a large directory of health and safety Internet resources. More detail about this site is available in the Top 50 Sites section.

Canadian General Standards Board

http://www.pwgsc.gc.ca/cgsb/

View CGSB's online catalogue and read about the standards development process.

Canadian Reflection

http://www.reflection.gc.ca

This site describes a study launched by the Federal Labour Minister to examine how the Canadian workplace is changing.

CDC Travel Information

http://www.cdc.gov/travel/index.htm

Information provided includes lists of reference materials, information about disease outbreaks, and specific health hazard information for different geographical locations. This site is maintained by the U.S. Centers for Disease Control and Prevention.

CDC Wonder

http://wonder.cdc.gov

CDC Wonder provides access to text and numeric databases guidelines and reports from the U.S. Centers for Disease Control and Prevention.

Centers for Disease Control and Prevention (CDC)

http://www.cdc.gov

This U.S. government site contains a wealth of information, including occupational and environmental health, and travellers' health. Numerous publications and reports are available, including the Morbidity and Mortality Weekly Report (MMWR). More detail about this site is available in the Top 50 Sites section.

Central Institute for Labour Protection (Poland)

http://www.ciop.waw.pl

Information is provided about the institute and its work, along with subscription information for the institute's journal, *International Journal of Occupational Safety and Ergonomics*.

Centre de Toxicologie du Québec

http://www.ctq.qc.ca

The CTQ's mission is to provide quality services in the area of human toxicology, primarily for the population of Quebec. Read about the organization and its services.

Chemical Reactivity Worksheet

http://response.restoration.noaa.gov/chemaids/react.html

This free, downloadable program was developed by the CAMEO team at the Hazardous Materials Response and Assessment Division, ORCA/NOS/NOAA, and the Chemical Emergency Prevention and Preparedness Office of the EPA. It includes a database of reactivity information for over 4000 common hazardous materials. It also allows you to "virtually mix" chemicals to find out what dangers could arise. More detail about this site is available in the Top 50 Sites section.

Child Health Programs, Agency for Toxic Substances and Disease Registry

http://atsdr1.atsdr.cdc.gov:8080/child/

"This site was created by the ATSDR Office of Children's Health to share information on a variety of programs that emphasize the vulnerabilities of infants and children in communities faced with the contamination of their water, soil, air, or food.

City of Alexandria, Public Safety

http://ci.alexandria.va.us/kyc2/kyc06.htm

Descriptions of the public safety-related departments in Alexandria, Virginia are provided.

City of Toronto

http://www.city.toronto.on.ca

Access public health documents and information about emergency services.

Colorado Department of Public Health and Environment

http://www.state.co.us/gov_dir/cdphe_dir/cdphehom.html

The site provides lots of contact information, frequently asked questions (FAQ), plus downloadable copies of Colorado public health and environment regulations.

Commission de la santé et de la sécurité du travail du Québec (CSST)

http://www.csst.qc.ca/

The site describes the role and services of the commission and the laws it administers. Other information includes CSST news releases, summaries of their magazine *Prévention au travail* and a publications list. Most information is in French, but some is also in English.

Consumer Information Center, Pueblo, CO

http://www.pueblo.gsa.gov/

The site provides access to the full text of hundreds of U.S. government consumer publications, including occupationally related material.

Consumer Product Safety Commission (CPSC)

http://www.cpsc.gov

Lots of product safety information is available, including press releases and many full text publications from the U.S. CPSC. More detail is available in the Top 50 Sites section.

CRISP—Computer Retrieval of Information on Scientific Projects

gopher://gopher.nih.gov:70/11/res/crisp

CRISP system is a major biomedical database containing information on research ventures supported by the United States Public Health Service.

Dealing with Workplace Violence: A Guide for Agency Planners

http://www.opm.gov/workplac/

This handbook, developed by the U.S. Office of Personnel Management and the Interagency Working Group on Violence in the Workplace, "is intended to assist those who are responsible for establishing workplace violence initiatives at their agencies."

Department of Energy, Environment Safety and Health, International Health Programs

http://www.eh.doe.gov/ihp/

View information about health effects of radiation and related environmental hazards. Three programs are described: those in Europe, Japan, and the Marshall Islands.

Dept. of Energy, Office of Human Radiation Experiments Home Page

http://tis.eh.doe.gov/ohre/

"The Office of Human Radiation Experiments, established in March 1994, leads the Department of Energy's efforts to tell the agency's Cold War story of radiation research using human subjects.

Department of Energy, Technical Information Service

http://www.tis.eh.doe.gov

This site provides access to a wealth of publications, alerts, fact sheets, and other health and safety-related information. It was designed for use by U.S. DOE staff, and as such, has an orientation towards electrical generating facilities, but has many useful resources for non-DOE people. More detail about this site is available in the Top 50 Sites section.

Department of Health (U.K.)

http://www.doh.gov.uk/dhhome.htm

The site provides information about the department itself, news releases, departmental programs, and publications (some full text).

Department of Transportation

http://www.dot.gov

As well as containing U.S. regulatory and policy information, the site provides access to a variety of transportation safety resources: research and reports on transportation safety from various sources, hazardous materials safety, etc.

Dept. of Transportation, Office of Hazardous Materials Safety (OHM)

http://www.volpe.dot.gov/ohm/

This office is responsible for coordinating a national (U.S.) safety program for the transportation of hazardous materials by air, rail, highway, and water. Highlights include an electronic version of the Emergency Response Guidebook, regulatory info., and OHM publications.

Dept. of Veterans Affairs, Office of Occupational Safety and Health

http://www.va.gov/vasafety/index.htm

As well as providing information about the office itself, the site provides access to information on a variety of OSH topics, such as OSH program documents, violent behavior prevention, needle safety, asbestos, and latex allergy prevention.

Division of Occupational Safety and Health (California OSHA)

http://www.dir.ca.gov/DIR/OS&H/DOSH/dosh1.html

Information is available about Cal/OSHA's services, personnel, etc. The full text of the division's publications and its policies and procedures manual is also included.

Documentation for Immediately Dangerous to Life or Health Concentrations (IDLHs)

http://www.cdc.gov/niosh/idlh/idlh-1.html

This online publication documents the criteria and information sources that have been used by the National Institute for Occupational Safety and Health (NIOSH) to determine concentrations that are immediately dangerous to life or health. The IDLHs for the chemicals themselves are also provided. More detail is available in the Top 50 Sites section.

EHIS Publications

http://ehpnet1.niehs.nih.gov/docs/publications.html

Access is provided to various publications of the U.S. National Institute of Environmental Health Sciences: Environmental Health Perspectives, National Toxicology Program Technical Reports, etc.

Emergency Information Infrastructure Project (EII)

gopher://oes1.oes.ca.gov:5555/11/eii

Read about EII, "an interorganizational effort to foster the development and application of advanced telecommunications networks in support of the emergency management community as a whole."

Emergency Management Australia

http://www.ema.gov.au/

EMA's mission is to promote and support comprehensive, integrated, and effective emergency management in Australia and its region of interest. Information includes publications and manuals, training information, and links to related Australian sites.

Emergency Preparedness Information Exchange (EPIX)

http://hoshi.cic.sfu.ca:80/~anderson/index.html

Content includes links and documents about various emergency topics, related emergency organizations, upcoming conferences, alerts, etc.

Emergency Response Notification System (ERNS) from U.S. EPA

http://www.epa.gov/ERNS

Information on the releases of oil and hazardous substances is provided.

EMF Research Activities Completed under the Energy Policy Act of 1992 (1995)

http://www.nap.edu/readingroom/enter2.cgi?NX006605.html

Read a summary of this report from the Committee to Review the Research Activities Completed under the Energy Policy Act of 1992, Board on Radiation Effects Research, Commission on Life Sciences, National Research Council (U.S.A.).

Envirofacts Warehouse, Master Chemical Integrator

http://www.epa.gov/enviro/html/emci/emci_query.html

This page allows you to obtain the acronyms, chemical identification numbers, and chemical names reported by the U.S. Environmental Protection Agency's Envirofacts program system databases (AFS, PCS, RCRIS, and TRIS). Links to factsheets containing information regarding health risks, exposure limits, handling regulations, and chemical properties.

Environment Canada

http://www.ec.gc.ca

A Canadian federal government department, its initiatives, structure, and legislation.

Environmental Chemicals Data and Information Network (ECDIN)

http://ecdin.etomep.net/

"ECDIN is a factual databank, created under the Environmental Research Programme of the Joint Research Centre (JRC) of the Commission of European Communities at the Ispra Establishment." "ECDIN deals with the whole spectrum of parameters and properties that might help the user to evaluate real or potential risk in the use of a chemical and its economical and ecological impact."

Environmental Health Information Service

http://ehis.niehs.nih.gov/

This is "a full-service site that produces, maintains, and disseminates information on the environment in the form of a searchable directory." Contents include the monthly journal, Environmental Health Perspectives, and the National Toxicology Program's 8th Report on Carcinogens.

Environmental Protection Agency (EPA)

http://www.epa.gov

This well-developed U.S. government site contains a huge amount of environmental information, including environmental regulations, a variety of database systems, and numerous indoor air quality documents. More detail is available in the Top 50 Sites section.

Environmental Protection Agency, Indoor Air Quality Home Page

http://www.epa.gov/iaq/

This is a very content-rich site provided by the U.S. EPA. It contains documents on common indoor air pollutants, sources of indoor air pollution, and problems in different types of buildings. More detail about this site is available in the Top 50 Sites section.

Environmental Protection Agency, Office of Pesticide Programs

http://www.epa.gov/pesticides/

Access is provided to detailed U.S. EPA pesticide information.

EPA—Technical Information Packages (TIPS)

http://www.epa.gov/oia/tips/

This page contains descriptions of eleven Technical Information Packages (TIPs). Aimed at the international community, the packages focus on key environmental and public health issues being investigated by the U.S. Environmental Protection Agency (EPA).

ERDEC Safety Office

http://www.apgea.army.mil/RDA/erdec/risk/safety/

The Edgewood Research Development and Engineering Center (ERDEC) provides health and safety information and services to its clientele, primarily the U.S. Armed Forces.

ErgoEASER Software

http://tis-hq.eh.doe.gov:80/others/ergoeaser/download.html

Download a copy of this software, "developed to aid in identifying, evaluating, and preventing work-related musculoskeletal disorders."

Federal Aviation Administration, Office of System Safety

http://nasdac.faa.gov/

This U.S. government site provides access to aviation safety-related databases, safety reports and publications, and information on the Global Analysis Information Network (GAIN) project.

Federal Emergency Management Agency (FEMA)

http://www.fema.gov

This U.S. agency's mission is "to reduce loss of life and property and protect our nation's critical infrastructure from all types of hazards through a comprehensive, risk-based, emergency management program of mitigation, preparedness, response and recovery."

FedWorld

http://www.fedworld.gov

This site is a gateway to U.S. federal government information. From here, you can find and order recent U.S. government reports from all agencies and search the Web sites of hundreds of federal government agencies. More detail is available in the Top 50 Sites section.

Finnish Institute of Occupational Health

http://www.occuphealth.fi

Information is provided about the institute, its purpose, and its work. Lists of institute publications and journals are provided. There is also a list of OH&S conferences.

Food and Drug Administration (FDA)

http://www.fda.gov

Information is provided about various U.S. FDA program areas, including radiological health, toxicology, and drugs.

Hazardous Materials Information Exchange

telnet://hmix.dis.anl.gov

This telnet-accessible bulletin board system was created to provide a centralized database for federal, state, local, and private-sector personnel to share information pertaining to hazardous materials emergency management, training, resources, technical assistance, and regulations.

HazDat—Hazardous Substance Release/Health Effects Database

http://atsdr1.atsdr.cdc.gov:8080/hazdat.html

This database, from the U.S. Agency for Toxic Substances and Disease Registry (ATSDR), provides information on the release of hazardous substances from Superfund sites or from emergency events and on effects of hazardous substances on the health of human populations.

HazMat Safety

http://hazmat.dot.gov/

This site is managed by the U.S. Department of Transportation's Office of Hazardous Materials Safety. It provides access to HazMat related regulations, DOT forms and documents, the North American Emergency Response Guidebook and much more.

Health and Safety Division

http://www.dedni.gov.uk/hsd/

This is the Health and Safety Division of the Northern Ireland Department of Economic Development. Information is provided about the division itself and its responsibilities.

Health and Safety Authority

http://www.hsa.ie/osh/welcome.htm

Information is provided about the Health and Safety Authority (Ireland), a list of publications, a list of acts and regulations, and information about the authority's programs.

Health and Safety Executive (U.K.)

http://www.open.gov.uk/hse/hsehome.htm

Information is provided about this U.K. regulatory agency, its publications and videos, and research.

Health Canada, Environmental Health Program

http://www.hc-sc.gc.ca/ehp/

The Environmental Health Program consists of four bureaus: Chemical Hazards, Product Safety, Tobacco Control, and Radiation Protection. The site provides access to numerous technical reports and publications and general interest fact sheets.

Health Canada, Health Information Network

http://www.hc-sc.gc.ca/

This Canadian federal government department presents its health alerts, public health documents, and news releases, along with links to its various directorates—Laboratory Centre for Disease Control, Environmental Health Directorate, etc. More detail about this site is available in the Top 50 Sites section.

Health Canada, Laboratory Centre for Disease Control

http://www.hc-sc.gc.ca/hpb/lcdc/

The site contains information about the centre itself, public health documents, travel health advisories, and disease prevention guidelines. More detail about this site is available in the Top 50 Sites section.

Health Canada, Laboratory Centre for Disease Control, Laboratory Biosafety Guidelines

http://hwcweb.hwc.ca/hpb/lcdc/bmb/biosafty/

This site provides the full text of LCDC's guidelines document. More detail about this site is available in the Top 50 Sites section.

HEBSWeb

http://www.hebs.scot.nhs.uk/

This is the site of the Health Education Board for Scotland. Bibliographies and assorted documents on a variety of health topics are provided.

Hill Air Force Base, Utah, Environment Safety and Health

http://137.241.179.15/esoh/index.htm

Read about their ESH programs and find links to related sites.

Hong Kong Education and Manpower Bureau

http://www.info.gov.hk/emb/welcome.htm

This government agency's site provides information about occupational health and safety policies and labor relations.

Human Resources Development Canada—Labour Program

http://labour-travail.hrdc-drhc.gc.ca/eng/

"The mission of the Labour Program of Human Resources Development Canada (HRDC) is to promote a fair, safe, healthy, stable, cooperative and productive work environment that contributes to the social and economic well-being of all Canadians."

Human Resources Development Canada, Labour Operations, Occupational Safety and Health

http://info.load-otea.hrdc-drhc.gc.ca/~oshweb/homeen.shtml

Information is provided to allow Canadian federally regulated organizations to meet the requirements under the Canada Labour Code Part II. There is an overview of the code and information about setting up an OSH program in your workplace. Information of general interest includes occupational injury statistics.

Institut national de recherche et de sécurité (INRS)

http://www.inrs.fr/

The role of the INRS in France "is to contribute technically, using appropriate means, to the prevention of occupational accidents and diseases to ensure the protection of human health and safetyat work." The site provides information about INRS' research, consulting, and publications.

Institute for Environment, Safety and Occupational Health Risk Analysis HSW

http://www.brooks.af.mil/ESOH/

The institute, at Brooks Air Force Base in Texas, is the Air Force consult and testing center for human health risks in daily systems operations, the U.S. Air Force's jurisdiction.

Institute of Occupational Safety and Health, Taiwan

http://192.192.46.66/INDEXE0.HTM

The missions of this research institute include "application of scientific technology, surveys and analyses of various risk factors in the working environment, as well as development of countermeasures." The site describes the institute and the research areas of its various divisions.

Internet Grateful Med (MEDLINE database)

http://igm.nlm.nih.gov/

Free access is provided to this U.S. National Library of Medicine database of over 9 million bibliographic citations from 3900 journals. More detail about this site is available in the Top 50 Sites section.

Investigation of the Safety Implications of Wireless Communications in Vehicles

http://www.nhtsa.dot.gov/people/injury/research/wireless/

This report, available in its entirety, was written by the U.S. National Highway Traffic Safety Administration. The report addresses four specific issues: does cellular telephone use while driving increase crash risk; magnitude of the traffic safety problem as a result; will crashes increase with increased cell phone use; options for safer cell phone use in vehicles.

Jefferson Lab EH&S

http://www.cebaf.gov/ehs/

The site provides this nuclear research facility's environmental health and safety manual and emergency management documents.

Job Accommodation Network

http://janweb.icdi.wvu.edu/

JAN is an international consulting service that provides information about job accommodations and the employability of people with disabilities. Provides a large directory of related sites, information about job accommodation in the U.S., and Americans With Disabilities Act.

Justice Canada

http://canada.justice.gc.ca/Loireg/index_en.html

This site provides access to the full text of many Canadian federal acts and regulations.

Lawrence Berkeley Lab, Environmental Health and Safety Home Page

http://ehssun.lbl.gov/ehsdiv/ehswww.html

Access is provided to a number of Berkeley Lab's health and safety documents: its lab safety manual, site environmental report, and "lessons learned" documents. There are also links to a number of public MSDS collections and a few other sites.

Lawrence Livermore National Lab (LLNL), Environment, Safety and Health

http://www.llnl.gov/es_and_h/

A number of documents are provided, including LLNL's health and safety manual, environmental guidelines documents, and an environmental compliance manual.

Manitoba Workplace Safety and Health Division

http://www.gov.mb.ca/labour/safety/index.html

Information about the various branches of this provincial government division (Workplace Safety and Health, Mining Inspection and Occupational Health), plus access to the full text of a number of its WorkSafepublications. More detail about this site is available in the Top 50 Sites section.

MediBel-Net—the Belgian Information Kiosk for Health and Environment

http://www.health.fgov.be/

View information about MediBel-Net, the responsibilities and work of the Belgian Ministry of Social Affairs, Health and Environment. Most information is in French and Dutch.

MICOM Safety Office Home Page

http://www.redstone.army.mil/safety/home.html

The Army Missile Command (MICOM) safety resources include a DRAFT AMC Regulation 385-100 and a DRAFT Employee Handbook and Guide to Occupational Health Services.

Mine Safety and Health Administration (MSHA)

http://www.msha.gov

This U.S. government site provides access to mining-related regulatory information, health and safety documents, and information about various programs.

Ministry of Health, Peru

http://www.digesa.sld.pe

This site provides information about the Peruvian Ministry of Health, plus links to the Directorate of Environmental Health, General Office of Epidemiology, and Peruvian Association of Public Health. All information is in Spanish.

Minnesota Department of Public Safety

http://www.dps.state.mn.us/

Information about the department's various divisions and public safety issues (floods, tornadoes).

Missouri Department Of Labor and Industrial Relations

http://www.dolir.state.mo.us/

Information is provided about the department's divisions, including labor standards, workers compensation, and the Governor's Council on Disability.

Mortality and Morbidity Weekly Report (MMWR)

http://www.cdc.gov/epo/mmwr/mmwr.html

Current and previous issues of this weekly report from the U.S. Centers for Disease Control and Prevention are presented in their entirety.

NASA/LeRC Office of Safety and Mission Assurance

http://www-osma.lerc.nasa.gov

Information about the office itself is provided. Links to many related NASA resources, such as the Spaceflight Safety Handbook.

National Institute for Occcupational Safety and Health (NIOSH)

http://www.cdc.gov/niosh/homepage.html

This home page has been established to provide information about NIOSH and related activities. Content includes NIOSH publications, respirator information, and meetings and symposia. More detail about this site is available in the Top 50 Sites section.

National Institute of Environmental Health Sciences

http://www.niehs.nih.gov/

The site provides access to detailed information about this U.S. government agency, plus links to NIEHS-sponsored sites.

National Institute of Health Sciences—Japan

http://www.nihs.go.jp

Information about the various divisions within NIHS, plus links to NIHS-sponsored sites.

National Institutes of Health

http://www.nih.gov

A wide variety of health information is available from this U.S. government site. Highlights include the NIH Health Information Index, HealthFinder, Consumer Health Information, along with links to other NIH sites such as the National Library of Medicine and the National Cancer Institute.

National Library of Medicine

http://www.nlm.nih.gov

This U.S. government site is one of the best starting points for finding medically-related information. Highlights include free access to the MEDLINE database, newsletters, fact sheets, and information about NLM programs.

National Occupational Health and Safety Commission (WorkSafe Australia)

http://www.worksafe.gov.au

Read about this national OH&S agency, view statistical information, Australian regulatory requirements, fact sheets, and alerts.

National Technical Information Service (NTIS)

http://www.ntis.gov/health/health.htm

The site provides information about this U.S. Government agency's health and safety publications, databases, and other information.

National Toxicology Program

http://ntp-server.niehs.nih.gov

This program is associated with the U.S. National Institute of Environmental Health Sciences. Access abstracts of NTP studies and chemical health and safety data. More detail about this site is available in the Top 50 Sites section.

National Toxicology Program, 7th Annual Report on Carcinogens

http://ntp-server.niehs.nih.gov/Main_Pages/NTP_ARC_PG.html

This site provides information about known and suspected carcinogens to which a significant number of people in the United States are exposed.

Natural Disaster Reference Database

http://ltpwww.gsfc.nasa.gov/ndrd/cgi/ndrd.cgi

A joint U.S.-Japan project, this is a bibliographic database on research, programs, and results which relate to the use of satellite remote sensing for disaster mitgation.

Nebraska Dept. of Administrative Services, Risk Management Health Benefit Program

http://www.das.state.ne.us/das_dorm/index.html

View information about this state government program.

New Brunswick Workplace Health, Safety, and Compensation Commission

http://www.gov.nb.ca/whscc/

This site provides information about the commission and its programs, calculating WCB claims in New Brunswick, industry assessment rates, etc.

New Jersey Dept. of Health and Senior Services, Occupational Disease Epidemiology and Surveillance Program

http://www.state.nj.us/health/eoh/survweb/survhome.htm

"The Occupational Disease Epidemiology and Surveillance Program targets its activities to workers who are exposed to hazardous chemical, physical or biological agents."

New Jersey Dept. of Health and Senior Services, Right-to-Know Program

http://www.state.nj.us/health/eoh/rtkweb/rtkhome.htm

The site provides information about the program, training courses, and commonly asked questions about NJ Right-to-Know, publications. A number of Hazardous Substance Fact Sheets are available as PDF files. (Adobe Acrobat reader software required.)

New Mexico Department of Public Safety

http://www.dps.state.nm.us

Provides information about the department itself, press releases, upcoming events.

New United Kingdom Official Publications Online

http://www.soton.ac.uk/~nukop/index.html

NUKOP provides bibliographic and ordering information for thousands of U.K. government publications in a variety of subject areas, including health and safety.

New York State Dept. of Health, Environmental and Occupational Health

http://www.health.state.ny.us/nysdoh/consumer/environ/homeenvi.htm

This site provides access to a series of basic documents on environmental and occupational health. The content is geared to the public.

New Zealand's Health and Safety Net

http://www.OSH.dol.govt.nz/

This is the home page of the Occupational Safety and Health Service of the Department of Labour, New Zealand. The site provides information about the organization, work hazards, health and safety law, and training programs.

Newfoundland Environment and Labour

http://www.gov.nf.ca/env/default.asp

The site contains information about this Canadian provincial government department, press releases, etc.

NIOSH Agricultural Health and Safety Centers

http://agcenter.ucdavis.edu/agcenter/niosh/niosh.html

The site contains the full text of back issues of *NIOSH AgCenter News*, information about specific centers, and links to related sites

NIOSH Educational Research Centers (ERCs)

http://rocky.utah.edu/erc.html

This page, provided by the Rocky Mountain Center for Occupational and Environmental Health, provides links to all 14 U.S. NIOSH ERCs.

NIOSH Manual of Analytical Methods (NMAM)

http://www.cdc.gov/niosh/nmam/nmammenu.html

NMAM is the result of part of the research activities of the U.S. National Institute for Occupational Safety and Health (NIOSH) relating to the determination of workplace contaminants. NMAM is a collection of methods for sampling and analysis of contaminants in workplace air and in the blood and urine of workers who are occupationally exposed.

NIOSH Pocket Guide to Chemical Hazards

http://www.cdc.gov/niosh/npg/pgdstart.html

Access the full text of this popular publication from the U.S. National Institute for Occupational Safety and Health (NIOSH).

North American Emergency Response Guidebook (1996)

http://www.tc.gc.ca/canutec/english/guide/menug_e.htm

This is an online version of the *Guidebook*, developed jointly by Transport Canada, U.S. Department of Transportation, and the Secretariat of Communications and Transportation of Mexico. More detail about this site is available in the Top 50 Sites section.

Nova Scotia Department of Labour

http://www.gov.ns.ca/labr/

Information is provided about the department, the minister, departmental publications, and news.

Nova Scotia Workers Compensation Board

http://www.gov.ns.ca/bacs/acns/paal/ndxwork.htm

Information is provided about registration with the WCB.

Nuclear Regulatory Commission

http://www.nrc.gov

Publications, news, and other information produced by the commission are accessible.

Occupational Disease Panel Electronic Library

http://www.ccohs.ca/odp/

The Occupational Disease Panel (ODP) was created under the Workers' Compensation Act of Ontario. In its 10 years of operation, ending in 1997, it produced numerous publications on issues relating to workplace/occupational diseases and their compensation. These publications are available in their entirety on this site.

Occupational Safety and Health Administration (OSHA)

http://www.osha.gov

This is one of the best starting points for U.S. regulatory compliance information. View OSHA standard interpretive documents, statistics, technical data, and OSHA publications. Many other useful documents and data collections are available. More detail about this site is available in the Top 50 Sites section.

Occupational Safety and Health Admin., Index to Construction Regulations

http://www.osha-slc.gov/OCIS/Construction.html

Access is provided to the full text of U.S. OSHA's construction regulations, 29 CFR—1926.

Office of the Assistant Deputy Under Secretary of Defense for Safety and Occupational Health — ADU.S.D(SH)

http://www.acq.osd.mil:80/ens/sh

Information about safety in the U.S. military: programs and policy documents, statistics.

Office of the Employer Advisor, Ontario

http://www.gov.on.ca/lab/oea/

The Office of the Employer Advisor (OEA), an independent agency of the Ontario Ministry of Labour, is in business to provide employers with advisory services, education, and information on Workplace Safety Insurance. Read about OEA's services.

Office of the Fire Marshal (Ontario, Canada)

http://www.gov.on.ca/OFM/

The site provides information about the office, public fire safety information, fire statistics, legislation (including the Ontario Fire Code), guidelines, and technical papers.

Ogden Air Logistics Center ESOH

http://137.241.179.15/

The site provides information about programs, contact offices, and databases available at Hill Air Force Base, Ogden Air Logistics Center.

Ontario Ministry of Labour

http://www.gov.on.ca/LAB/main.htm

The site contains information about the ministry, provincial government OSH discussion papers, and a list of ministry publications.

Ontario Statutes and Regulations

http://204.191.119.105/

This site presents the full text of Ontario statutes and regulations. At the time of writing, it was current to January 1, 1998.

Oregon Emergency Management Office

http://www.osp.state.or.us/oem/

Information provided includes descriptions of this state agency itself, its programs, and reports of recent incidents.

Oregon Occupational Safety and Health Division (OR-OSHA)

http://www.cbs.state.or.us/external/osha/index.html

The site contains a directory of OR-OSHA's services, news releases, an employer's toolkit, information about codes, and publications and videos for sale. The *Oregon Health and Safety Resource*, OR-OSHA's newsletter, is also accessible.

OSHA Analytical Methods

http://www.osha-slc.gov/SLTC/analytical_methods/index.html

Provides access to analytical methods from the U.S. Occupational Safety and Health Administration: fully validated methods for organic and inorganic chemicals and partially validated methods. Some methods are provided in Adobe Acrobat (PDF) format.

OSHA Permit Spaces Advisor

http://www.osha.gov/oshasoft/psa.html

OSHA's Permit Spaces Advisor is Windows software that gives users interactive expert help to apply OSHA's Permit Required Confined Spaces Standard (29 CFR 1910.146). Download it from this page.

OSHA Regulations (Standards—29 CFR)

http://www.osha-slc.gov/OshStd_toc/OSHA_Std_toc.html

Access is provided to the full text of U.S. Occupational Safety and Health Administration's regulations.

OSHNET—The Occupational Safety and Health Network of Western Australia

http://sage.wt.com.au/safetyline/oshnet/osh_ind.htm

"Western Australian occupational safety and health service and product providers, who have a common interest in export markets, have formed a network called 'OSHNET'."

Pennsylvania Department of Labor and Industry

http://www.state.pa.us/PA_Exec/Labor_Industry/

Read about the department's various divisions and programs.

Pest Management Regulatory Agency (PMRA)

http://www.hc-sc.gc.ca/pmra-arla/

This agency part of Health Canada has as its mandate "to protect human health, safety, and the environment by minimizing risks associated with pesticides while enabling access to pest management tools, namely pest control products and pest management strategies." PMRA

Pittsburgh and Spokane Research Centers, Mining Health and Safety Research

http://www.usbm.gov

This NIOSH site provides information about its health and safety research, worker health, worker safety, and disaster prevention. Search its database of publications and download MADSS (Mine Accident Decision Support System).

Plain English Guide to the Clean Air Act (U.S. EPA)

http://www.epa.gov/oar/oaqps/peg_caa/pegcaain.html

As the title indicates, this guide offers an explanation of the Clean Air Act and its various provisions in simple, straightforward language.

Prevention and Control of Legionnaires' Disease

http://www.wt.com.au/safetyline/codes/legion.htm

This detailed code of practice is presented in its entirety by WorkSafe Western Australia.

Published International Literature on Traumatic Stress

http://dciswww.dartmouth.edu/cgi-bin/dcis/
wdi?&Alexandria.Dartmouth.EDU&51001&PILOTS%20Catalog&s

PILOTS is a bibliographic database produced at the U.S. National Center for Post-Traumatic Stress Disorder.

Radiation Protection Division, U.S. Environmental Protection Agency

http://www.epa.gov/radiation/

The site provides information about the division and also provides basic radiation protection information in the form of a booklet and fact sheets. There is also a kids' page, "Perfect for book reports or other homework!"

Radiological and Environmental Sciences Laboratory (RESL) at the Idaho National Engineering Laboratory

http://www.inel.gov/resl/

Read about this U.S. federal laboratory's ongoing programs.

RCRA Online

http://www.epa.gov/rcraonline/

Provided by the U.S. Environmental Protection Agency (EPA), RCRA Online is a "database of thousands of selected letters, memoranda and questions and answers written by the EPA's Office of Solid Waste since 1980."

Risk Excellence Notes

http://www.doe.gov/riskcenter/ren.html

Produced by the U.S. Department of Energy (DOE), this newsletter contains digests on exemplary projects, new risk technologies, who's who in the world of risk, notices of upcoming meetings, and news from other U.S. federal and state agencies, academia, and industry.

Rådet för arbetslivsforskning

http://www.ralf.se

The Swedish Council for Work Life Research is a government agency which plans and funds Swedish research on subjects related to the work environment and working life. The site provides contact information, descriptions of its programmes, a newsletter, and its publications. Information is in Swedish and English.

SafetyLine

http://www.safetyline.wa.gov.au

An online information service provided by WorkSafe Western Australia. A content-rich site, with safety and health solutions in many subject areas, training materials, Australian federal and state legislation. More detail is available in the Top 50 Sites section.

SafetyLine Institute

http://www.safetyline.wa.gov.au/institute/

"The SafetyLine Institute is an online education and training facility established by WorkSafe Western Australia to provide high quality education and training materials in occupational safety and health." Available courses include OSH Management, OSH Information Sources, Safe Systems of Work, and OSH Promotion.

Saskatchewan Department of Labour

http://www.gov.sk.ca/govt/labour/

Describes the department's mandate and provides information about the minister.

South Carolina Department of Health and Environmental Control

http://www.state.sc.us/dhec/

Read about this state department's functions, programs, and publications.

Standards Council of Canada

http://www.scc.ca

Access the database of Canadian standards, find out about ISO 9000 and ISO 14000, read a concept paper on occupational health and safety systems standards, and view the full contents of *Consensus, Canada's News Magazine of Standardization*.

State of Ohio Environmental Protection Agency

http://www.epa.ohio.gov

Information is provided about this U.S. state environment agency. Content includes rules and regulations, press releases, policy documents, publications, etc.

State of Washington, Military Department, Emergency Management Div.

http://www.wa.gov/mil/wsem/

"The mission of the Emergency Management Division in Washington State is to coordinate and facilitate resources to minimize the impacts of disasters and emergencies on people, property, and the environment."

Statutes and Regulations of Ontario

http://204.191.119.105/

Provides access to the full text of Acts and Regulations from the Government of Ontario.

Strategis

http://strategis.ic.gc.ca

This Industry Canada site provides information about many aspects of Canadian industry: industry profiles, technology, etc.

Swedish Occupational Safety and Health Administration

http://www.arbsky.se/arbskeng.htm

View Swedish work environment legislation, selected documents produced by the organization, and information about the organization itself. Content is in Swedish and English.

Transport Canada

http://www.tc.gc.ca

The site contains current transportation of dangerous goods (TDG) regulations, draft regulations, and other transportation safety information: aviation safety, marine, railway, road, etc.

Transportation Safety Board of Canada

http://bst-tsb.gc.ca/

This site contains the full text of TSB's latest occurrence reports, occurrence statistics, and significant safety issues, as well as information about the board itself.

U.S. Chemical Safety and Hazard Investigation Board (CSB)

http://www.chemsafety.gov/

CSB was set up as an independent federal agency "to serve as a new resource in the effort to enhance industrial safety." Its mission is "to provide industries that manufacture, use or otherwise handle chemicals with information to enable identification and mitigation of operational conditions that compromise safety."

U.S. Code of Federal Regulations (CFRs)

http://www.access.gpo.gov/nara/cfr/cfr-table-search.html

This is the U.S. Code of Federal Regulations, provided by the National Archives and Records Administration

---

U.S. House of Representatives Internet Law Library: *Federal Register*

http://law.house.gov/7.htm

View proposed U.S. regulations and regulations adopted so recently that they are not yet in the Code of Federal Regulations.

Utah Department of Public Safety

http://www.ps.ex.state.ut.us/

Information about the department and its programs is presented.

Washington State Department of Health

http://www.doh.wa.gov/

The site provides a number of public health fact sheets, monographs, and reports.

West Virginia Bureau of Employment Programs, Workers' Compensation

http://www.state.wv.us/bep/WC/default.HTM

The site provides access to the bureau's publications, frequently asked questions about compensation in West Virginia, along with links to other compensation-related sites.

Westray Mine Public Inquiry

http://www.gov.ns.ca/legi/inquiry/westray/

The site presents the findings of a public inquiry into the 9 May 1992 underground explosion at the Westray coal mine in Nova Scotia that killed 26 miners.

Workers Compensation Board of British Columbia

http://www.wcb.bc.ca/

This site provides access to information about the WCB itself, answers a series of frequently asked questions (FAQ), information about what's new (initiatives, legislation, etc.), and online publications.

Workers Compensation Board of Manitoba

http://www.wcb.mb.ca/

The site provides information about the WCB, claims, and employer and employee responsibilities. The WCB Act, Policy Manual, and fact sheets are provided.

Workers' Compensation Board of Alberta

http://www.wcb.ab.ca/

The site provides information about the WCB, claims, and employer and employee responsibilities. The WCB Publications section contains the full text of the WCB Policies and Information Manual, the Workers' Compensation Act and Regulation, various booklets, information and reports, and forms.

Workers' Compensation Board of Nova Scotia

http://www.wcb.ns.ca/

The site provides information about the WCB, claims, and employer and employee responsibilities. There is also a "library," including the Workers' Compensation Act, regulations, a WCB policy manual, and fact sheets.

Workplace Safety and Insurance Appeals Tribunal (Ontario)

http://www.wsiat.on.ca

The tribunal is the final level of appeal to which workers and employers may bring disputes concerning workers' compensation matters in Ontario. The site provides a description of the appeal process, a list of recent decisions, and research publication. (Formerly the Workers' Compensation Appeals Tribunal)

Workplace Safety and Insurance Board, Ontario, Canada

http://www.wsib.on.ca

This content-rich site contains information about the board itself, information for employers and workers, news releases, facts and figures, and the Ontario Workers' Compensation Act. This board has replaced theWorkers' Compensation Board of Ontario.

Worplace Health, Safety, and Compensation Com. (Newfoundland)

http://www.whscc.nf.ca/

The site provides information about the commission, claims, and employer and employee responsibilities. The Workers' Compensation Act and Regulations, Policy Manual, and other publications are available on the site.

Young Worker Awareness, Workplace Health and Safety Agency (Ontario Canada)

http://www.whsa.on.ca

The site contains information for young workers, their parents, teachers, principals, employers, and others. Information categories include what are the risks, the law (Ontario), your rights, what if I get hurt, resources, and true health and safety stories. More detail about this site is available in the Top 50 Sites section.

7. Regulatory Information

3M Occupational Health and Safety

http://www.mmm.com/occsafety/

The site provides several categories of information: regulatory updates (including a version of 3M's Respiratory Protection Regulations Handbook for downloading), product information, newsletters, and information about 3M's training courses.

ADA and Disability Information

http://www.public.iastate.edu/~sbilling/ada.html

This site, maintained by Iowa State University, provides an extensive list of resources related to the Americans with Disabilities Act (ADA), plus many other disability resource links.

Agricultural Safety

http://hammock.ifas.ufl.edu/txt/fairs/31538

This page, part of the Florida Agricultural Information Retrieval System, provides access to a number of documents, information about training materials, and links to relevant legislation.

AINnet

http://www.ain.es/basedat.htm

The site contains a Spanish and European Union legislative database. All information is in Spanish.

Alberta Labour

http://www.gov.ab.ca/lab/index1.html

This Alberta government department's site provides access to the full text of acts and regulations, accident and injury statistics, and publications.

American Crop Protection Association

http://www.acpa.org/

ACPA represents "the companies that produce, sell and distribute virtually all the active compounds used in crop protection chemicals registered for use in the United States." Read about the association.

American Institute of Chemical Engineers (AIChE), Consultant and Expert Witness Electronic Directory

http://www.resume-link.com/consultants/

This searchable database references chemical engineering consultants and expert witnesses from various backgrounds and expertise including environmental engineering, process design, and hazardous waste management.

American Society of Safety Engineers (ASSE)

http://www.asse.org

View information about the society, membership, and continuing education. Information is also provided about professional accreditation and relevant U.S. regulatory developments. More detail about this site is available in the Top 50 Sites section.

Americans with Disabilities Act Document Center

http://janweb.icdi.wvu.edu/kinder/

This site contains copies of the Americans with Disabilities Act (ADA) of 1990, ADA regulations, and technical assistance manuals.

Arizona State Mine Inspector

http://www.indirect.com/www/mining/

The site contains information about this state government agency, its annual report, and upcoming conferences.

Bara Environmental Compliance

http://www.bara.com

Bara Environmental Solutions is a Texas-based consulting firm that "creates advanced environmental safety solutions through computer tracking of hazardous materials and safety equipment." Read about the company's services.

Belgian Biosafety Server

http://biosafety.ihe.be

This server, located at the Institute of Hygiene and Epidemiology (IHE) in Brussels, provides detailed information about biosafety, particularly in the European Union.

California Department of Industrial Relations (DIR)

http://www.dir.ca.gov/

This site contains information about workers' compensation, occupational safety and health, and California labor law.

Canadian Centre for Occupational Health and Safety (CCOHS)

http://www.ccohs.ca

CCOHS's site provides access to Canadian environmental and OSH legislation, plus major OSH databases (*e.g.*, NIOSHTIC, RTECS, HSELine) on a subscription basis. Other content includes information and hazard alerts, a large directory of health and safety Internet resources. More detail about this site is available in the Top 50 Sites section.

Canadian Environmental Law Association

http://www.web.net/cela/

The CELA is a nonprofit, public interest organization established in 1970 to use existing laws to protect the environment and to advocate environmental law reforms. Provides information about CELA, a list of publications, subscription information for its newsletter, and descriptions of current cases and causes.

CCINFOWeb

http://ccinfoweb.ccohs.ca

This is the home page for the commercial OHS database service on the Web, provided by the Canadian Centre for Occupational Health and Safety (CCOHS). Databases include MSDS, CHEMINFO, RTECS, NIOSHTIC, and HSELine, plus the full text of all Cdn. environmental and OSH legislation. Other databases will be added on an ongoing basis.

CELEX

http://europa.eu.int/celex/

"CELEX is a comprehensive and authoritative information source on European Community law. It offers multilingual, full text coverage of a wide range of legal acts including the founding treaties, binding and non-binding legislation, opinions and resolutions issued by the EU institutions and consultative bodies and the case law of the European Court of Justice."

Ceske pracovni zdravi a bezpecnost

http://www.markl.cz

Czech occupational health and safety. The site contains a description of this consulting company's products, a directory of OSH organizations in the Czech republic, and Czech OSH legislation. All information is available in Czech, and some in English.

Christie Communications—WHMIS Course

http://www.mrg.ab.ca/course/whindex.htm

This inexpensive Web-based course provides a general introduction to WHMIS (Workplace Hazardous Materials Information System).

Colorado Department of Public Health and Environment

http://www.state.co.us/gov_dir/cdphe_dir/cdphehom.html

The site provides lots of contact information, frequently asked questions (FAQ), and downloadable copies of Colorado public health and environment regulations.

Commission de la santé et de la sécurité du travail du Québec (CSST)

http://www.csst.qc.ca/

The site describes the role and services of the commission and the laws it administers. Other information includes CSST news releases, summaries of their magazine *Prévention au travail* and a publications list. Most information is in French, but some is also in English.

Compliance Online

http://www.ieti.com/taylor/compliance.html

This is an "Internet newsletter for finding your way through the maze of environmental regulations. This site contains an archive of newsletter issues back to Jan. 1996."

Controlled Risk International

http://www5.electriciti.com/risk/

Read about the company's risk management services, courses, and OSHA and CAL/OSHA's most frequently cited serious violations.

Counterpoint Publishing

http://www.counterpoint.com

Read about the publications, databases, and data collections available from this publisher of regulatory, legal, and chemical information. A lot of the information is available on a subscription basis.

CTDNews Online

http://ctdnews.com/

CTDNEWS is a monthly newsletter related to cumulative trauma disorders. The full text of current and past issues is presented, along with CTD-related information about products and regulatory initiatives.

Dakota Decision Support Software

http://www.dakotasoft.com

Read about and view sample screens from Dakota's EH&S auditing software.

Department of Transportation

http://www.dot.gov

As well as containing U.S. regulatory and policy information, the site provides access to a variety of transportation safety resources: research and reports on transportation safety from various sources, hazardous materials safety, etc.

Dept. of Transportation, Office of Hazardous Materials Safety (OHM)

http://www.volpe.dot.gov/ohm/

Responsible for coordinating a national (U.S.) safety program for the transportation of hazardous materials by air, rail, highway, and water. Highlights include an electronic version of the Emergency Response Guidebook, regulatory info., and OHM publications.

Division of Occupational Safety and Health (California OSHA)

http://www.dir.ca.gov/DIR/OS&H/DOSH/dosh1.html

Information is available about Cal/OSHA's services, personnel etc. The full text of the division's publications and its policies and procedures manual is also included.

Domestic Substances List/ Non-Domestic Substances List (DSL/NDSL)

http://www2.ec.gc.ca/cceb1/cas_e.html

This site provides access to the Canadian DSL/ NDSL. Searching is by CAS number only.

Donald W. Fohrman and Associates, Ltd.

http://www.chicagolegalnet.com/

The site "provides victims of Carpal Tunnel Syndrome and other types of Repetitive Stress injuries, with over 100 pages of straight-forward legal information about the Illinois Workers' Compensation Act."

EC Chemical Regulation Reporter

http://www.incadinc.com/edr/ecchemicalregulationreporter.html

This site provides basic information about European Community chemical regulations and subscription information about "EC Chemical Regulation," a monthly notification of EC actions affecting the chemical industry.

EnviroInfo Research Service

http://members.aol.com/elsevierhr/index.htm

Read about Elsevier's Enviroinfo Research Service, which "specializes in finding information for industry and environmental professionals." Environmental compliance information is their specialty.

Environmental Chemicals Data and Information Network (ECDIN)
http://ecdin.etomep.net/

"ECDIN is a factual databank, created under the Environmental Research Programme of the Joint Research Centre (JRC) of the Commission of European Communities at the Ispra Establishment." "ECDIN deals with the whole spectrum of parameters and properties that might help the user to evaluate real or potential risk in the use of a chemical and its economical and ecological impact."

Environmental Protection Agency (EPA)
http://www.epa.gov

This well-developed U.S. government site contains a huge amount of environmental information, including environmental regulations, a variety of database systems, and numerous indoor air quality documents. More detail is available in the Top 50 Sites section.

Environmental Protection Agency, Office of Pesticide Programs
http://www.epa.gov/pesticides/

Access is provided to detailed U.S. EPA pesticide information.

EPA Pesticide Registration Kit
http://www.epa.gov/pesticides/registrationkit/

Online registration kit contains "pertinent forms and information needed to register a pesticide with the U.S. Environmental Protection Agency's Office of Pesticide Programs.

Europa—Governments online
http://europa.eu.int/en/gonline.html

This is a directory to government pages for European Union member countries.

European Agency for Safety and Health at Work
http://www.eu-osha.es/

This site (will) contain(s) European information in the following areas: legislation and standards, research, practice, strategies and programmes, statistics, information, and news. More detail about this site is available in the Top 50 Sites section.

European Chemicals Bureau
http://ecb.ei.jrc.it/

"The principal task of the ECB is to carry out and coordinate the scientific/technical work needed for the implementation of EU-legislation (directives and regulations) in the area of chemical control." Read about ECB's projects and activities and publications.

FAQ: Sources of EMC and Safety Compliance Information

http://world.std.com/~techbook/compliance_faq.html

Frequently-asked-questions document (FAQ) for the sci.engr.electrical.compliance newsgroup.

FedWorld

http://www.fedworld.gov

A gateway to U.S. federal government information. From here, you can find and order recent U.S. Government Reports from all agencies and search the Web sites of hundreds of federal government agencies. More detail is available in the Top 50 Sites section.

First Report—Workers' Compensation and OSHA Recordkeeping Software

http://www.firstreport.com/

Really Useful Software provides information about this software package. They also provide "commentary and information" about workers' compensation and safety issues.

Food and Drug Administration (FDA)

http://www.fda.gov

Information is provided about various U.S. FDA program areas, including radiological health, toxicology, and drugs.

Harvard Environmental Resources Online

http://environment.harvard.edu/

The site provides information about environmental resources at Harvard University (U.S.A.), an international environmental policy resource guide, plus archives for the ENVCONFS-L and ENVREFLIB-L mailing lists.

Hazardous Materials Information Exchange

telnet://hmix.dis.anl.gov

This telnet-accessible bulletin board system was created to provide a centralized database for federal, state, local, and private-sector personnel to share information pertaining to hazardous materials emergency management, training, resources, technical assistance, and regulations.

HazMat Safety

http://hazmat.dot.gov/

This site is managed by the U.S. Department of Transportation's Office of Hazardous Materials Safety. It provides access to HazMat related regulations, DOT forms and documents, the North American Emergency Response Guidebook, and much more.

Health and Safety Authority

http://www.hsa.ie/osh/welcome.htm

Information is provided about the Health and Safety Authority (Ireland), a list of publications, a list of acts and regulations, and information about the authority's programs.

Health and Safety Executive (U.K.)

http://www.open.gov.uk/hse/hsehome.htm

Information is provided about this U.K. regulatory agency, its publications and videos, and its research.

Health and Safety Promotion in the European Union

http://www.hsa.ie/hspro/index1.html

HSPro—EU will set up a remotely accessible telematics system providing a "one-stop-shop" for occupational health and safety information. Categories of information include news and events, guidance, OSH information, OSH databases, bibliographies, library catalogs, and publications. More detail is available in the Top 50 Sites section.

Hong Kong Education and Manpower Bureau

http://www.info.gov.hk/emb/welcome.htm

This government agency's site provides information about occupational health and safety policies and labor relations.

Hong Kong Occupational Safety and Health Association

http://www.hk.super.net/~hkosha/

Information is provided about the association, Hong Kong safety and health legislation and standards, and the association's publications, along with links to other sites.

House of Representatives Internet Law Library

http://law.house.gov

This site provides access to the United States Code, Code of Federal Regulations, the *Federal Register*, and numerous other law sites.

Human Resources Development Canada, Labour Operations, Occupational Safety and Health

http://info.load-otea.hrdc-drhc.gc.ca/~oshweb/homeen.shtml

Information is provided to allow Canadian federally regulated organizations to meet the requirements under the Canada Labour Code Part II. There is an overview of the code and information about setting up an OSH program in your workplace. Information of general interest includes occupational injury statistics.

InteractiveWare Inc.—HAZCOM training software

http://www.interactiveware.com

Download a demo of this Windows-based hazard communication training software.

Job Accommodation Network

http://janweb.icdi.wvu.edu/

An international consulting service that provides information about job accommodations and the employability of people with disabilities. Provides a large directory of sites, information about job accommodation in the U.S., Americans with Disabilities Act, etc.

Justice Canada

http://canada.justice.gc.ca/Loireg/index_en.html

This site provides access to the full text of many Canadian federal acts and regulations.

Lab XL: Performance Oriented Environmental Regulation for Labs

http://esf.uvm.edu/labxl/

The purpose of this Web page is to provide a clearinghouse for information related to performance-oriented laboratory environmental regulation reform. The site is maintained by Ralph Stuart.

LegiFrance

http://www.legifrance.gouv.fr

This site contains the entire legislative code of France. All information is in French.

Mathews, Dinsdale & Clark

http://www.mdclabourlaw.com/

This is a Toronto, Canada-based law practice "restricting its practice to labour and employment law." The site provides a number of interpretive documents about Ontario law, along with quizzes to test your knowledge.

MGMT Alliances Inc.

http://www.mgmt14k.com

MGMT Alliances develops and presents training courses on Environmental Management Systems, ISO 14000, ISO 9000, and Integrated Management Systems.

Mine Safety and Health Administration (MSHA)

http://www.msha.gov

This U.S. government site provides access to mining-related regulatory information, health and safety documents, and information about various programs.

Missouri Department Of Labor and Industrial Relations

http://www.dolir.state.mo.us/

Information is provided about the department's divisions, including labor standards, workers' compensation, and the Governor's Council on Disability.

National Institute for Occcupational Safety and Health (NIOSH)

http://www.cdc.gov/niosh/homepage.html

This home page has been established to provide information about NIOSH and related activities. Content includes NIOSH publications, respirator information, meetings and symposia, etc. More detail about this site is available in the Top 50 Sites section.

National Institute of Environmental Health Sciences

http://www.niehs.nih.gov/

The site provides access to detailed information about this U.S. government agency, along with links to NIEHS-sponsored sites.

National Occupational Health and Safety Commission (WorkSafe Australia)

http://www.worksafe.gov.au

Read about this national OH&S agency, view statistical information, Australian regulatory requirements, fact sheets, and alerts.

Nebraska Dept. of Administrative Services, Risk Management Health Benefit Program

http://www.das.state.ne.us/das_dorm/index.html

View information about this state government program.

New Brunswick Workplace Health, Safety and Compensation Commission

http://www.gov.nb.ca/whscc/

This site provides information about the commission and its programs, calculating WCB claims in New Brunswick, and industry assessment rates.

New Jersey Dept. of Health and Senior Services, Right-to-Know Program

http://www.state.nj.us/health/eoh/rtkweb/rtkhome.htm

The site provides information about the program, training courses, commonly asked questions about NJ Right-to-Know, and publications. Hazardous Substance Fact Sheets are available as PDF files. (Adobe Acrobat reader software required)

New Substances Notification Regulations

http://www2.ec.gc.ca/cceb1/eng/nsnr_cp.htm

"Environment Canada and Health Canada have prepared this brochure to alert persons who manufacture or import chemical substances and polymers of their obligation to submit a notification package before they import or manufacture a substance that is considered new to Canada." The brochure is presented in its entirety.

New Zealand's Health and Safety Net

http://www.OSH.dol.govt.nz/

This is the home page of the Occupational Safety and Health Service of the Department of Labour, New Zealand. The site provides information about the organization, work hazards, health and safety law, and training programs.

Newfoundland Environment and Labour

http://www.gov.nf.ca/env/default.asp

The site contains information about this Canadian provincial government department, press releases, etc.

NIOSH Criteria Documents

http://www.cdc.gov/niosh/critdoc2.html

View Criteria Documents, Occupational Hazard Assessments, Special Hazard Reviews, and Joint Occupational Health Documents from the U.S. National Institute for Occupational Safety and Health (NIOSH). More detail about this site is available in the Top 50 Sites section.

Nova Scotia Department of Labour

http://www.gov.ns.ca/labr/

Information is provided about the department, the minister, departmental publications, news, etc.

Nova Scotia Workers Compensation Board

http://www.gov.ns.ca/bacs/acns/paal/ndxwork.htm

Information is provided about registration with the WCB.

Nuclear Regulatory Commission

http://www.nrc.gov

Publications, news, and other information produced by the commission are accessible.

Occupational Noise Exposure

http://hammock.ifas.ufl.edu/txt/fairs/oa/19401.html

This is a summary of U.S. OSHA Standard 1910.95, written primarily for agricultural business owners and managers.

Occupational Safety and Health Administration (OSHA)

http://www.osha.gov

This is one of the best starting points for U.S. regulatory compliance information. View OSHA standard interpretive documents, statistics, technical data, and OSHA publications. Many other useful documents and data collections are available. More detail about this site is available in the Top 50 Sites section.

Occupational Safety and Health Admin., Index to Construction Regulations

http://www.osha-slc.gov/OCIS/Construction.html

Access is provided to the full text of U.S. OSHA's construction regulations, 29 CFR—1926.

Office of the Fire Marshal (Ontario Canada)

http://www.gov.on.ca/OFM/

The site provides information about the office, public fire safety information, fire statistics, legislation (including the Ontario Fire Code), guidelines, and technical papers.

Ontario Ministry of Labour

http://www.gov.on.ca/LAB/main.htm

The site contains information about the mnistry, provincial government OSH discussion papers, and a list of ministry publications.

Ontario Statutes and Regulations

http://204.191.119.105/

This site presents the full text of Ontario statutes and regulations. At the time of writing, it was current to January 1, 1998.

Oregon Occupational Safety and Health Division (OR-OSHA)

http://www.cbs.state.or.us/external/osha/index.html

The site contains a directory of OR-OSHA's services, news releases, an employer's toolkit, information about codes, publications, and videos for sale. The *Oregon Health and Safety Resource*, OR-OSHA's newsletter, is also accessible.

OSHA Permit Spaces Advisor

http://www.osha.gov/oshasoft/psa.html

OSHA's Permit Spaces Advisor is Windows software that gives users interactive expert help to apply OSHA's Permit Required Confined Spaces Standard (29 CFR 1910.146). Download it from this page.

OSHA Regulations (Standards—29 CFR)

http://www.osha-slc.gov/OshStd_toc/OSHA_Std_toc.html

Access is provided to the full text of U.S. Occupational Safety and Health Administration's regulations.

OSHA-Data

http://www.oshadata.com

OSHA-Data's site describes the company's OSHA compliance services, including searching its database of OSHA violation and enforcement records.

Pathfinder Associates

http://www.webcom.com/pathfndr/welcome.html

The company's Web pages provide current information on OSHA regulations and citations. There is also company product and service information, along with various health and safety documents.

Pennsylvania Department of Labor and Industry

http://www.state.pa.us/PA_Exec/Labor_Industry/

Read about the department's various divisions and programs.

Performance Safety Home Page

http://www.uow.edu.au/crearts/theater/safehome.html

The Performance Safety Home Page is an expanding site dedicated to peforming arts safety. It includes a range of information including Australian OH&S legislation, fire safety, and hazardous materials.

Pest Management Regulatory Agency (PMRA)

http://www.hc-sc.gc.ca/pmra-arla/

This agency, a part of Health Canada, has as its mandate "to protect human health safety and the environment by minimizing risks associated with pesticides while enabling access to pest management tools, namely pest control products and pest management strategies." PMRA

Pine Lane Resources

http://www.1pinelane.com

This site is primarily a directory to sites that are useful for regulatory support. The sponsoring company also provides consulting services in the regulatory support field.

Plain English Guide to the Clean Air Act (U.S. EPA)

http://www.epa.gov/oar/oaqps/peg_caa/pegcaain.html

As the title indicates, this guide offers an explanation of the Clean Air Act and its various provisions in simple, straightforward language.

PSM Report

http://www.infoassets.com/kbi/psmrpt.html

Read about the *Process Safety Management Report*, a subscription service that provides information about OSHA's Process Safety Management (29 CFR 1910.119) and EPA's rule for Risk Management Programs (58 FR 54190).

Radiation related rules, regulations, and laws

http://www.physics.isu.edu/radinf/law.htm

This is an extensive list of links to primarily U.S. resources.

RCRA Online

http://www.epa.gov/rcraonline/

Provided by the U.S. Environmental Protection Agency (EPA), RCRA Online is a "database of thousands of selected letters, memoranda and questions and answers written by the EPA's Office of Solid Waste since 1980."

Regulatory Compliance Information Center

http://www.rcic.com

This site has a search engine with only compliance-related sites indexed, compliance Web chatrooms, etc. Subject areas include radiation, telecommunications, and electricity.

SafetyLine

http://www.safetyline.wa.gov.au

SafetyLine is an online information service provided by WorkSafe Western Australia. This is a very content-rich site, with safety and health solutions in a variety of subject areas, training materials, and Australian federal and state legislation. More detail about this site is available in the Top 50 Sites section.

Saskatchewan Department of Labour

http://www.gov.sk.ca/govt/labour/

Describes the department's mandate and provides information about the minister.

SASSI Training Systems

http://www.mindspring.com/~sassitng/index.html

Read about the company's environmental counseling, compliance training, and ISO 14000 training.

Seguranca e Saude no Trabalho

http://www.geocities.com/CapeCanaveral/2616/

This Brazilian occupational health and safety site presents general information, Brazilian legislation, articles, upcoming events, and links to Brazilian and international organizations. All information on this site is in Portuguese.

Seton's Technical and Regulatory Assistance Center (SmartTrac)

http://www3.seton.com/smarttrac.html

To quote from the site itself, Seton's "Technical and Regulatory Assistance Center will help you get the answers you need to your important safety and regulatory questions." This appears to be a free service, available to the general Web-surfing public.

Solutions Software Corporation

http://www.env-sol.com/

This site contains descriptions of the company's CD-ROM products, many of them health and safety related, plus links to various sites.

Sortor Occupational Safety Consulting

http://www.safestyle.com/

Sortor is a management consulting firm specializing in occupational safety and health. It serves attorneys, insurance companies, government, and private industry. The site describes the company's services.

South Carolina Department of Health and Environmental Control

http://www.state.sc.us/dhec/

Read about this state department's functions, programs, and publications.

State of Ohio Environmental Protection Agency

http://www.epa.ohio.gov

Information is provided about this U.S. state environment agency. Content includes rules and regulations, press releases, policy documents, and publications.

Statutes and Regulations of Ontario

http://204.191.119.105/

This site provides access to the full text of acts and regulations of the government of Ontario.

Swedish Occupational Safety and Health Administration

http://www.arbsky.se/arbskeng.htm

View Swedish work environment legislation, selected documents produced by the organization, and information about the organization itself. Content is in Swedish and English.

Text-Trieve Environmental Health and Safety Page

http://www.halcyon.com/ttrieve/

Read about the company's electronic text-management systems that "present government regulations in easy-to-use electronically published formats." A number of U.S. regulatory subject areas are described.

Transport Canada

http://www.tc.gc.ca

The site contains current regulations on the transportation of dangerous goods (TDG), draft regulations, and other transportation safety information, such as aviation safety, marine, railway, road etc.

Transportation Safety Board of Canada

http://bst-tsb.gc.ca/

This site contains the full text of TSB's latest occurrence reports, occurrence statistics, significant safety issues, as well as information about the board itself.

U.K. Maximum Exposure Limits

http://physchem.ox.ac.uk/MSDS/mels.html

This site provides maximum exposure limits in the U.K. for a number of chemicals, based on Schedule 1 of the COSHH regulations.

U.S. Code of Federal Regulations (CFRs)

http://www.access.gpo.gov/nara/cfr/cfr-table-search.html

This is the U.S. Code of Federal Regulations, provided by the National Archives and Records Administration

---

U.S. House of Representatives Internet Law Library: *Federal Register*

http://law.house.gov/7.htm

View proposed U.S. regulations and regulations adopted so recently that they are not yet in the Code of Federal Regulations.

Vermont SIRI

http://siri.org

The site contains the archives of the SAFETY and Occ-Env-Med-L mailing lists, a large MSDS database, safety clip art, various safety-related documents, and regulatory information. More detail about this site is available in the Top 50 Sites section.

WHMIS 96

http://www.bbsi.net/50para/p_web.html-ssi

Read about this CD-ROM-based WHMIS (Canadian right-to-know legislation) training package, produced by C.D. 50e PARALLELE INC.

WHMIS Software Training

http://www.niagara.com/blmc/

This site contains a brief description of BLMC's WHMIS training software, with contact information.

Workers' Compensation Board of British Columbia

http://www.wcb.bc.ca/

This site provides access to information about the WCB itself, answers to a series of frequently asked questions (FAQ), information about what's new (initiatives, legislation, etc.), and online publications.

Workplace Safety and Insurance Appeals Tribunal (Ontario)

http://www.wsiat.on.ca

The t ribunal is the final level of appeal to which workers and employers may bring disputes concerning workers' compensation matters in Ontario. The site provides a description of the appeal process, a list of recent decisions, and research publications. (Formerly the Workers' Compensation Appeals Tribunal)

Workplace Safety and Insurance Board, Ontario, Canada

http://www.wsib.on.ca

This content-rich site contains information about the board itself, information for employers and workers, news releases, facts and figures, and the Ontario Workers' Compensation Act. This board has replaced the Workers Compensation Board of Ontario.

Wright's PestLaw—Pesticide News and Regulations

http://www.pestlaw.com/

This is a free public service that provides full text regulatory information and other resources of interest to crop-protection and antimicrobial pesticide registrants, regulatory agencies, academia, and pesticide users.

Young Worker Awareness, Workplace Health and Safety Agency (Ontario Canada)

http://www.whsa.on.ca

The site contains information for young workers, their parents, teachers, principals, employers, and others. Information categories include what are the risks, the law (Ontario), your rights, what if I get hurt, resources, and true health and safety stories. More detail about this site is available in the Top 50 Sites section.

---

8. Material Safety Data Sheet Sites and Software

CCINFOWeb

http://ccinfoweb.ccohs.ca

This is the home page for the commercial OHS database service on the Web, provided by the Canadian Centre for Occupational Health and Safety (CCOHS). Databases include MSDS, CHEMINFO, RTECS, NIOSHTIC, and HSELine, as well as the full text of all Cdn. environmental and OSH legislation. Other databases will be added on an ongoing basis.

ChemFinder Chemical Searching and Information Integration

http://www.chemfinder.com

Comprehensive access to chemical information from a variety of sources is provided: MSDS collections, toxicological profiles, etc. The site is provided by CambridgeSoft Corporation. More detail about this site is available in the Top 50 Sites section.

Chemical Safety Associates

http://www.chemical-safety.com

The site provides information about this consulting company's services, including site remediation, MSDS preparation, and assistance with safe handling of hazardous materials.

Cornell Planning Design and Construction, Material Safety Data Sheets

http://msds.pdc.cornell.edu/issearch/msdssrch.htm

This is a large MSDS database, containing approximately 325,000 MSDSs from the U.S. Dept. of Defense MSDS database, the Vermont SIRI MSDS collection, and data sheets maintained by Cornell's Environmental Health and Safety Dept.

Eclipse Software Technologies

http://www.eclipsesoft.com

Information about and downloadable demonstrations of "MSDS Wizard" and "MSDS Scan Wizard," MSDS management software.

Euroware Associates Software

http://www.euroware.com/index.html

Provides information about the company's MSDS and chemical-related software.

Fisher Scientific

http://www.fisher1.com

View Fisher's product catalogs and MSDS and order products online.

Genium Publishing

http://www.genium.com

This site provides information about Genium's safety and health software programs, MSDS collection, etc. An online safety and health glossary and a database of hazard labelling are also provided.

Information Services for Agriculture (ISA)

http://www.aginfo.com/isa.html

This service provides access to the company's label and MSDS management system. View MSDSs for agricultural products from Dupont, Bayer, Monsanto, and Rhone Poulenc. ISA also creates Web sites for agricultural companies.

Material Safety Data Sheets (MSDSs) for Infectious Substances

http://www.hc-sc.gc.ca/hpb/lcdc/biosafty/msds/index.html

Health Canada, Health Protection Branch, Laboratory Centre for Disease Control, presents these MSDSs "for personnel working in the life sciences as quick safety reference material relating to infectious micro-organisms." Data sheets available in English and French.

MSDS links from U.S. Dept Commerce/NOAA/NMFS/NWFSC

http://research.nwfsc.noaa.gov/msds.html

This fill-in search form provides access to the University of Utah and Vermont SIRI MSDS collections, along with links to a number of other MSDS collections.

MSDS Pro

http://www.msdspro.com

Download a demo of this MSDS management software by Aurora Data Systems.

MSDS-Search

http://www.msdssearch.com

According to Envirocare, the company that runs the site, "Our mission is the electronic connection of manufacturers, distributors, customers, government agencies transporters, and emergency responders in a single, international data base with all MSDSs available in text format at no cost to users." At the time of writing, sample MSDSs were available.

MSDS2

http://www.whmis.com

Read about and "test drive" BC Hydro's material safety data sheet management system.

Oxford University, MSDS

http://physchem.ox.ac.uk/MSDS/

The site provides access to a collection of publicly accessible MSDSs. It appears to be derived from the University of Utah collection.

Safety Management Services

http://www.cyberg8t.com/~jprice/

This consulting company's site provides access to its Online Safety Program, including an Injury and Illness Prevention Program, Hazardous Communications Program, MSDSs, and links to other sites.

Safety Officer II

http://www.safetyofficer.com/

Safety Officer II is a chemical information software system for converting data associated with chemicals into end-user specific MSDSs, custom Avery labels, and management reports. Detailed descriptions of the software are provided.

Seton Identification Products

http://www.seton.com

This extensive, well-organized site provides detailed information about Seton's safety products, MSDSs for its chemical products, and a health and safety resources directory.

The Wercs(r)

http://www.thewercs.com/

Read about and view a demo of this MSDS management software.

University of Utah, MSDS

http://www.chem.utah.edu/MSDS/msds.html

This site contains a public collection of approximately 1500 material safety data sheets. Many appear to be J. T. Baker data sheets from the mid-1980's.

Vermont SIRI

http://siri.org

The site contains the archives of the SAFETY and Occ-Env-Med-L mailing lists, a large MSDS database, safety clip art, various safety-related documents, and regulatory information. More detail about this site is available in the Top 50 Sites section.

Where to Find MSDS on the Internet

http://www.ilpi.com/msds/index.chtml

This site, maintained by Rob Toreki, a chemistry professor at the University of Kentucky, contains descriptive information about MSDSs themselves, as well as links to MSDS collections.

Xerox MSDS

http://www.xerox.com/ehs/msds/index.htm

The site provides a searchable database of MSDSs for Xerox products sold in the Western Hemisphere.

Section B

Health and Safety Specialties

1. Agricultural Safety

Agricultural Safety

http://hammock.ifas.ufl.edu/txt/fairs/31538

Part of the Florida Agricultural Information Retrieval System, this provides access to a number of documents, information about training materials, and links to relevant legislation.

American Crop Protection Association

http://www.acpa.org/

ACPA represents "the companies that produce, sell and distribute virtually all the active compounds used in crop protection chemicals registered for use in the United States." Read about the association.

Environmental Protection Agency, Office of Pesticide Programs

http://www.epa.gov/pesticides/

Access is provided to detailed U.S. EPA pesticide information.

EPA Pesticide Registration Kit

http://www.epa.gov/pesticides/registrationkit/

This online registration kit contains "pertinent forms and information needed to register a pesticide with the U.S. Environmental Protection Agency's Office of Pesticide Programs."

Extoxnet—The EXtention TOXicology NETwork

http://ace.orst.edu/info/extoxnet/

EXTOXNET is a cooperative effort of University of California-Davis, Oregon State University, Michigan State University, and Cornell University. Various types of pesticide toxicology and environmental chemistry information are available.

Farm Safety Association of Ontario

http://www.fsai.on.ca

The site provides information about the association itself, along with safety information for agriculture/ horticulture and landscapers.

Information Services for Agriculture (ISA)

http://www.aginfo.com/isa.html

This service provides access to the company's label and an MSDS management system. View MSDSs for agricultural products from Dupont, Bayer, Monsanto, and Rhone Poulenc. ISA also creates Web sites for agricultural companies.

National Agricultural Safety Database

http://www.cdc.gov/niosh/nasd/nasdhome.html

NASD contains over 2,000 ag-health and safety publications from 25 states, 4 federal agencies, and 5 national organizations. The collection includes OSHA and EPA standards, extension publications, and a database of abstracts and ordering information for over 1,000 ag safety-related videos.

National Antimicrobial Information Network

http://ace.ace.orst.edu/info/nain/

NAIN provides information concerning antimicrobial toxicology, as well as pesticide manufacturer and distributor information.

National Pesticide Telecommunications Network

http://ace.ace.orst.edu/info/nptn/

NPTN is a joint initiative of Oregon State University and the U.S. Environmental Protection Agency. The site provides links to various pesticide-related sites: pesticide companies, sources for pest control information, pesticide toxicology, and poison control centers.

NIOSH Agricultural Health and Safety Centers

http://agcenter.ucdavis.edu/agcenter/niosh/niosh.html

The site contains the full text of back issues of *NIOSH Ag Center News*, information about specific centers, and links to other related sites.

Occupational Noise Exposure

http://hammock.ifas.ufl.edu/txt/fairs/oa/19401.html

This is a summary of U.S. OSHA Standard 1910.95, written primarily for agricultural business owners and managers.

Pesticide Information Fact Sheets

http://fshn.ifas.ufl.edu/pest/default.htm#FactSheets

This is a series of documents describing various environmental and occupational aspects of pesticides. The site is maintained by the Pesticide Information Office, Institute of Food and Agricultural Sciences, University of Florida.

Pesticide Information Fact Sheets

http://gnv.ifas.ufl.edu/~foodweb/links.htm#FactSheets

This is a series of documents describing various environmental and occupational aspects of pesticides. The site is maintained by the Pesticide Information Office, Institute of Food and Agricultural Sciences, University of Florida

Pesticide Poisoning Handbook

http://hammock.ifas.ufl.edu/txt/fairs/19729.html

This is the full text of a handbook maintained by the Institute of Food and Agricultural Sciences, University of Florida.

Pesticide Poisoning Handbook

http://hammock.ifas.ufl.edu/txt/fairs/pp/19729.html

This is the full text of a handbook maintained by the Institute of Food and Agricultural Sciences, University of Florida.

Prairie Agricultural Machinery Institute, "Be Seen, Be Safe!"

http://www.pami.ca/beseennf/beseennf.htm

This site has been adapted from the guidebook of the same name, originally produced by the Prairie Agricultural Machinery Institute (PAMI) under the direction of Manitoba Highways and Transportation.

Southwest Rural and Agricultural Safety

http://ag.arizona.edu/agsafety/

This site, an extension of University of Arizona, provides numerous links and documents related to a variety of aspects of agricultural safety.

Timber Falling Consultants

http://www.forestindustry.com/ddouglasdent/index.html

View information about the company's films and books on safety in the lumber industry.

Wright's PestLaw—Pesticide News and Regulations

http://www.pestlaw.com/

This is a free public service that provides full text regulatory information and other resources of interest to crop-protection and antimicrobial pesticide registrants, regulatory agencies, academia, and pesticide users.

---

2. Art Safety

Arts, Crafts, and Theater Safety (ACTS)

http://www.caseweb.com/acts/

Information is provided about this free art/theater hazard information service. A list of its publications and further contact information is available.

Center for Safety in the Arts

http://www.artswire.org:70/1/csa/

Full text documents describing numerous arts-related hazards are presented, along with a catalog of books, videos, etc. More detail is available in the Top 50 Sites section.

Performance Safety Home Page

http://www.uow.edu.au/crearts/theater/safehome.html

The Performance Safety Home Page is an expanding site dedicated to peforming arts safety. It provides a range of information including Australian OH&S legislation, fire safety, and hazardous materials.

---

3. Biosafety

ABSA Canada

http://www.nanook.com/absa_canada/

ABSA Canada is a Canadian affiliate of the American Biological Safety Association. Its objective is to establish a Canadian network of individuals interested in a wide variety of biological safety issues.

Aerobiological Engineering—Airborne Pathogen Control Systems

http://www.engr.psu.edu/www/dept/arc/server/wjkaerob.html

Provided by the Graduate School of Architectural Engineering and the Department of Biology at Penn State University, this content-rich site provides information about numerous aerobiological engineering research topics: airborne pathogens, fungi and bacteria in ventilation systems, legionnaires disease, and the spread of disease in office buildings.

American Biological Safety Association

http://www.absa.org

Information is provided about the association itself, the biosafety profession, the BIOSAFTY mailing list, and an extensive directory of biosafety resources on the Internet.

Ass'n. for Professionals in Infection Control and Epidemiology, Inc. (APIC)

http://www.apic.org

This international organization's site lists resources for infection control professionals, a list of publications for sale, U.S. government information, and courses offered.

Belgian Biosafety Server

http://biosafety.ihe.be

This server, located at the Institute of Hygiene and Epidemiology (IHE) in Brussels, provides detailed information about biosafety, particularly in the European Union.

Biosafety Information Network and Advisory Service (BINAS)

http://binas.unido.org/binas/binas.html

BINAS, a service of the United Nations Industrial Development Organization, monitors global developments in regulatory issues in biotechnology. Site highlights include links to regulations in many countries, UNIDO publications, the BINAS newsletter, and OECD Environment Monographs.

Biosafety Resource Page

http://www.orcbs.msu.edu/absa/resource.html

This is an extensive, high quality list of biosafety resources selected and maintained by Stefan Wagener, a biosafety professional at Michigan State University. More detail about this site is available in the Top 50 Sites section.

ChinaHawk Enterprises Ltd.

http://home.hkstar.com/~chinahwk/home.html

Information about this Beijing-based company's disposable safety products for the health care, industrial and food services industries. Offices are in Beijing and Hong Kong.

Dept. of Veterans Affairs, Office of Occupational Safety and Health

http://www.va.gov/vasafety/index.htm

As well as providing information about the office itself, the site provides access to information on a variety of OSH topics, such as OSH program documents, violent behavior prevention, needle safety, asbestos, and latex allergy prevention.

Directory of Equipment and Services for Controlling Airborne Microorganisms

http://www.bio.psu.edu/faculty/whittam/research/products.html

This site provides names, contact information, and Web links for providers of a wide range of products and services related to controlling airborne microorganisms.

Effective Prevention

http://www.interlog.com/~effprev/

Information is provided about Effective Prevention, a skin cream that "repels bodily fluids and many biological hazards in the medical and food services profession."

European Biological Safety Association

http://biosafety.ihe.be/EBSA/HomeEBSA.html

EBSA was formed in 1996 "to promote biosafety as a scientific discipline and serve the growing needs of the biosafety professionals throughout Europe."

European Federation, Biotechnology Working Party on Safety in Biotechnology

http://www.boku.ac.at/iam/efb/efb_wp.htm

Information available includes a list of publications from and membership information for the Working Party, biosafety links, and a bibliography on laboratory-acquired infections.

Howard Hughes Medical Institute, Office of Laboratory Safety

http://www.hhmi.org/science/labsafe

This site contains information about laboratory safety materials available from the institute, laboratory safety summaries, training videos, emergency response guidelines, and help with laboratory hazard recognition.

International Centre for Genetic Engineering and Biotechnology, Biosafety Web Pages

http://www.icgeb.trieste.it/biosafety/

ICGEB's site provides the full text of numerous biosafety documents, a bibliographic database of biosafety papers, European and U.S. biosafety regulations, a biosafety mailing list, and links to other biosafety resources. More detail is available in the Top 50 Sites section.

International Register on Biosafety

http://irptc.unep.ch/biodiv/

This site, provided by the United Nations Environment Programme, presents a number of biosafety resources. Highlights include a series of international directories, guidelines and fact sheets, and descriptions of international biosafety activities.

Legionnaires' Disease

http://www.hcinfo.com/legionella/

This site, provided by HC Information Resources, provides a frequently-asked-questions document (FAQ) about Legionnaires' Disease, along with excerpts from and details about Matthew Freije's book, *Legionellae Control in Healthcare Facilities: A Guide for Minimizing Risk*.

M.O.S.T. Consulting Services

http://www.mostsafety.com/

Medical and Occupational Safety and Training (M.O.S.T.) provides consulting services for safety programs in medical facilities. Read about the company's services.

Material Safety Data Sheets (MSDS) for Infectious Substances

http://www.hc-sc.gc.ca/hpb/lcdc/biosafty/msds/index.html

Health Canada, Health Protection Branch, Laboratory Centre for Disease Control, presents these MSDSs "for personnel working in the life sciences as quick safety reference material relating to infectious micro-organisms." Data sheets available in English and French.

Michigan State Univ. Office of Radiation, Chemical and Biological Safety

http://www.orcbs.msu.edu

Various documents are presented, including MSU's chemical hygiene program, their chemical lab safety checklist, radiation safety manual, and *Safe Science* newsletter.

National Sanitation Foundation

http://www.nsf.com

The site describes NSF International and provides a database of NSF-certified products (biohazard cabinets, water purification systems, etc.) and online newsletters.

Portland Community College Distance Learning—Bloodborne Pathogens

http://www.online.pcc.edu/courses/bloodborne/

This distance learning course is presented periodically and may be taken through the Web site. Paid registration is required.

Prevention and Control of Legionnaires' Disease

http://www.wt.com.au/safetyline/codes/legion.htm

This detailed code of practice is presented in its entirety by WorkSafe Western Australia.

ProtectAide Inc.

http://www.protectaide.com

This site provides an illustrated catalog of the company's personal protective equipment products and information about its bloodborne pathogens training program on CD-ROM.

Safe-tec Canada

http://www.safetec.com

This is an online catalog of the company's infection control and personal safety products.

University of Alberta, Medical Laboratory Science Program

http://www.ualberta.ca/~medlabsc/

Read about their undergraduate and graduate programs and Internet resources available to students.

---

4. Chemical Safety

Consulting Services

Advanced Chemical Safety

http://www.chemical-safety.com

The site provides information about this consulting company's services, including chemical safety training and U.S. regulatory compliance documentation.

Advanced Waste Management Systems Inc.

http://www.awm.net/

AWMS is an environmental consulting firm offering services which include HAZWOPER Worker Right-to-Know Emergency Planning and Community Right-to-Know Act. This site describes their services and qualifications.

Chemical Safety Associates

http://www.chemical-safety.com

Provides information about this consulting company's services, including site remediation, MSDS preparation, and assistance with safe handling of hazardous materials.

DuPont Safety and Environmental Management Services (SEMS)

http://www.dupont.com/safety/

This site describes DuPont SEMS' services. They provide safety seminars, training materials, and consulting services.

Water Technology International

http://www.wti.cciw.ca

WTI is an environmental technology and services company. Read about their products and services (hazardous waste, site remediation, etc.). Of particular interest is their 40-hour health and safety course, which satisfies general Canadian (federal and provincial) and U.S. (29 CFR1910.120, "HAZWOPER") health and safety training requirements for contaminated-site workers.

Databases

Carcinogenic Potency Database

http://potency.berkeley.edu/cpdb.html

Provides the results of chronic long-term animal cancer tests. A single database includes sufficient information on each experiment to permit investigations into many research areas of carcinogenesis. Analyses of 5152 experiments on 1298 chemicals are presented

ChemFinder Chemical Searching and Information Integration

http://www.chemfinder.com

Comprehensive access to chemical information from a variety of sources is provided: MSDS collections, toxicological profiles, etc. The site is provided by CambridgeSoft Corporation. More detail about this site is available in the Top 50 Sites section.

Cornell Planning Design and Construction, Material Safety Data Sheets

http://msds.pdc.cornell.edu/issearch/msdssrch.htm

This is a large MSDS database, containing approximately 325,000 MSDSs from the U.S. Dept. of Defense MSDS database, the Vermont SIRI MSDS collection, and data sheets maintained by Cornell's Environmental Health and Safety Dept.

Envirofacts Warehouse, Master Chemical Integrator

http://www.epa.gov/enviro/html/emci/emci_query.html

Obtain the acronyms, chemical identification numbers, and chemical names reported by the U.S. Environmental Protection Agency's Envirofacts program system databases (AFS, PCS, RCRIS, and TRIS). There are links to factsheets, containing information regarding health risks, exposure limits, handling regulations, and chemical properties.

Genium Publishing

http://www.genium.com

This site provides information about Genium's safety and health software programs, MSDS collection, etc. An online safety and health glossary and database of hazard labelling are also provided.

Global Information Network for Chemicals (GINC)

http://www.nihs.go.jp/GINC

GINC is a worldwide information network for safe use of chemicals. Read about the GINC project and access its database of integrated chemical databases.

HazDat—Hazardous Substance Release/Health Effects Database

http://atsdr1.atsdr.cdc.gov:8080/hazdat.html

This database, from the U.S. Agency for Toxic Substances and Disease Registry (ATSDR), provides information on the release of hazardous substances from Superfund sites or from emergency events and on effects of hazardous substances on the health of human populations.

Information Services for Agriculture (ISA)

http://www.aginfo.com/isa.html

This service provides access to the company's label and MSDS management system. View MSDSs for agricultural products from Dupont, Bayer, Monsanto, and Rhone Poulenc. ISA also creates Web sites for agricultural companies.

Integrated Risk Information System (IRIS)

http://www.epa.gov/ngispgm3/iris/

"The Integrated Risk Information System (IRIS), prepared and maintained by the U.S. Environmental Protection Agency (U.S. EPA), is an electronic database containing information on human health effects that may result from exposure to various chemicals in the environment."

Projet Regetox—Programme "Protection du travailleur en matiäre de santé"

http://www.regetox.med.ulg.ac.be/

The object of the study is to develop and install a server equipped with a physico-chemical and toxicological database and interactive risk evaluation software.

ToxNet

http://toxnet.nlm.nih.gov

This U.S. National Library of Medicine site provides "access to the TOXNET system of databases on toxicology, hazardous chemicals, and related areas."

UNEP Chemicals (International Registry of Potentially Toxic Chemicals—IRPTC)

http://irptc.unep.ch/irptc/

UNEP Chemicals is the center for all chemicals-related activities of the United Nations Environment Programme. Highlights include an inventory of information sources on chemicals, the UNEP Chemicals database, and information about UNEP programs.

Xerox MSDS

http://www.xerox.com/ehs/msds/index.htm

Provides a searchable database of MSDSs for Xerox products sold in the Western Hemisphere.

Directories

American Association of Poison Control Centers

http://198.79.220.3/aapcc/

This site provides a directory of centers, educational materials, and a newsletter.

Poisons Information Database

http://vhp.nus.sg/PID/

The site provides information about poisonous plants and animals, a directory of antivenoms, a directory of toxinologists, and a directory of poison control centers around the world.

Educational Resources / Training

Advanced Chemical Safety

http://www.chemical-safety.com

The site provides information about this consulting company's services, including chemical safety training and U.S. regulatory compliance documentation.

Chemical Safety Associates

http://www.chemical-safety.com

Provides information about this consulting company's services, including site remediation, MSDS preparation, and assistance with safe handling of hazardous materials.

DuPont Safety and Environmental Management Services (SEMS)

http://www.dupont.com/safety/

This site describes DuPont SEMS' services. They provide safety seminars, training materials, and consulting services.

Huntsville Training Associates

http://www.hsv.tis.net/~hta/

The site contains descriptions of the company's hazardous materials training courses.

Introduction to Applied Toxicology

http://www.bio.hw.ac.uk/edintox/page1.htm

This is an online course covering basic aspects of chemistry and health effects of chemicals. It is intended as a primer for those needing a grounding in chemical hazard and risk assessment.

Petroleum Industry Training Service

http://www.pits.ca/

Calgary, Alberta-based organization provides training services for the petroleum industry.

Seagull Environmental Training

http://www.seagulltraining.com/

This U.S.-wide company provides training on subjects such as asbestos, lead paint, hazardous materials, and environmental site assessment. Read about the company's courses.

Water Technology International

http://www.wti.cciw.ca

WTI is an environmental technology and services company. Read about their products and services (hazardous waste, site remediation, etc.). Of particular interest is their 40-hour health and safety course, which satisfies general Canadian (federal and provincial) and U.S. (29 CFR1910.120, "HAZWOPER") health and safety training requirements for contaminated site workers.

WHMIS 96

http://www.bbsi.net/50para/p_web.html-ssi

Read about this CD-ROM-based WHMIS (Canadian right-to-know legislation) training package, produced by C.D. 50e PARALLELE INC.

WHMIS Software Training

http://www.niagara.com/blmc/

Contains a brief description of BLMC's WHMIS training software with contact information.

Journals and Newsletters

American Association of Poison Control Centers

http://198.79.220.3/aapcc/

This site provides a directory of centers, educational materials, and a newsletter.

Hazardous Materials Management magazine

http://www.hazmatmag.com

The publication, plus selected articles and features from current and past issues.

Society of Environmental Toxicology and Chemistry (SETAC)

http://www.setac.org/

Available information includes details about membership in this professional society, a newsletter, activities and meetings, and publications.

Policy/Procedures Documents

Denison University Safety

http://www.denison.edu/sec-safe/safety.html

The university's chemical hygiene plan, MSDS links and information, and training information are provided. Check out Madame Curie in protective clothing.

Michigan State Univ. Office of Radiation, Chemical and Biological Safety

http://www.orcbs.msu.edu

Various documents are presented, including MSU's chemical hygiene program, their chemical lab safety checklist, radiation safety manual, and *Safe Science* newsletter.

Portland State University, Environmental Health and Safety

http://www.pp.pdx.edu/FAC/Safety/

The site provides access to a variety of university documents, such as their chemical hygiene program, lockout-tagout, and pesticide/herbicide use.

Univ. of Illinois (Urbana), Division of Environmental Health and Safety

http://phantom.ehs.uiuc.edu/dehs.html

Provides links to a number of documents and other resources from the division's chemical safety and waste management, occupational safety and health, and radiation safety sections.

University of Maryland at College Park, Environmental Safety

http://www.inform.umd.edu/DES/

Information at this site includes the university's Chemical Hygiene Plan and Laboratory Safety Guide, along with search forms for various MSDS collections.

Products

3M Environmental Safety and Energy Performance Products

http://www.mmm.com/market/government/env/envindex.html

Information about 3M's products.

Accu*Aire Controls, Inc.

http://www.accuaire.com/

Information is presented about the company's laboratory fume hoods and exhaust products, selected articles, and free software.

E H Lynn Industries

http://www.ehlynn.com/

The company distributes liquid and dry bulk handling equipment, tank truck equipment, hoses, valves, gaskets, etc., for the chemical, petroleum, and dry bulk markets. Read about the company and its products.

Fisher Scientific

http://www.fisher1.com

View Fisher's product catalogs and MSDSs, and order products online.

Flinn Scientific

http://www.flinnsci.com

This science supply company site provides a lot of useful information, including laboratory design tips and numerous documents on chemical safety.

Pioneer Products Inc.

http://www.pioneers.com

The site contains information about products from this specialty chemical company.

Publications

Baylor College of Medicine, Environmental Safety

http://www.bcm.tmc.edu/envirosafety/

Biosafety, chemical safety, and radiation safety policy documents and guidelines at Baylor.

Chemistry and Industry Magazine

http://ci.mond.org

Chemistry and Industry is an international magazine that provides news and features on chemistry and related sciences, as well as on the commercial and political aspects of these subjects. The site provides access to selected articles and subscription information.

Department of Transportation, Office of Hazardous Materials Safety (OHM)

http://www.volpe.dot.gov/ohm/

This office is responsible for coordinating a national (U.S.) safety program for the transportation of hazardous materials by air, rail, highway, and water. Highlights include an electronic version of the Emergency Response Guidebook, regulatory info., and OHM publications.

Documentation for Immediately Dangerous to Life or Health Concentrations (IDLHs)

http://www.cdc.gov/niosh/idlh/idlh-1.html

This online publication documents the criteria and information sources that have been used by the National Institute for Occupational Safety and Health (NIOSH) to determine concentrations that are immediately dangerous to life or health. The IDLHs for the chemicals themselves are also provided. More detail about this site is available in the Top 50 Sites section.

EC Chemical Regulation Reporter

http://www.incadinc.com/edr/ecchemicalregulationreporter.html

This site provides basic information about European Community chemical regulations and subscription information about "EC Chemical Regulation," a monthly notification of EC actions affecting the chemical industry.

Environmental Protection Agency, Office of Pesticide Programs

http://www.epa.gov/pesticides/

Access is provided to detailed U.S. EPA pesticide information.

Errata Notice—1998 ACGIH TLVs and BEIs

http://dspace.dial.pipex.com/hhsc/errtlv.html

Provides the text of ACGIH's errata notice for its 1998 booklet of TLVs and BEIs.

Extoxnet—The EXtention TOXicology NETwork

http://ace.orst.edu/info/extoxnet/

EXTOXNET is a cooperative effort of University of California-Davis, Oregon State University, Michigan State University, and Cornell University. Various types of pesticide toxicology and environmental chemistry information are available.

Hazardous Materials and Safety: Guide to Reference Sources

http://www.lib.lsu.edu/sci/chem/guides/srs103.html

Bibliography of HazMat and Safety publications compiled by Louisiana State U. libraries.

Hazardous Materials Management magazine

http://www.hazmatmag.com

The publication, along with selected articles and features from current and past issues.

Merck Publications

http://www.merck.com/!!rkoAi1qZlrkoBb0zrR/pubs/

Information about the *Merck Manual* and *Merck Index*.

National Toxicology Program

http://ntp-server.niehs.nih.gov

This program is associated with the U.S. National Institute of Environmental Health Sciences. Access abstracts of NTP studies, chemical health and safety data, etc. More detail about this site is available in the Top 50 Sites section.

NIOSH Pocket Guide to Chemical Hazards

http://www.cdc.gov/niosh/npg/pgdstart.html

Access the full text of this popular publication from the U.S. National Institute for Occupational Safety and Health (NIOSH).

Pesticide Information Fact Sheets

http://fshn.ifas.ufl.edu/pest/default.htm#FactSheets

This is a series of documents describing various environmental and occupational aspects of pesticides. The site is maintained by the Pesticide Information Office, Institute of Food and Agricultural Sciences, University of Florida.

Pesticide Poisoning Handbook

http://hammock.ifas.ufl.edu/txt/fairs/19729.html

This is the full text of a handbook maintained by the Institute of Food and Agricultural Sciences, University of Florida.

Society of Chemical Industry

http://sci.mond.org

Information is provided about this U.K.-based learned society, its membership, meetings, and publications. There are details about their health and safety group.

Society of Environmental Toxicology and Chemistry (SETAC)

http://www.setac.org/

Available information includes details about membership in this professional society, a newsletter, activities and meetings, and publications.

Vermont SIRI

http://siri.org

The site contains the archives of the SAFETY and Occ-Env-Med-L mailing lists, a large MSDS database, safety clip art, various safety-related documents, and regulatory information. More detail about this site is available in the Top 50 Sites section.

Volvo Group—Environment

http://www.volvo.se/environment/index.html

Volvo's Environmental policies, programs, and environmental report are provided. Volvo's "blacklist" of chemicals is included in the 1995 environmental report.

Programs

American Chemical Society

http://www.acs.org

Information is provided about the society itself, membership, programs, technical divisions, and local sections.

Canadian Chemical Producers' Association (CCPA)

http://www.ccpa.ca

Information about CCPA and its responsible care program is provided.

Dept. of Transportation, Office of Hazardous Materials Safety (OHM)

http://www.volpe.dot.gov/ohm/

This office is responsible for coordinating a national (U.S.) safety program for the transportation of hazardous materials by air, rail, highway, and water. Includes an electronic version of the Emergency Response Guidebook, regulatory info., and OHM publications.

Global Information Network for Chemicals (GINC)

http://www.nihs.go.jp/GINC

GINC is a worldwide information network for safe use of chemicals. Read about the GINC project and access its database of integrated chemical databases.

Society of Chemical Industry

http://sci.mond.org

Information is provided about this U.K.-based learned society, its membership, meetings, and publications. There are details about their health and safety group.

Volvo Group—Environment

http://www.volvo.se/environment/index.html

Volvo's Environmental policies, programs, and environmental report are provided. Volvo's "blacklist" of chemicals is included in the 1995 environmental report.

Software

Accu*Aire Controls, Inc.

http://www.accuaire.com/

Information is presented about the company's laboratory fume hoods and exhaust products, selected articles, and free software.

Aurora Data Systems

http://www.alaska.net/~aurora/

Read about the company's database applications: Chemical Information Management System, Equipment Maintenance System, and Training Records Management System.

Bara Environmental Compliance

http://www.bara.com

Bara Environmental Solutions is a Texas-based consulting firm that "creates advanced environmental safety solutions through computer tracking of hazardous materials and safety equipment." Read about the company's services.

Chemical Abstracts Service

http://info.cas.org

Read about CAS, its products, and its services. Find out about the STN database service and view a copy of the CAS registry structured searching manual.

Chempute Software

http://www.chempute.com

This South African company produces and provides a variety of software packages for chemical process industries. Plant safety and quality control programs are available. Download demos and tutorials.

Eclipse Software Technologies

http://www.eclipsesoft.com

Information about and downloadable demonstrations of "MSDS Wizard" and "MSDS Scan Wizard," MSDS management software.

Euroware Associates Software

http://www.euroware.com/index.html

Information is provided about the company's MSDS and chemical-related software packages.

Genium Publishing

http://www.genium.com

This site provides information about Genium's safety and health software programs, MSDS collection, etc. An online safety and health glossary and database of hazard labelling are also provided.

InteractiveWare Inc.—HAZCOM training software

http://www.interactiveware.com

Download a demo of this Windows-based hazard communication training software.

Map 80 Systems Ltd.

http://www.map80.co.uk/

Read about this U.K. company's labelling software for chemical, pharmaceutical, industrial, and retail settings.

MSDS Pro

http://www.msdspro.com

Download a demo of this MSDS management software by Aurora Data Systems.

PlantSafe

http://www.plantsafe.com

PlantSafe is a computerized chemical emergency response system. Read a detailed description and see ordering information.

Safety Officer II

http://www.safetyofficer.com/

Safety Officer II is a chemical information software system for converting data associated with chemicals into end-user specific MSDSs, custom Avery labels, and management reports. Detailed descriptions of the software are provided.

Safety4

http://www.safety4.com/

Safety4, provider of chemical protection wear, hosts a free online chemical protection guide, complete with hundreds of chemicals and their associated permeation rate.

Solutions Software Corporation

http://www.env-sol.com/

This site contains descriptions of the company's CD-ROM products, many of them health and safety related, along with links to various sites.

WHMIS 96

http://www.bbsi.net/50para/p_web.html-ssi

Read about this CD-ROM based WHMIS (Canadian right-to-know legislation) training package, produced by C.D. 50e PARALLELE INC.

WHMIS Software Training

http://www.niagara.com/blmc/

Contains a brief description of BLMC's WHMIS training software, with contact information.

5. Construction Safety

ABIH/BCSP Joint Committee for Certification of Occupational Health and Safety Technologists

http://www.ABIH.org/JCOHST_roster.htm

Search for the name, city, or state (or country—U.S.A. is assumed) of an Occupational Health and Safety Technologist (OHST), Construction Health and Safety Technician (CHST), or Safety Trained Supervisor in Construction (STS-Construction).

Alberta Construction Safety Association

http://www.compusmart.ab.ca/acsa/

The site provides information about this industry association, its training courses, a description of resources available to members, and products.

Building Industry Exchange

http://www.building.org

Provides a directory of products and services for the construction and building industry.

Comité National d'Action pour la Sécurité et l'Hygiène dans la Construction

http://www.cobonet.be/navb-cnac.be/

National Action Committee for Safety and Health in Construction. Read about its services, publications, and training.

Consortium for Construction Health and Safety

http://info.pmeh.uiowa.edu/construc/construc.htm

This consortium is an organization of three U.S. universities committed to the reduction of work-related musculoskeletal disorders in the construction industry. The site provides publications, bibliographies, and newsletters.

Construction Industry Research and Information Association

http://www.ciria.org.uk/

CIRIA is a nonprofit U.K. organization that provides best practice guidance to professionals. Read about the organization, its research, and publications.

Construction Safety Association of Ontario

http://www.csao.org/

As well as providing information about the association itself, the site provides a number of basic safety awareness documents. The full text of their magazine *Construction Safety* and their newsletter is also available.

Gravitec Systems Inc.

http://www.gravitec.com/

This U.S. company specializes in engineering, training, and consulting related to fall protection. Read about the company's services.

HG & Associates, Inc.

http://www.hgassociates.com

The site provides a description of this Florida construction industry consulting company.

Occupational Safety and Health Administration, Index to Construction Regulations

http://www.osha-slc.gov/OCIS/Construction.html

Provides access to the full text of U.S. OSHA's construction regulations, 29 CFR 1926.

Online Resources for the Construction Industry

http://www.copywriter.com/ab/constr.html

This is a detailed directory of construction-related resources on the Internet.

Operation Safe Site

http://www.opsafesite.com

This is a commercial site from the above-named consulting company. The site provides information about their products and services, along with links to sites and files of interest to the construction industry.

PearlWeave Safety Netting

http://www.pearlweave.com

View information about fall protection debris containment and OSHA compliance, plus PearlWeave products for the construction and safety industries.

Plant and Machinery Assessing Services

http://www.cyberstrategies.net/pamas/

This Australian consultancy develops safety programmes for the building industry. Services include preparation of safety plans, risk assessment for personal protective equipment, certificates of competency, and training programmes.

Professional Services Listing Directory for the Construction Industry— InterPRO Resources Inc.

http://www.ipr.com

This a directory of construction-related individuals and companies: architects, engineers, materials suppliers, and product manufacturers.

Regulatory Compliance Information Center

http://www.rcic.com

This site has a search engine with only compliance-related sites indexed, compliance Web chatrooms, etc. Subject areas include radiation, telecommunications, and electricity.

Safety Boot

http://members.aol.com/safetmaker/index.htm

Read a description of Safety Boot, "a reusable base for constructing free-standing temporary guardrails that exceed OSHA standards."

Safety Management Services

http://www.cyberg8t.com/~jprice/

This consulting company's site provides access to its Online Safety Program, including an Injury and Illness Prevention Program, Hazardous Communications Program, MSDSs, and links to other sites.

Sulowski Fall Protection Inc.

http://ourworld.compuserve.com/homepages/AndrewSulowski/sulowski.htm

This consulting company's site provides information about its services, fall protection courses, publications, and upcoming events.

Surety Manufacturing & Testing Ltd.

http://www.Suretyman.com/

This Edmonton, Canada-based company manufactures and sells equipment for fall protection and confined space entry. View the company's catalogue and read about training videos.

Trench Safety

http://www.bsc.auburn.edu/research/trench/

This is an online tutorial for constructors, provided by Auburn University.

Western Hard Hats

http://www.westernhardhat.com/

Information is provided about the western-style hard hats this company distributes.

---

6. Emergency Response, Fire and Rescue

Conferences

Emergency Preparedness Information Exchange (EPIX)

http://hoshi.cic.sfu.ca:80/~anderson/index.html

Content includes links and documents about various emergency topics, emergency related organizations, upcoming conferences, and alerts.

Consulting Services

Advanced Waste Management Systems Inc.

http://www.awm.net/

AWMS is an environmental consulting firm offering services which include HAZWOPER Worker Right-to-Know Emergency Planning and Community Right-to-Know Act. This site describes their services and qualifications.

Continental Hoisting Consultants

http://www.chcelevator.com/

Safety training for elevators: entrapment and fire emergency. Read about their services.

National Access and Rescue Centre

http://www.narc.co.uk/

This U.K. organization "is a training, development and equipment supply centre for rope access, line rescue, confined space entry, mast climbing, fall prevention and all work involving high places or other difficult access." Describes training programmes and services.

Rebernigg Associates

http://www.sulphurcanyon.com/magiers/default.htm

Read about the company's safety engineering, fire protection, and laser safety services.

Turning Point Group Inc.

http://www.turningpointgroup.com/

This company specializes in emergency preparedness, crisis management, and business recovery. Read about their services and participate in their emergency management bulletin board.

West Arm Consulting Services

http://www.westarm.bc.ca/

This company specializes in WWW page development for EMS. Their page has extensive links to various EMS and police sites.

Databases

Published International Literature on Traumatic Stress

http://dciswww.dartmouth.edu/cgi-bin/dcis/
wdi?&Alexandria.Dartmouth.EDU&51001&PILOTS%20Catalog&s

PILOTS is a bibliographic database produced at the U.S. National Center for Post-Traumatic Stress Disorder.

Directories

Emergency Services Registry and Search Site

http://www.district.north-van.bc.ca/admin/depart/fire/ffsearch/mainmenu.cfm

This is a directory of emergency services professionals.

Emergency Services WWW Site List (Dean Tabor)

http://pplant.uafadm.alaska.edu/www-911.htm

A list of Fire/Rescue/EMS/Emergency Services sites that can be found on the Internet.

Fire Departments on the Web

http://www.ctlnet.com/users/jdboyd/fdw/fire2.html

This site presents an extensive directory of fire departments and related resources accessible on the Web.

Educational Resources / Training

Affirmed Medical Inc.

http://www.affirmed.com

The company provides first aid and safety products. Find out about these and their U.S. regulatory compliance products and training videos.

American Safety and Health Institute

http://www.ashinstitute.com/

The site describes the organization's first-aid training programs.

Continental Hoisting Consultants

http://www.chcelevator.com/

Safety training for elevators: entrapment and fire emergency. Read about their services.

Crisis Care Education

http://www.hypersurf.com/~gnarles/frames/cce1.html

CCE is a small group of professional corporate trainers who specialize in emergency awareness and preparedness. Read about the company's services.

Emergency Management Australia

http://www.ema.gov.au/

EMA's mission is to promote and support comprehensive, integrated, and effective emergency management in Australia and its region of interest. Information includes publications and manuals, training information, and links to related Australian sites.

Huntsville Training Associates

http://www.hsv.tis.net/~hta/

The site contains descriptions of the company's hazardous materials training courses.

Idea Bank

http://www.theideabank.com/

The company's catalog of fire videos for public education is presented.

International Society of Fire Service Instructors

http://www.isfsi.org/

ISFSI develops and implements training and educational programs for the public and industrial fire and emergency response communities. Resources on the Web site include an extensive reference center (books, videos, training materials, etc.), information about and excerpts from its monthly publication, and details about upcoming event.

Life Safety Associates

http://www.lifesafety.com/

Life Safety Associates develops and provides onsite emergency, safety, and disaster training. Information is provided about the company and its products and services.

National Access and Rescue Centre

http://www.narc.co.uk/

This U.K. organization "is a training, development and equipment supply centre for rope access, line rescue, confined space entry, mast climbing, fall prevention and all work involving high places or other difficult access." Read about the organization's training programmes and services.

National Fire Protection Association (NFPA)

http://www.nfpa.org

The site provides information about the NFPA, its departments, publications, seminars, and educational programs.

Policy / Procedures Documents

California Office of Emergency Services

http://www.oes.ca.gov

Provides information about the Governor's Office of Emergency Services, the California Emergency Plan, current conditions (weather, roads, etc.), the office's earthquake program.

Products

Affirmed Medical Inc.

http://www.affirmed.com

The company provides first aid and safety products. Find out about these, plus their U.S. regulatory compliance products and training videos.

Crouch Fire and Safety Products

http://www.jtcrouch.com/

Find out about the company and its fire protection and public safety lighting products and services.

E. D. Bullard Co.

http://www.bullard.com

The site contains this company's catalogs of personal protective equipment: respiratory protection, head protection, and fire and rescue. A list of distributors is also provided.

First Aid 1st

http://www.seadrive.com/first.aid.1st/

This Alberta-based company provides first aid supplies and services to business, sports groups, nonprofit organizations, and individuals. View product descriptions and locate distributors.

Grace Sales, Inc.

http://www.gracesales.com/

Read about the company's safety products for the firefighting community.

HBA International

http://www.hbainternational.com/

This company is a supplier of medical equipment for hospitals and emergency medical services. Read about HBA's various product lines.

Inter Consulting Systems

http://www.wionline.com/ics/

Jerry E. Smith, former Los Angeles City Fire Captain and retired California OES Fire-rescue Assistant Chief has written a number of books for firefighters. This site describes his publications. This is also home to the "Emergency Grapevine," a Web-based discussion group for emergency management topics.

Mac's Fire and Safety

http://www.macsfire.com

This site provides a catalog of rescue tools, fire fighting equipment, and safety gear.

Onscene.com

http://www.onscene.com/

This site has been "created as a point of information exchange for public safety professionals." It provides links to fire and EMS products and services, news, and pictures.

Pre-ventronics

http://www.primenet.com/~prevent/

Read about products and services provided by this commercial alarm and security company.

VST Chemical Corp.

http://ecki.com/vst/

Information about this company's firefighting chemicals is available. There are also links to other fire sites.

Publications

Colorado Hazards Center at the University of Colorado, Boulder

http://adder.colorado.edu/~hazctr/Home.html

The center is a "national and international clearinghouse for information on natural hazards and human adjustments to hazards and disasters."

Disaster Connection

http://www.itn.is/~gro/disaster/

This is an extensive list of Internet disaster-related resources and documents.

Emergency Nursing World!

http://www.hooked.net/~ttrimble/enw/index.html

This site is intended to be the world's first comprehensive practice-based resource for emergency nursing and the allied professions.

Emergency Preparedness Information Exchange (EPIX)

http://hoshi.cic.sfu.ca:80/~anderson/index.html

Content includes links and documents about various emergency topics, emergency related organizations, upcoming conferences, and alerts.

Emergency Response and Research Institute (ERRI)

http://www.emergency.com

There is a tremendous amount of information on a variety of emergency-related subjects: documents, links to other sites, EmergencyNet News, etc.

Global Health Disaster Network

http://hypnos.m.ehime-u.ac.jp/GHDNet/index.html

GHDNet is maintained by Department of Emergency Medicine, Ehime University, Japan. Contains documents, bibliographies, conference information, links to related sites.

National Fire Protection Association (NFPA)

http://www.nfpa.org

The site provides information about the NFPA, its departments, publications, seminars, and educational programs.

Programs

Federal Emergency Management Agency (FEMA)

http://www.fema.gov

This U.S. agency's mission is "to reduce loss of life and property and protect our nation's critical infrastructure from all types of hazards through a comprehensive, risk-based, emergency management program of mitigation, preparedness, response and recovery."

Oregon Emergency Management Office

http://www.osp.state.or.us/oem/

Information provided includes descriptions of this state agency itself, its programs, and reports of recent incidents.

Software

Fire Safety Engineering Group

http://fseg.gre.ac.uk/

This Greenwich, U.K. consultancy group describes itself as "one of the largest research groups in the world dedicated to the development and application of mathematical modelling tools suitable for the simulation of fire related phenomena." Read about the company's services.

PlantSafe

http://www.plantsafe.com

PlantSafe is a computerized chemical emergency response system. Read a detailed description and see ordering information.

7. Ergonomics

Consulting Services

Advanced Ergonomics

http://www.advergo.com

AEI provides a variety of ergonomic services to corporations and individuals. Their products and services are described.

Comprehensive Loss Management, Inc.

http://www.clmi-training.com

This company produces safety and ergonomic training programs to help companies achieve OSHA compliance.

Ergonomic Central

http://www.ergocentral.com/

This site is an electronic mall of ergonomics products and services.

Usernomics

http://www.usernomics.com/

Ergonomics for hardware, software, and training. Find out about the company's services, or access links related to various aspects of ergonomics.

VDT Solution

http://home.earthlink.net/~ergo1/

The site provides a variety of types of information related to computer workstations and ergonomics: preferred equipment and services, a PC and laptop ergonomic guide, and ergonomic product reviews.

Databases

ErgoWeb

http://www.ergoweb.com

ErgoWeb is an extensive Web site, containing a variety of types of ergonomics-related information, such as workplace evaluation and design software, a database of ergonomics-related products and services, and case studies. More detail about this site is available in the Top 50 Sites section.

Educational Resources / Training

Center for Ergonomics, University of Michigan

http://www.engin.umich.edu/dept/ioe/C4E/

Information about the center's education and research in ergonomics, programs, and publications.

Usernomics

http://www.usernomics.com/

Ergonomics for hardware, software, and training. Find out about the company's services, or access links related to various aspects of ergonomics.

Journals and Newsletters

CTDNews Online

http://ctdnews.com/

CTDNEWS is a monthly newsletter related to Cumulative Trauma Disorders. The full text of current and past issues is presented, along with CTD-related information about products and regulatory initiatives.

Policy / Procedures Documents

University of Waikato (New Zealand), Health and Safety

http://www.waikato.ac.nz/fmd/hsc/

A number of documents are available, including the university's Occupational Health and Safety Policy and information about Occupational Overuse Syndrome.

Products

Advanced Ergonomics

http://www.advergo.com

AEI provides a variety of ergonomic services to corporations and individuals. Their products and services are described.

Back Be Nimble

http://www.backbenimble.com/

The company's online catalog provides information and products for back and body care. The site also includes basic information on ergonomics, back supports, and back surgery.

BackHealth U.S.A

http://www.backhealth.com

Some information is provided about back pain causes and treatments. Most of the information is about BackHealth 2000, the company's back exercise equipment.

Carpal Tunnel Syndrome

http://www.netaxs.com/~iris/cts/

Access summary information about carpal tunnel syndrome, along with links to other carpal tunnel resources and ergonomic products. Information is also provided about Metro Smallwares' Compfort keyboard extender.

Combo Ergonomic and Contract Furnishings

http://www.combo.com

View the company's ergonomic furniture catalog and a "collection of information and solutions pertaining to office ergonomics."

---

comp.speech FAQ

http://www.speech.cs.cmu.edu/comp.speech/

This frequently-asked-questions (FAQ) document originated with the comp.speech news-group, and covers information on speech technology, including speech synthesis, speech recognition, and related material. The speech recognition section is of particular interest.

Conveyerail Systems Ltd.

http://www.conveyerail.com/

View descriptive information about VacuMove, a vacuum lifting system.

Entropy Software, OOS Software

http://www.albatross.co.nz/~miker/main.htm

Describes this New Zealand company's OOS software, a program that tells you when to take a break from your computer to help prevent/alleviate OOS (occupational overuse syndrome or repetitive strain injuries). Download a copy of the software from this site.

Ergonomic Central

http://www.ergocentral.com/

This site is an electronic mall of ergonomics products and services.

Ergonomic Design Inc.

http://www.ergodesign.com/

View information about the company's computer-related ergonomic products: wrist rests, keyboard holders, and document holders.

Ergonomic Sciences Corporation

http://www.ergosci.com

The goal of this Web site is to provide up-to-date information about the science of workspace design and function and present information that will help you choose the most beneficial ergonomic products to address specific ergonomic issues.

LeMitt

http://www.lemitt.com

Commercial information about this temperature-regulated therapeutic cold pack for hands and wrists. There is also a link to some repetitive strain injury information.

Micronite

http://www.micronite.com

This commercial site provides information about the company's repetitive strain injury prevention software and publications.

Mouse Escalator

http://www.ergonomic-pad.com/

Description of this ergonomic product. It is a "positive inclined work surface" for a computer mouse.

Proformix

http://www.proformix.com

The company provides ergonomic keyboard platforms and related products, plus a suite of ergonomic management software.

Questec Input Devices

http://www.questecmouse.com

Read about this company's computer related products: mice, mice pens, trackballs, ergonomic keyboards, and antiglare filters.

Safe Computing

http://www.safecomputing.com

Access the company's extensive catalog of computer ergonomics-related products.

Publications

Boletin Argentino de Ergonomia

http://www.geocities.com/CapeCanaveral/6616/

News about Argentine ergonomists and their laboratories. (In Spanish)

Center for Ergonomics, University of Michigan

http://www.engin.umich.edu/dept/ioe/C4E/

Information about the center's education and research in ergonomics, programs, and publications.

Combo Ergonomic and Contract Furnishings

http://www.combo.com

View the company's ergonomic furniture catalog and a "collection of information and solutions pertaining to office ergonomics."

comp.speech FAQ

http://www.speech.cs.cmu.edu/comp.speech/

This frequently-asked-questions (FAQ) document originated with the comp.speech newsgroup and covers information on speech technology, including speech synthesis, speech recognition, and related material. The speech recognition section is of particular interest.

Computer Related RSI Primer

http://engr-www.unl.edu/ee/eeshop/rsi.html

This is a well presented summary document about repetitive strain injuries, with links to related documents, and an RSI bibliography.

Cornell Ergonomics Web

http://ergo.human.cornell.edu/

This site contains the full text of a number of studies related to ergonomics and human factors: keyboard systems, cumulative trauma disorders (CTDs), low back injuries, lighting, and indoor air quality (IAQ). More detail is available in the Top 50 Sites section.

CybErg 1996

http://www.curtin.edu.au/conference/cyberg/

Proceedings of the first ergonomics conference in CyberSpace, presented by the Ergonomics Society of Australia and the International Ergonomics Association. Preliminary information about CybErg 1999 is also available.

Ergonomics Task Force—Online Ergonomics Resources

http://www.lib.utexas.edu/Pubs/etf/

Sponsored by the Ergonomics Task Force at the General Libraries, University of Texas at Austin. Numerous documents and links to other ergonomics sites are available.

ErgoWeb

http://www.ergoweb.com

ErgoWeb is an extensive Web site, containing a variety of types of ergonomics-related information, such as workplace evaluation and design software, a database of ergonomics-related products and services, and case studies. More detail about this site is available in the Top 50 Sites section.

Human Factors and Ergonomics Society

http://hfes.org/

Read about the society itself, its news, publications, and meetings.

International Journal of Industrial Ergonomics

http://www.elsevier.nl/locate/ergon

View the tables of contents of printed issues, information about special issues, an author index, and subject index for the journal.

Occupational Overuse Syndrome/RSI resources

http://www.mcs.vuw.ac.nz/comp/General/OOS/

The site contains articles and bulletins from New Zealand Occupational Safety and Health, the typing injuries FAQ, and links to other RSI information.

Patient's Guide to Cumulative Trauma Disorder

http://www.sechrest.com/mmg/ctd/

Fairly detailed information is provided about cumulative trauma disorders or repetitive motion injuries. Information about Medical Media Group and their software.

Pointing Device Summary

http://www.ME.berkeley.edu/ergo/tips/pdtips.html

This is an online document about keyboard injuries related to pointing devices (mice, trackballs, etc.). It was produced by the University of California San Francisco/University of California Berkeley Ergonomics Program.

RSI Network—Newsletter

http://www.tifaq.com/rsinet/index.html

"The RSI Network newsletter is provided by the Cumulative Trauma Disorder (CTD) Resource Network for dissemination of information to the repetitive strain injury (RSI) community." Individuals and not-for-profit organizations may receive free electronic copies of the newsletter by email.

The Mouse List—Ergonomic, Adaptive and Alternate Pointing Devices

http://www.setbc.org/mouselist/

This online document by Diana Carroll provides information about specific mice, trackballs, and other computer pointing devices. There are descriptions, reviews, and company contact information.

Typing Injury FAQ

http://www.tifaq.com

This is a frequently-asked-questions document (FAQ) about typing injuries. Sections include general information, keyboards, voice recognition, mice, software, furniture, and links to other resources. More detail is available in the Top 50 Sites section.

VDT Solution

http://home.earthlink.net/~ergo1/

Provides information related to computer workstations and ergonomics: preferred equipment and services, a PC and laptop ergonomic guide, and ergonomic product reviews.

Programs

Center for Ergonomics, University of Michigan

http://www.engin.umich.edu/dept/ioe/C4E/

Information about the center's education and research in ergonomics, programs, and publications.

Center for Industrial Ergonomics, University of Louisville (Kentucky)

http://www.spd.louisville.edu/~ergonomics/

Read about the center's research and educational activities, which focus on integrating people, organization, and technology at work, and improving quality and productivity through ergonomics, safety, and health management.

Software

Center for Ergonomics, University of Michigan

http://www.engin.umich.edu/dept/ioe/C4E/

Information about the center's education and research in ergonomics, programs, and publications.

DataChem Software

http://www.datachemsoftware.com

The company's site provides information about its various software packages "that help you study and practice for your professional certification examination."

Entropy Software, OOS Software

http://www.albatross.co.nz/~miker/main.htm

The site describes this New Zealand company's OOS software, a program that tells you when to take a break from your computer to help prevent/alleviate OOS (occupational overuse syndrome or repetitive strain injuries). Download a copy of the software from this site.

ErgoEASER Software

http://tis-hq.eh.doe.gov:80/others/ergoeaser/download.html

Download a copy of this software, "developed to aid in identifying, evaluating, and preventing work-related musculoskeletal disorders."

Micronite

http://www.micronite.com

This commercial site provides information about the company's repetitive strain injury prevention software and publications.

Occupational Overuse Syndrome/RSI resources

http://www.mcs.vuw.ac.nz/comp/General/OOS/

The site contains articles and bulletins from New Zealand Occupational Safety and Health, the typing injuries FAQ, and links to other RSI information.

Patient's Guide to Cumulative Trauma Disorder

http://www.sechrest.com/mmg/ctd/

Fairly detailed information is provided about cumulative trauma disorders or repetitive motion injuries. Information about Medical Media Group and their software is also provided.

Proformix

http://www.proformix.com

The company provides ergonomic keyboard platforms and related products, plus a suite of ergonomic management software.

---

8. Industrial/Occupational Hygiene

Consulting Services

Emilcott-dga Inc.

http://www.emilcott-dga.com

Find out about the company's services, including industrial hygiene, safety and environmental consulting, and training.

Environmental Safety and Health of Alaska

http://www.alaska.net/~esha/

The site contains descriptions of this company's safety training courses, Industrial Hygiene and Marine Chemist Services, plus an industrial hygiene news and current events section, and news from the Midnight Sun Section of AIHA.

Safety Advantage

http://www.SafetyAdvantage.com

This safety and industrial hygiene consultants' home page provides information about their services and training packages.

SAREC

http://www.sarec.ca

The expertise and services of this Canadian occupational hygiene consulting company.

W. H. Interscience

http://www.gj.net/~whicotom/

View information about the company's safety and industrial hygiene services: consulting, publications, and software.

Databases

Patty's Industrial Hygiene and Toxicology CD-ROM

http://www.wiley.com/products/subject/chemistry/pattys/

Information is provided about the CD-ROM version of Patty's. View the table of contents, see sample screens, and download a CD-ROM demo.

Directories

Pro-Am Industrial Safety and Hygiene

http://www.pro-am.com

This is one of the best developed "safety malls" on the Web. Find products, services, links to OSHA standards, and a good directory of Internet resources.

Educational Resources / Training

American Board of Industrial Hygiene (ABIH)

http://www.abih.org

Information is provided about ABIH, CIH certification, examinations, and ABIH publications. More detail about this site is available in the Top 50 Sites section.

Applications of Phase-contrast Microscopy to NIOSH Method 7400 and OSHA ID-160

http://www.steve-o.com/public/pdcftu/pdcftu.html

An online Professional Development Course. When last visited, it had been awarded one Certification Maintenance point by the American Board of Industrial Hygiene.

Canadian Registration Board of Occupational Hygienists

http://www.crboh.ca/

The CRBOH is a national, not-for-profit organization, which sets standards of professional competence for occupational hygienists and occupational hygiene technologists in Canada. Provides information about ROH and ROHT certification, exams, maintenance.

Environmental Safety and Health of Alaska

http://www.alaska.net/~esha/

The site contains descriptions of this company's safety training courses, Industrial Hygiene and Marine Chemist Services, plus an industrial hygiene news and current events section, and news from the Midnight Sun Section of AIHA.

Safety Advantage

http://www.SafetyAdvantage.com

This safety and industrial hygiene consultants' home page provides information about their services and training packages.

Journals and Newsletters

Annals of Occupational Hygiene

http://www.elsevier.nl:80/inca/publications/store/2/0/1/

Get tables of contents and ordering information for this British Occupational Hygiene Society publication. A free table of contents service via email is available.

Applied Occupational and Environmental Hygiene

http://www.acgih.org/applied/welcome.htm

The site provides ordering information for this ACGIH journal, plus information for authors and contents of upcoming issues.

Industrial Hygiene News

http://www.rimbach.com/ihnpage/ihn.htm

Read past articles from this publication, view subscription information, descriptions of IH-related products, trade shows, and courses.

Industrial Safety and Hygiene News

http://SafetyOnline.net/ishn/

Information about and the full text of selected portions of this journal are presented.

Products

American Conference of Government Industrial Hygienists (ACGIH)

http://www.acgih.org/

View information about ACGIH, a catalog of its publications, demos of the organization's electronic products, upcoming events, and an "OH Talk" Web chat section. More detail about this site is available in the Top 50 Sites section.

ENMET Corporation

http://www.enmet.com

Read about this company's industrial gas and vapor monitoring instrumentation.

Major Safety Service

http://www.pilot.infi.net/~majsaf/

Products and services provided by this environmental instrument and safety equipment supplier are described.

Pro-Am Industrial Safety and Hygiene

http://www.pro-am.com

This is one of the best developed "safety malls" on the Web. Find products, services, links to OSHA standards, and a good directory of Internet resources.

SKC Online

http://www.skcinc.com/

View air sampling equipment information from the manufacturer. The site also contains frequently asked questions about the company's air sampling instrumentation and fill-in forms for requesting more information.

Technika Scientific and Industrial Hygiene Measurement Instruments

http://www.technika.com/

Technika is a supplier of health, safety, lab, and field meters and detectors. "Our site also has general information on various areas of safety, how metering in that particular area is done, and what specifications you should look for."

Publications

American Board of Industrial Hygiene (ABIH)

http://www.abih.org

Information is provided about ABIH, CIH certification, examinations, and ABIH publications. More detail about this site is available in the Top 50 Sites section.

American Conference of Government Industrial Hygienists (ACGIH)

http://www.acgih.org/

View information about ACGIH, a catalog of its publications, demos of the organization's electronic products, upcoming events, and an "OH Talk" Web chat section. More detail about this site is available in the Top 50 Sites section.

American Industrial Hygiene Association (AIHA)

http://www.aiha.org

Provides access to information about AIHA, its conferences, publications, and lists of accredited laboratories. More detail about this site is available in the Top 50 Sites section.

Army Industrial Hygiene Program

http://chppm-www.apgea.army.mil/Armyih/

Provides access to Army industrial hygiene documents, publications and policies, plus links to a number of other documents and sites of interest to industrial hygienists.

Australian Institute of Occupational Hygienists (AIOH)

http://www.curtin.edu.au/org/aioh/

This is the official home page of the AIOH. It contains information about the institute's publications, membership, office holders, and conferences.

British Occupational Hygiene Society (BOHS)

http://www.bohs.org/

This page describes the society, its mission, members, publications, conferences, and meetings. There are also links to related sites.

Corporate Health and Safety—Managing Environmental Issues in the Workplace

http://www.boxelectronics.com/

View a description and ordering information for this industrial hygiene guide.

Environmental Safety and Health of Alaska

http://www.alaska.net/~esha/

The site contains descriptions of this company's safety training courses, industrial hygiene and marine chemist services, plus an industrial hygiene news and current events section, and news from the Midnight Sun Section of AIHA.

Industrial Safety and Hygiene News

http://SafetyOnline.net/ishn/

Information about and the full text of selected portions of this journal are presented.

Western Kentucky University Industrial Hygiene Student Association

http://www2.wku.edu/www/stuorgs/ihsa/index.html

Access information about the students, the industrial hygiene program, the full text of documents on a number of IH-related subjects, and industrial hygiene calculation programs written in Javascript.

Programs

American Board of Industrial Hygiene (ABIH)

http://www.abih.org

Information is provided about ABIH, CIH certification, examinations, and ABIH publications. More detail about this site is available in the Top 50 Sites section.

Army Industrial Hygiene Program

http://chppm-www.apgea.army.mil/Armyih/

Provides access to Army industrial hygiene documents, publications and policies, plus links to a number of other documents and sites of interest to industrial hygienists.

Western Kentucky University Industrial Hygiene Student Association

http://www2.wku.edu/www/stuorgs/ihsa/index.html

Access information about the students, the industrial hygiene program, the full text of documents on a number of IH-related subjects, and industrial hygiene calculation programs written in Javascript.

Software

American Conference of Government Industrial Hygienists (ACGIH)

http://www.acgih.org/

View information about ACGIH, a catalog of its publications, demos of the organization's electronic products, upcoming events, and an "OH Talk" Web chat section. More detail about this site is available in the Top 50 Sites section.

DataChem Software

http://www.datachemsoftware.com

The company's site provides information about its various software packages "that help you study and practice for your professional certification examination."

Hyginist

http://home.wxs.nl/~ihpc/

Produced by Scheffers Industrial Hygiene Publishing and Consultancy in the Netherlands, HYGINIST is an "industrial hygiene statistical tool. It evaluates the exposure data collected with the standard exposure assessment strategies."

Industrial Hygiene on the World Wide Web

http://www.industrialhygiene.com

A growing number of industrial hygiene calculation programs are available at this site. Available programs include the Revised NIOSH Lifting Equation, Gaussian Plume Dispersion Model, a Safety and Health Program Assessment Worksheet, Convert % Dose to 8 hour TWA, and Determine ventilation rate required to maintain a concentration. More detail about this site is available in the Top 50 Sites section.

---

9. Industrial Safety

Consulting Services

Dynamic Scientific Controls, Inc.

http://www.fallsafety.com

They are consultants in occupational fall protection safety planning and commercial egress systems. Site provides a list of the company's publications, courses, and videos.

Gravitec Systems Inc.

http://www.gravitec.com/

This U.S. company specializes in engineering, training, and consulting related to fall protection. Read about the company's services.

Shiftwork Services

http://host.mwk.co.nz/shiftwork/

Shiftwork Services is a division of the Auckland (New Zealand) Sleep Management Centre. The site provides access to the full text of the organization's newsletter, sleep tips, and guidelines for managers.

Sulowski Fall Protection Inc.

http://ourworld.compuserve.com/homepages/AndrewSulowski/sulowski.htm

This consulting company's site provides information about its services, fall protection courses, publications, and upcoming events.

SynchroTech

http://www.schedule-masters.com/

SyncroTech is a consulting firm specializing in shiftwork, sleep, and schedule management. The company provides customized educational programs and work scheduling systems to all levels of personnel in a wide range of operational settings.

Educational Resources / Training

Dynamic Scientific Controls, Inc.

http://www.fallsafety.com

They are consultants in occupational fall protection safety planning and commercial egress systems. Site provides a list of the company's publications, courses, and videos.

SASSI Training Systems

http://www.mindspring.com/~sassitng/index.html

The company's environmental counseling, compliance training, and ISO 14000 training.

Journals and Newsletters

Canadian Auto Workers

http://www.caw.ca/caw/

Information about the union, its work, and studies undertaken. Its health, safety, and environment newsletter and health and safety fact sheets are also available.

Circadian Information

http://www.shiftwork.com/

Circadian Information publishes the monthly *Shiftwork Alert* newsletter, the *Shiftwork Family Calendar*, the new monthly *Working Nights* newsletter, and other books and pamphlets about different aspects of shiftwork and fatigue management.

Industrial Safety and Hygiene News

http://SafetyOnline.net/ishn/

Information about and the full text of selected portions of this journal are presented.

Saf-T-Gard International

http://www.saftgard.com/

The site provides access to the company's catalog of industrial safety products, its newsletter, and selected links to other sites.

Uni-Hoist

http://www.uni-hoist.com

Find out about the company's confined space products. The site also includes a photo gallery, links to related articles and sites, OSHA regulations, and a newsletter.

Products

3M Environmental Safety and Energy Performance Products

http://www.mmm.com/market/government/env/envindex.html

Information about 3M's products.

Ansell Edmont

http://www.anselledmont.com/

Information about the company's safety gloves, the company itself, and its distributors.

Best Gloves

http://www.bestglove.com

Information about the company's safety gloves is presented. There is also a free downloadable program, "Comprehensive Guide to Chemical-Resistant Best Gloves."

Bouton Eye Protection

http://www.hlbouton.com/

This is a searchable catalog of the company's safety eyewear.

ChinaHawk Enterprises Ltd.

http://home.hkstar.com/~chinahwk/home.html

Information about this Beijing-based company's disposable safety products for the health care, industrial, and food services industries. Offices are in Beijing and Hong Kong.

Conveyerail Systems Ltd.

http://www.conveyerail.com/

View descriptive information about VacuMove, a vacuum lifting system.

ENMET Corporation

http://www.enmet.com

Read about this company's industrial gas and vapor monitoring instrumentation.

Major Safety Service

http://www.pilot.infi.net/~majsaf/

Products and services provided by this environmental instrument and safety equipment supplier are described.

Nikos International

http://www.antinfortunistica.it

The site describe's the company's industrial cleaning products and personal protective equipment. Most information is in Italian.

Northern Industrial Supply Company

http://www.nisco.net/

This Ontario, Canada company sells industrial heating, ventilation, and air conditioning equipment. Read about the products they provide.

Pro Grip Safety Sole

http://www.uleth.ca/~gadd/index.htm

Information about Pro-Grip nonslip safety soles.

Protecta Fall Arrest Systems

http://www.protecta.com

Detailed information is provided about the company's fall protection, confined space entry and retrieval equipment, rescue devices, and heavy-load fall arrestors.

Saf-T-Gard International

http://www.saftgard.com/

The site provides access to the company's catalog of industrial safety products, its newsletter, and selected links to other sites.

Safe Shop Tools

http://www.safeshop.com/

Provides information about the company's safety products for the truck repair industry.

Safemate International

http://www.safemate.com

Read about this company's antislip products.

Safety Boot

http://members.aol.com/safetmaker/index.htm

Read a description of Safety Boot, "a reusable base for constructing free-standing temporary guardrails that exceed OSHA standards."

Shiftwork Systems, Inc.

http://members.tripod.com/~Shiftwork/

Read about the company's "Circadian Lighting Technology," along with links to a number of documents and sites containing shiftwork-related information.

Surety Manufacturing and Testing Ltd.

http://www.Suretyman.com/

This Edmonton, Canada-based company manufactures and sells equipment for fall protection and confined space entry. View the catalogue and read about training videos.

Uni-Hoist

http://www.uni-hoist.com

Find out about the company's confined space products. The site also includes a photo gallery, links to related articles and sites, OSHA regulations, and a newsletter.

Uvex—Serving Safety

http://www.uvex.com

The company's various lines of protective eyewear are highlighted.

Waldmann Lichttechnik

http://www.waldmann.de/

View information about this German company's range of workplace lighting products.

Western Hard Hats

http://www.westernhardhat.com/

Information is provided about the western-style hard hats this company distributes.

Publications

Canadian Auto Workers

http://www.caw.ca/caw/

Information about the union, its work, and studies undertaken. Its health, safety, and environment newsletter and health and safety fact sheets are also available.

DieselNet

http://www.dieselnet.com/

The Diesel Emissions Evaluation Program (DEEP) is an industry-sponsored initiative to facilitate research and transfer of technology on diesel emissions, their health effects, measurement, and control. Provides documents, information about research, and links.

Dynamic Scientific Controls, Inc.

http://www.fallsafety.com

They are consultants in occupational fall protection, safety planning, and commercial egress systems. Provides a list of the company's publications, courses, and videos.

Industrial Safety and Hygiene News

http://SafetyOnline.net/ishn/

Information about and the full text of selected portions of this journal are presented.

Software

ErgoEASER Software

http://tis-hq.eh.doe.gov:80/others/ergoeaser/download.html

Download a copy of this software, "developed to aid in identifying, evaluating, and preventing work-related musculoskeletal disorders."

Map 80 Systems Ltd.

http://www.map80.co.uk/

Read about this U.K. company's labelling software for chemical, pharmaceutical, industrial, and retail settings.

---

10. Laboratory Safety

Accu*Aire Controls, Inc.

http://www.accuaire.com/

Information is presented about the company's laboratory fume hoods and exhaust products, selected articles, and free software.

Best Gloves

http://www.bestglove.com

Information about the company's safety gloves is presented. There is also a free downloadable program, "Comprehensive Guide to Chemical-Resistant Best Gloves."

Bouton Eye Protection

http://www.hlbouton.com/

This is a searchable catalog of the company's safety eyewear.

Brookhaven National Lab, Safety and Environmental Protection Div.

http://sun10.sep.bnl.gov/seproot.html

Information at this site includes health and safety questions and answers, EHS policies, standards and notices, and BNL's employee safety handbook.

DuPont Protective Apparel Information Service

http://www.dupont.com/tyvek/protective-apparel/

At this site, users can search for appropriate protective apparel by end use chemical name, class, or CAS number.

Flinn Scientific

http://www.flinnsci.com

This science supply company site provides a lot of useful information, including laboratory design tips and numerous documents on chemical safety.

Health Canada, Laboratory Centre for Disease Control

http://www.hc-sc.gc.ca/hpb/lcdc/

The site contains information about the centre itself, public health documents, travel health advisories, and disease prevention guidelines. More detail about this site is available in the Top 50 Sites section.

Health Canada, Laboratory Centre for Disease Control, Laboratory Biosafety Guidelines

http://hwcweb.hwc.ca/hpb/lcdc/bmb/biosafty/

This site provides the full text of LCDC's guidelines document. More detail about this site is available in the Top 50 Sites section.

Howard Hughes Medical Institute, Office of Laboratory Safety

http://www.hhmi.org/science/labsafe

This site contains information about laboratory safety materials available from the institute, laboratory safety summaries, training videos, emergency response guidelines, and help with laboratory hazard recognition.

Lab Safety Supply

http://www.labsafety.com

As well as providing access to the company's catalog of lab safety equipment, the site presents a number of useful documents in a variety of health and safety categories.

Lawrence Livermore National Lab (LLNL), Environment, Safety and Health

http://www.llnl.gov/es_and_h/

Various documents are provided, including LLNL's health and safety manual, environmental guidelines documents, and an environmental compliance manual.

University of Alberta, Medical Laboratory Science Program

http://www.ualberta.ca/~medlabsc/

Read about their undergraduate and graduate programs and Internet resources available to students.

University of Maryland at College Park, Environmental Safety

http://www.inform.umd.edu/DES/

Information at this site includes the University's Chemical Hygiene Plan and Laboratory Safety Guide, search forms for various MSDS collections, etc.

University of Virginia, Office of Environmental Health and Safety

http://www.virginia.edu/~enhealth/

This well-developed site provides access to a number of documents, information about the office's programs, UVA's Laboratory Survival Manual, and a Laboratory Safety Checklist.

Uvex—Serving Safety

http://www.uvex.com

The company's various lines of protective eyewear are highlighted.

---

11. Mine Safety

Arizona State Mine Inspector

http://www.indirect.com/www/mining/

The site contains information about this state government agency, its annual report, and upcoming conferences.

DieselNet

http://www.dieselnet.com/

The Diesel Emissions Evaluation Program (DEEP) is an industry-sponsored initiative to facilitate research and transfer of technology on diesel emissions, their health effects, measurement, and control. Provides documents, information about research, and links.

Mine Safety and Health Administration (MSHA)

http://www.msha.gov

This U.S. government site provides access to mining-related regulatory information, health and safety documents, and information about various programs.

Mines and Aggregates Safety and Health Association

http://www.masha.on.ca/

MASHA is a designated "Safe Workplace Association" in Ontario, Canada. Provides information about the association, its publication catalogue, and a calendar of events.

Pittsburgh and Spokane Research Centers, Mining H&S Research

http://www.usbm.gov

This NIOSH site provides information about its health and safety research, worker health, worker safety, and disaster prevention. Search its database of publications and download MADSS (Mine Accident Decision Support System).

Safety Mining and Eng.

http://www.instanet.com/~pfc/SME/safety_1.html

Safety Mining and Eng. is a group of safety professionals who provide safety consulting and training to both mining and industry.

Tensor Technologies—MineNet

http://www.microserve.net:80/~doug/

MineNet is a directory of mining-related products, services, technology, and news.

Westray Mine Public Inquiry

http://www.gov.ns.ca/legi/inquiry/westray/

The site presents the findings of a public inquiry into the 9 May 1992 underground explosion at the Westray coal mine in Nova Scotia that killed 26 miners.

---

12. Noise and Hearing

Occupational Noise Exposure

http://hammock.ifas.ufl.edu/txt/fairs/oa/19401.html

This is a summary of U.S. OSHA Standard 1910.95, written primarily for agricultural business owners and managers.

OSHEX/ESA

http://www.oshex.com

A State of New York safety equipment provider of noise control and equipment safety solutions. Environmental Safety Associates (ESA) provides safety and noise-related consulting services, including litigation. Information about the company's services.

Safe-At-Work.com

http://www.safe-at-work.com/

Site is the host for HearSaf 2000 and James Anderson and Associates. Various documents related to hearing conservation and demonstration software are available for download.

Thorburn Associates

http://www.ta-inc.com/

Read about this company's acoustical consulting and audiovisual system design consulting services.

---

13. Occupational Medicine

Conferences

4th ICOH International Conference on Occupational Health for Health Care Workers

http://www.santepub-mtl.qc.ca/icoh1999/icoh1999.html

Read about this conference in Montreal, Canada, September 28 to October 1, 1999.

Consulting Services

Alex Hartov Consulting

http://www.newcc.com/alex/

Read about his consulting services, including medical and biomedical instrumentation.

Assure Health Management

http://www.assure.ca

This company designs and delivers occupational health and disability management programs to Canadian business and industry. Read about the company and its services.

HealthGate

http://www.healthgate.com/HealthGate/MEDLINE/search.shtml

HealthGate provides free access to Medline, plus information about their commercial health, wellness, and biomedical information services.

IDEWE Occupational Health and Safety Services on the Internet

http://www.idewe.be/

Idewe is a nonprofit occupational health service in Flanders, Belgium. Read about the organization's services.

Life in Motion

http://www.rockies.net/~eberry/

This company provides occupational health nurses and registered nurses with programs for use in both the workplace and the community.

McMaster Univ., Occupational and Environmental Health Laboratory

http://www-fhs.mcmaster.ca/oehl/index.html

This AIHA accredited laboratory in Hamilton Canada provides occupational hygiene services to industry, labour, and the general public. Read about the lab's research projects, services, etc.

Medical Device Link—Safety Consultants

http://www.devicelink.com/consult/Safety.html

This directory provides descriptions of companies that provide safety services.

Medical Horizons Unlimited (TM)

http://www.medhorizons.com

This company is a medical educational consulting firm. Read about its educational services in a variety of medically related fields, including occupational health and safety.

MedTox Health Services

http://home.earthlink.net/~medtox/

Provides information about this California occupational medicine services company.

Occupational Medicine Center of Tuscarawas County (Ohio, U.S.A)

http://web1.tusco.net/omctc/index.htm

Describes the center's occupational health care services for businesses and their employees.

Span Corporation Occupational Health Services

http://www.spancorp.com/

The site describes this company's occupational medicine, compliance, prevention, hazard recognition, and information services.

Databases

Carcinogenic Potency Database

http://potency.berkeley.edu/cpdb.html

CPDB provides the results of chronic long-term animal cancer tests. It provides a single database that includes sufficient information on each experiment to permit investigations into many research areas of carcinogenesis. Analyses of 5152 experiments on 1298 chemicals are presented

CRISP—Computer Retrieval of Information on Scientific Projects

gopher://gopher.nih.gov:70/11/res/crisp

CRISP (Computer Retrieval of Information on Scientific Projects) System is a major biomedical database containing information on research ventures supported by the United States Public Health Service.

HealthGate

http://www.healthgate.com/HealthGate/MEDLINE/search.shtml

HealthGate provides free access to Medline, plus information about their commercial health, wellness, and biomedical information services.

Internet Grateful Med (MEDLINE database)

http://igm.nlm.nih.gov/

Free access is provided to this U.S. National Library of Medicine database of over 9 million bibliographic citations from 3900 journals. More detail about this site is available in the Top 50 Sites section.

MedScape

http://www.medscape.com

This site provides free access to MEDLINE, TOXLINE, and AIDSLINE from the U.S. National Library of Medicine, plus a document delivery service for a small fee; the full text of thousands of health related articles; the full text of various journals; and more. Registration required (free). More detail is available in the Top 50 Sites section.

National Library of Medicine

http://www.nlm.nih.gov

This U.S. government site is one of the best starting points for finding medically related information. Highlights include free access to the MEDLINE database, newsletters, fact sheets, and information about NLM programs.

SilverPlatter

http://www.silverplatter.com

The site provides information about databases commercially available from SilverPlatter, both on CD-ROM and via the Internet.

Directories

Hardin Meta Directory—Public, Occupational and Environmental Health

http://www.lib.uiowa.edu/hardin/md/publ.html

This is a "list of lists," providing links to a variety of health-related directories.

Medical Device Link—Safety Consultants

http://www.devicelink.com/consult/Safety.html

This directory provides descriptions of companies that provide safety services.

University of Edinburgh, Health Environment and Work

http://www.med.ed.ac.uk/hew/

The site provides a long list of "tutorials" (educational resources) on a variety of occupational and environmental health topics, plus information on Edinburgh's academic programs. There is also an excellent directory of primarily European Internet resources. More detail about this site is available in the Top 50 Sites section.

Your Business, Your Health

http://www.siu.edu/departments/bushea/

This Health Promotion directory is a joint project of the Southern Illinois University Health Education Program, the Center for Injury Control and Worksite Health Promotion (CICWHP), and the SIU Pontikes Center for Management of Information.

Educational Resources / Training

Center for Occupational and Environmental Health, Univ. of California

http://ehs.sph.berkeley.edu/coeh/

View information about this research center, its academic programs, continuing education courses, and research projects.

Cornell OSH Archive of Educational Materials

http://ext.ilr.cornell.edu/osh/edu/occindex/srchindex.html

Cornell University's School of Industrial and Labor Relations, OSH Extension is piloting an online archive of public domain and freeware occupational safety and health educational materials and policies. Find training materials on a varietyof OSH subjects.

Griffith University, Faculty of Health and Behavioural Sciences

http://www.ua.gu.edu.au/hbk/hbs_ind.htm

Information about courses provided by this Queensland, Australia university is available. The School of Occupational Health and Safety is within this faculty.

McGill University, Occupational Health Sciences

http://www.mcgill.ca/occh/

The site provides information about their academic programs, activities/news/seminars, research groups/labs, and publications.

Medical Horizons Unlimited (TM)

http://www.medhorizons.com

This company is a medical educational consulting firm. Read about its educational services in a variety of medically related fields, including occupational health and safety.

North Carolina Occupational Safety and Health Education and Research Center

http://www.sph.unc.edu/osherc/

"A partnership of the University of North Carolina at Chapel Hill, and the Duke University Medical C Center, NC OSHERC was established in 1977 through funding from the National Institute for Occupational Safety andHealth (NIOSH)." Read about the center's graduate and continuing education programs.

University of Birmingham (U.K.), Institute of Occupational Health

http://www.bham.ac.uk/IOH/

Read about the institute's postgraduate programs, upcoming local, national, and international conferences and meetings.

Univ. of California Davis, Dept. of Epidemiology and Preventive Medicine

http://www-oem.ucdavis.edu

Information is provided about the department, its programs, and research.

University of Edinburgh, Health Environment and Work

http://www.med.ed.ac.uk/hew/

The site provides a long list of "tutorials" (educational resources) on a variety of occupational and environmental health topics, plus information on Edinburgh's academic programs. There is also an excellent directory of primarily European Internet resources. More detail about this site is available in the Top 50 Sites section.

Journals and Newsletters

British Medical Journal

http://www.bmj.com/bmj/

Abstracts of papers from current and past issues of the journal are available.

Circadian Information

http://www.shiftwork.com/

Circadian Information publishes the monthly *Shiftwork Alert* newsletter, the *Shiftwork Family Calendar*, the new monthly *Working Nights* newsletter and other books and pamphlets about different aspects of shiftwork and fatigue management.

EHIS Publications

http://ehpnet1.niehs.nih.gov/docs/publications.html

Access is provided to various publications of the U.S. National Institute of Environmental Health Sciences: Environmental Health Perspectives, National Toxicology Program Technical Reports, etc.

Institute for Work and Health

http://www.iwh.on.ca

Information about this Toronto, Canada-based organization, its publications, programs, special events, newsletter, etc.

MedScape

http://www.medscape.com

This site provides free access to MEDLINE, TOXLINE, and AIDSLINE from the U.S. National Library of Medicine, plus a document delivery service for a small fee; the full text of thousands of health related articles; the full text of various journals; and more. Registration required (free). More detail is available in the Top 50 Sites section.

Scandinavian Journal of Work, Environment, and Health

http://www.occuphealth.fi/eng/dept/sjweh/

The site provides information about upcoming articles in the journal, subscribing to the paper version, etc. The journal appears bimonthly.

Policy/Procedures Documents

American College of Occupational and Environmental Medicine (ACOEM)

http://www.acoem.org

Categories of available information include information about ACOEM, membership information, courses/conferences, statement papers/guidelines, publications, press releases, a self-assessment exam, employment service, and other OEM links. More detail about this site is available in the Top 50 Sites section.

Canadian Medical Ass'n., Occupational Medicine, Clinical Practice Guidelines

http://www.cma.ca/cpgs/occup.htm

Links are provided to a series of documents describing specific guidelines.

Products

BackHealth U.S.A

http://www.backhealth.com

Some information is provided about back pain causes and treatments. Most of the information is about BackHealth 2000, the company's back exercise equipment.

Quick-Aid, Inc.

http://www.quick-aid.com

Read about products and services from this medical and safety supply company.

Stress Assessment Profile

http://www.opd.net/stress.html

Tthis tool for stress assessment is produced by Organizational Performance Dimensions.

Publications

American College of Occupational and Environmental Medicine (ACOEM)

http://www.acoem.org

Categories of available information include information about ACOEM, membership information, courses/conferences, statement papers/guidelines, publications, press releases, a self-assessment exam, employment service, and other OEM links. More detail about this site is available in the Top 50 Sites section.

Asian-Pacific Network on Occupational Safety and Health Information

http://www.ilo.org/public/english/270asie/asiaosh/

Presented by the ILO/FINNIDA Asian-Pacific Regional Programme on Occupational Safety and Health, this site was established with a view to sharing OSH information in and about the Asian-Pacific region.

Centers for Disease Control and Prevention (CDC)

http://www.cdc.gov

This U.S. government site contains a wealth of information, including occupational and environmental health, and travellers' health. Numerous publications and reports are available, including the Morbidity and Mortality Weekly Report (MMWR). More detail about this site is available in the Top 50 Sites section.

Cornell OSH Archive of Educational Materials

http://ext.ilr.cornell.edu/osh/edu/occindex/srchindex.html

Cornell University's School of Industrial and Labor Relations, OSH Extension is piloting an online archive of public domain and freeware occupational safety and health educational materials and policies. Find training materials on a variety of OSH subjects.

DieselNet

http://www.dieselnet.com/

The Diesel Emissions Evaluation Program (DEEP) is an industry sponsored initiative to facilitate research and transfer of technology on diesel emissions, their health effects, measurement, and control. Provides documents,information about research, and links.

Duke University Occupational and Environmental Medicine

http://occ-env-med.mc.duke.edu/oem

This site contains numerous documents and links to resources in occupational and environmental medicine. It is also the "home" of the occ-env-med-l mailing list. More detail about this site is available in the Top 50 Sites section.

Finnish Institute of Occupational Health

http://www.occuphealth.fi

Information is provided about the Institute, its purpose, its work, etc. Lists of Institute publications and journals are provided. There is also a list of OH&S conferences.

Global Health Disaster Network

http://hypnos.m.ehime-u.ac.jp/GHDNet/index.html

GHDNet is maintained by Department of Emergency Medicine, Ehime University, Japan. The site contains documents, bibliographies, conference information, and links to related sites.

Health Canada, Health Information Network

http://www.hc-sc.gc.ca/

This Canadian federal government department presents its health alerts, public health documents, news releases, etc., plus links to its various directorates: Laboratory Centre for Disease Control, Environmental Health Directorate, etc. More detail about this site is available in the Top 50 Sites section.

Health Promotion Center, University of California Irvine

http://www.seweb.uci.edu/users/dstokols/hpc.html

"UCIHPC is a research and consulting unit operating within the School of Social Ecology at the University of California, Irvine. We are interested in all aspects of health promotion. Our research focuses on comprehensive, integrated approaches to health promotion, especially on workplace-based health promotion."

HEBSWeb

http://www.hebs.scot.nhs.uk/

This is the site of the Health Education Board for Scotland. Bibliographies and assorted documents on a variety of health topics are provided.

Institute for Work and Health

http://www.iwh.on.ca

Information about this Toronto, Canada-based organization, its publications, programs, special events, newsletter, etc.

International Agency for Research on Cancer (IARC)

http://www.iarc.fr

IARC's mission is "to coordinate and conduct research on the causes of human cancer, and to develop scientific strategies for cancer control. The agency is involved in both epidemiological and laboratory research and disseminates scientific information through meetings, publications, courses and fellowships." The full text of some IARC publications is available on the site.

Joint Commission on Accreditation of Healthcare Organizations (JCAHO)

http://www.jcaho.org/

The Joint Commission evaluates and accredits health care organizations in the U.S.. The site provides information about JCAHO publications and standards for accreditation.

Journal of Occupational and Environmental Medicine

http://www.acoem.org/pubs/joem/joemgen.htm

Provides ordering information for this publication from the American College of Occupational and Environmental Medicine. It also provides abstracts of articles from past issues.

McGill University, Occupational Health Sciences

http://www.mcgill.ca/occh/

The site provides information about their academic programs, activities/news/seminars, research groups/labs, publications, etc.

MedScape

http://www.medscape.com

This site provides free access to MEDLINE, TOXLINE and AIDSLINE from the U.S. National Library of Medicine, plus a document delivery service for a small fee; the full text of thousands of health related articles; the full text of various journals; and more. Registration required (free). More detail is available in the Top 50 Sites section.

Mortality and Morbidity Weekly Report (MMWR)

http://www.cdc.gov/epo/mmwr/mmwr.html

Current and previous issues of this weekly report from the U.S. Centers for Disease Control and Prevention are presented in their entirety.

Mosby Publishing

http://www.mosby.com

Read about the company's health care-related books and periodicals. Tables of contents and abstracts are provided for the periodicals.

National Institutes of Health

http://www.nih.gov

A wide variety of health information is available from this U.S. government site. Highlights include the NIH Health Information Index, HealthFinder, Consumer Health Information, plus links to other NIH sites such as the National Library of Medicine and the National Cancer Institute.

National Library of Medicine

http://www.nlm.nih.gov

This U.S. government site is one of the best starting points for finding medically-related information. Highlights include free access to the MEDLINE database, newsletters, fact sheets, information about NLM programs, etc.

New York State Dept. of Health, Environmental and Occupational Health

http://www.health.state.ny.us/nysdoh/consumer/environ/homeenvi.htm

This site provides access to a series of basic documents on environmental and occupational health. The content is geared to the public.

Occupational Disease Panel Electronic Library

http://www.ccohs.ca/odp/

The Occupational Disease Panel (ODP) was created under the Workers' Compensation Act of Ontario. In its 10 years of operation, ending in 1997, it produced numerous publications on issues relating to workplace/occupational diseases and their compensation. These publications are available in their entirety on this site.

OncoLink

http://www.oncolink.org/

This site is probably the best place to start when looking for information about any type of cancer, whether you are a patient, caregiver, or health care professional. More detail about this site is available in the Top 50 Sites section.

Pan American Health Organization (PAHO)

http://www.paho.org

The Pan American Health Organization (PAHO) is an international public health agency working to improve health and living standards of the countries of the Americas. Read about PAHO's various programs and publications.

Scandinavian Journal of Work, Environment, and Health

http://paja.occuphealth.fi/sjweh/

The site provides information about upcoming articles in the journal, subscribing to the paper version, etc. The journal appears bimonthly.

Southern Africa Occupational Health Web

http://www.und.ac.za/und/med/comhlth/occ_hlth.html

Maintained by the Occupational Health Programme, Dept. of Community Health, University of Natal, the site describes the University's Occupational Health Programme, presents a discussion paper on women in the workplace, and provides links to various occupational health sites.

University of Washington, Department of Environmental Health Library

http://weber.u.washington.edu/~dehlib/

The site provides a series of links to environmental health publications, theses, and academic programs at the University of Washington and elsewhere.

Workers Health and Safety Centre

http://www.whsc.on.ca/

The centre is a worker-driven health and safety delivery organization in Ontario, Canada. The site provides information about the centre and its services, including the full text of publications and links to labor and health and safety resources.

World Health Organization

http://www.who.int

WHO's programs, its regional offices around the world, and international health information reports and publications. More detail is available in the Top 50 Sites section.

Programs

Asian-Pacific Network on Occupational Safety and Health Information

http://www.ilo.org/public/english/270asie/asiaosh/

Presented by the ILO/FINNIDA Asian-Pacific Regional Programme on Occupational Safety and Health, this site was established with a view to sharing OSH information in and about the Asian-Pacific region.

Association of Ontario Health Centres

http://www.aohc.org/

The Association of Ontario Health Centres (AOHC) is the nonprofit organization that represents community health centers (CHCs) and some health service organizations (HSOs) in the province of Ontario, Canada. The site contains information about AOHC, its position papers, and health promotion programs.

Assure Health Management

http://www.assure.ca

This company designs and delivers occupational health and disability management programs to Canadian business and industry. Read about the company and its services.

Center for Occupational and Environmental Health, Univ. of California

http://ehs.sph.berkeley.edu/coeh/

View information about this research center, its academic programs, continuing education courses, and research projects.

Griffith University, Faculty of Health and Behavioural Sciences

http://www.ua.gu.edu.au/hbk/hbs_ind.htm

Information about courses provided by this Queensland, Australia university is available. The School of Occupational Health and Safety is within this faculty.

McGill University, Occupational Health Sciences

http://www.mcgill.ca/occh/

The site provides information about their academic programs, activities/news/seminars, research groups/labs, and publications.

University of Birmingham (U.K.), Institute of Occupational Health

http://www.bham.ac.uk/IOH/

Read about the institute's postgraduate programs, upcoming local, national, and international conferences and meetings.

Univ. of California Davis, Dept. of Epidemiology and Preventive Medicine

http://www-oem.ucdavis.edu

Information is provided about the department, its programs, and research.

University of Washington, Department of Environmental Health Library

http://weber.u.washington.edu/~dehlib/

The site provides a series of links to environmental health publications, theses, and academic programs at the University of Washington and elsewhere.

World Health Organization

http://www.who.int

WHO's programs, its regional offices around the world, international health information reports, and publications. More detail is available in the Top 50 Sites section.

Software

EnviroDx

http://www.auhs.edu/envirodx/

"EnviroDx is a multimedia, case-focused, computer-based learning program on environmental-related diseases. The organizing metaphor for EnviroDx is an exploratory 'virtual clinic' affiliated with a busy medical school. The program user takes the part of a practicing physician faced with a patient with an unknown disease or condition that is possibly caused by exposure to environmental factors."

Environment and Safety Data Exchange (ESDX)

http://www.esdx.org/esdhome.html

Find out about ESDX, an industry association organized to participate in setting standards for more cost-effective management information technology in environmental, health and safety (EHS) activities.

Occupational Health Research

http://ohr.systoc.com/index.htm

Read about the company's software, SYSTOC, a program to manage patient care. There are also a lot of links to resources for occupational health professionals: U.S. regulatory info, hazardous materials, etc.

---

14. Office Safety

Consulting Services

National HVAC

http://www.nationalhvac.com/

The content of the site describes this Canadian company's services, including indoor environment consulting; building ventilation assessments; field investigations, air testing, and remediation management services.

Usernomics

http://www.usernomics.com/

Ergonomics for hardware, software, and training. Find out about the company's services, or access links related to various aspects of ergonomics.

VDT Solution

http://home.earthlink.net/~ergo1/

The site provides a variety of types of information related to computer workstations and ergonomics: preferred equipment and services, a PC and laptop ergonomic guide, and ergonomic product reviews.

Databases

ErgoWeb

http://www.ergoweb.com

ErgoWeb is an extensive Web site, containing a variety of types of ergonomics-related information, such as workplace evaluation and design software, a database of ergonomics-related products and services, and case studies. More detail about this site is available in the Top 50 Sites section.

Educational Resources / Training

Usernomics

http://www.usernomics.com/

Ergonomics for hardware, software, and training. Find out about the company's services, or access links related to various aspects of ergonomics.

Products

21st Century Eloquence

http://www.voicerecognition.com/

This Florida-based company is a reseller of voice recognition products for computers. View information about their products and services.

Combo Ergonomic and Contract Furnishings

http://www.combo.com

View the company's ergonomic furniture catalog, plus a "collection of information and solutions pertaining to office ergonomics."

---

comp.speech FAQ

http://www.speech.cs.cmu.edu/comp.speech/

This frequently-asked-questions (FAQ) document originated with the comp.speech newsgroup and covers information on speech technology, including speech synthesis, speech recognition, and related material. The speech recognition section is of particular interest.

DeskEx Body Awareness System

http://www.iftech.com/products/deskex/

The site describes this computer workstation wellness program, consisting of an online wellness manual, exercise reminders, break reminders, and an activity monitor.

Ergonomic Design Inc.

http://www.ergodesign.com/

View information about the company's computer-related ergonomic products: wrist rests, keyboard holders, document holders, etc.

Ergonomic Sciences Corporation

http://www.ergosci.com

The goal of this Web site is to provide up-to-date information about the science of workspace design and function and present information that will help you choose the most beneficial ergonomic products to address specific ergonomic issues.

Mouse Escalator

http://www.4amouse.com/

Read a description of this ergonomic product. It is a "positive inclined work surface" for a computer mouse.

Proformix

http://www.proformix.com

The company provides ergonomic keyboard platforms and related products, plus a suite of ergonomic management software.

Questec Input Devices

http://www.questecmouse.com

Read about this company's computer related products: mice, mice pens, trackballs, ergonomic keyboards, and antiglare filters.

Safe Computing

http://www.safecomputing.com

Access the company's extensive catalog of computer ergonomics-related products.

Surviving the Workplace Jungle

http://vvv.com/m2/swj/

The site presents detailed information about this workplace violence video and workbook.

Publications

Combo Ergonomic and Contract Furnishings

http://www.combo.com

View the company's ergonomic furniture catalog, plus a "collection of information and solutions pertaining to office ergonomics."

---

comp.speech FAQ

http://www.speech.cs.cmu.edu/comp.speech/

This frequently-asked-questions (FAQ) document originated with the comp.speech newsgroup, and covers information on speech technology, including speech synthesis, speech recognition, and related material. The speech recognition section is of particular interest.

ErgoWeb

http://www.ergoweb.com

ErgoWeb is an extensive Web site, containing a variety of types of ergonomics-related information, such as workplace evaluation and design software, a database of ergonomics-related products and services, and case studies. More detail about this site is available in the Top 50 Sites section.

Occupational Overuse Syndrome/RSI resources

http://www.mcs.vuw.ac.nz/comp/General/OOS/

The site contains articles and bulletins from New Zealand Occupational Safety and Health, the typing injuries FAQ, and links to other RSI information.

Pointing Device Summary

http://www.ME.berkeley.edu/ergo/tips/pdtips.html

This is an online document about keyboard injuries related to pointing devices (mice, trackballs, etc.). It was produced by the University of California San Francisco/University of California Berkeley Ergonomics Program.

The Mouse List—Ergonomic, Adaptive and Alternate Pointing Devices

http://www.setbc.org/mouselist/

This online document by Diana Carroll provides information about specific mice, trackballs, and other computer pointing devices. There are descriptions, reviews, and company contact information.

Typing Injury FAQ

http://www.tifaq.com

This is a frequently-asked-questions document (FAQ) about typing injuries. Sections include general information, keyboards, voice recognition, mice, software, furniture, and links to other resources. More detail is available in the Top 50 Sites section.

VDT Solution

http://home.earthlink.net/~ergo1/

The site provides a variety of types of information related to computer workstations and ergonomics: preferred equipment and services, a PC and laptop ergonomic guide, and ergonomic product reviews.

Software

ErgoEASER Software

http://tis-hq.eh.doe.gov:80/others/ergoeaser/download.html

Download a copy of this software, "developed to aid in identifying, evaluating, and preventing work-related musculoskeletal disorders."

Occupational Overuse Syndrome/RSI resources

http://www.mcs.vuw.ac.nz/comp/General/OOS/

The site contains articles and bulletins from New Zealand Occupational Safety and Health, the typing injuries FAQ, and links to other RSI information.

Proformix

http://www.proformix.com

The company provides ergonomic keyboard platforms and related products, plus a suite of ergonomic management software.

---

15. Product Safety

Consumer Information Center, Pueblo CO

http://www.pueblo.gsa.gov/

The site provides access to the full text of hundreds of U.S. government consumer publications, including occupationally related material.

Consumer Product Safety Commission (CPSC)

http://www.cpsc.gov

Lots of product safety information is available, including press releases and many full text publications from the U.S. CPSC. More detail is available in the Top 50 Sites section.

GPS Gas Protection Systems Inc.

http://www.GasProtection.com/

The site provides information about the company's automatic gas shutoff systems.

Health Canada, Environmental Health Program

http://www.hc-sc.gc.ca/ehp/

The Environmental Health Program consists of four bureaus: Chemical Hazards, Product Safety, Tobacco Control, and Radiation Protection. The site provides access to numerous technical reports and publications, general interest fact sheets, etc.

National Sanitation Foundation

http://www.nsf.com

The site describes NSF International, provides a database of NSF-certified products (biohazard cabinets, water purification systems, etc.), and online newsletters.

---

16. Public Health

Databases

CRISP—Computer Retrieval of Information on Scientific Projects

gopher://gopher.nih.gov:70/11/res/crisp

CRISP (Computer Retrieval of Information on Scientific Projects) System is a major biomedical database containing information on research ventures supported by the United States Public Health Service.

Health in Action

http://www.health-in-action.org/

Health In Action is the Alberta Information Clearinghouse on Prevention and Promotion. Find out about the organization's programs and access its databases.

National Library of Medicine

http://www.nlm.nih.gov

This U.S. government site is one of the best starting points for finding medically related information. Highlights include free access to the MEDLINE database, newsletters, fact sheets, information about NLM programs, etc.

National Sanitation Foundation

http://www.nsf.com

The site describes NSF International, provides a database of NSF-certified products (biohazard cabinets, water purification systems, etc.,) and online newsletters.

Outbreak

http://www.outbreak.org

"Outbreak is an online information service that addresses emerging diseases. It provides in-depth information for interested laypersons as well as medical and health professionals. It also provides a world-wide collaborative databasefor the collection of information about possible disease outbreaks."

Directories

Hardin Meta Directory—Public, Occupational, and Environmental Health
http://www.lib.uiowa.edu/hardin/md/publ.html

This is a "list of lists," providing links to a variety of health-related directories.

Poisons Information Database

http://vhp.nus.sg/PID/

The site provides information about poisonous plants and animals, a directory of antivenoms, a directory of toxinologists, and a directory of poison control centers around the world.

Your Business, Your Health

http://www.siu.edu/departments/bushea/

This Health Promotion directory is a joint project of the Southern Illinois University Health Education Program, the Center for Injury Control and Worksite Health Promotion (CICWHP), and the SIU Pontikes Center for Management of Information.

Educational Resources / Training

Center for Applied Environmental Public Health

http://caeph.tulane.edu/

Distance learning courses are provided through Tulane University School of Public Health and Tropical Medicine. Read about their degree programs—MSPH in Industrial Hygiene and MPH in Safety and Health Management.

Royal Environmental Health Institute of Scotland

http://www.ed.ac.uk/~tbell/rehis.html

"The main aims of the institute are to promote the advancement of all aspects of health and hygiene, to stimulate interest in public health, and to disseminate knowledge on health matters to the benefit of the community." Read about the institute's activities, including the courses it provides.

Journals and Newsletters

American Public Health Association

http://www.apha.org

Read about APHA, news releases, information about and abstracts from its journal, etc. Links are also provided to legislative, science policy, and other public health-related sites.

National Sanitation Foundation

http://www.nsf.com

The site describes NSF International, provides a database of NSF-certified products (biohazard cabinets, water purification systems, etc.,) and online newsletters.

Weekly Epidemiological Record

http://www.who.int/wer/

WER provides weekly information about outbreaks of diseases of public health importance. This World Health Organization publication is available in English and French.

Policy / Procedures Documents

Health Canada, Laboratory Centre for Disease Control

http://www.hc-sc.gc.ca/hpb/lcdc/

The site contains information about the centre itself, public health documents, travel health advisories, disease prevention guidelines, etc. More detail about this site is available in the Top 50 Sites section.

Publications

Centers for Disease Control and Prevention (CDC)

http://www.cdc.gov

This U.S. government site contains a wealth of information, including occupational and environmental health, and travellers' health. Numerous publications and reports are available, including the Morbidity and Mortality Weekly Report (MMWR). More detail about this site is available in the Top 50 Sites section.

Consumer Information Center, Pueblo CO

http://www.pueblo.gsa.gov/

The site provides access to the full text of hundreds of U.S. government consumer publications, including occupationally related material.

Department of Energy, Technical Information Service

http://www.tis.eh.doe.gov

This site provides access to a wealth of publications, alerts, fact sheets, and other health and safety related information. It was designed for use by U.S. DOE staff and as such has an orientation towards electrical generating facilities, but has many useful resources for non-DOE people. More detail about this site is available in the Top 50 Sites section.

Environmental Health Information Service

http://ehis.niehs.nih.gov/

This is "a full-service site that produces, maintains, and disseminates information on the environment in the form of a searchable directory." Contents include the monthly journal, *Environmental Health Perspectives*, and the National Toxicology Program's 8th Report on Carcinogens.

EPA—Technical Information Packages (TIPS)

http://www.epa.gov/oia/tips/

This page contains descriptions of eleven Technical Information Packages (TIPs). Aimed at the international community, the packages focus on key environmental and public health issues being investigated by the U.S. Environmental Protection Agency (EPA).

Health Canada, Health Information Network

http://www.hc-sc.gc.ca/

This Canadian federal government department presents its health alerts, public health documents, news releases, etc., plus links to its various directorates—Laboratory Centre for Disease Control, Environmental Health Directorate, etc. More detail about this site is available in the Top 50 Sites section.

HEBSWeb

http://www.hebs.scot.nhs.uk/

This is the site of the Health Education Board for Scotland. Bibliographies and assorted documents on a variety of health topics are provided.

Mortality and Morbidity Weekly Report (MMWR)

http://www.cdc.gov/epo/mmwr/mmwr.html

Current and previous issues of this weekly report from the U.S. Centers for Disease Control and Prevention are presented in their entirety.

National Institutes of Health

http://www.nih.gov

A wide variety of health information is available from this U.S. government site. Highlights include the NIH Health Information Index, HealthFinder, Consumer Health Information, plus links to other NIH sites such as the National Library of Medicine and the National Cancer Institute.

National Library of Medicine

http://www.nlm.nih.gov

This U.S. government site is one of the best starting points for finding medically related information. Highlights include free access to the MEDLINE database, newsletters, fact sheets, information about NLM programs, etc.

New York State Department of Health, Environmental and Occupational Health

http://www.health.state.ny.us/nysdoh/consumer/environ/homeenvi.htm

This site provides access to a series of basic documents on environmental and occupational health. The content is geared to the public.

Pan American Health Organization (PAHO)

http://www.paho.org

The Pan American Health Organization (PAHO) is an international public health agency working to improve health and living standards of the countries of the Americas. Read about PAHO's various programs, publications, etc.

World Health Organization

http://www.who.int

WHO's programs, its regional offices around the world, international health information reports and publications, etc. More detail is available in the Top 50 Sites section.

Programs

Food and Drug Administration (FDA)

http://www.fda.gov

Information is provided about various U.S. FDA program areas, including radiological health, toxicology, and drugs.

World Health Organization

http://www.who.int

WHO's programs, its regional offices around the world, international health information reports and publications, etc. More detail is available in the Top 50 Sites section.

Software

Epi Info

http://www.cdc.gov/epo/epi/downepi6.htm

Epi Info, a series of computer programs produced by the U.S. Centers for Disease Control and Prevention and the World Health Organization, provides public-domain software for word processing, database, and statistics work in public health. Download a copy from here.

---

17. Public Safety

Databases

Federal Aviation Administration, Office of System Safety

http://nasdac.faa.gov/

This U.S. government site provides access to aviation safety databases, safety reports and publications, and information on the Global Analysis Information Network (GAIN) project.

Educational Resources / Training

Canada Safety Council (CSC)

http://www.safety-council.org

CSC is an independent, national, nonprofit membership safety organization. The Web site offers information about the organization, its training programs, publications, etc.

Idea Bank

http://www.theideabank.com/

The company's catalog of fire videos for public education is presented.

Justice Institute of British Columbia

http://www.jibc.bc.ca/

"The Justice Institute of British Columbia develops and delivers training in all areas of justice, public safety, and human services." Read about its various programs.

Journals and Newsletters

Safe Communities Foundation

http://www.safecommunities.ca

"The Safe Communities Foundation is a unique partnership between the private and public sectors that is dedicated to making Canada the safest place to work, live, and play in the world." The site includes safety tips, lists of safe communities, the foundation's newsletter, etc.

Products

Crouch Fire and Safety Products

http://www.jtcrouch.com/

Find out about the company and its fire protection and public safety lighting products and services.

___

North American Detectors, Inc.

http://www.nadi.com/

A description of the company's carbon monoxide detector is provided. There is also a frequently-asked-questions (FAQ) document about carbon monoxide.

___

Onscene.com

http://www.onscene.com/

This site has been "created as a point of information exchange for public safety professionals." It provides links to fire and EMS products and services, news, pictures, etc.

___

SafeSun Personal UV Meter

http://www.SafeSun.com/

Read about this device that "lets you know how much UV you've absorbed throughout the day." The site also provides basic UV information and technical scientific data.

___

Publications

Canada Safety Council (CSC)

http://www.safety-council.org

CSC is an independent, national, nonprofit membership safety organization. The Web site offers information about the organization, its training programs, publications, etc.

___

Consumer Information Center, Pueblo CO

http://www.pueblo.gsa.gov/

The site provides access to the full text of hundreds of U.S. government consumer publications, including occupationally related material.

___

Consumer Product Safety Commission (CPSC)

http://www.cpsc.gov

Lots of product safety information is available, including press releases and many full text publications from the U.S. CPSC. More detail is available in the Top 50 Sites section.

___

Department of Transportation

http://www.dot.gov

As well as containing U.S. regulatory and policy information, the site provides access to a variety of transportation safety resources: research and reports on transportation safety from various sources, hazardous materials safety, etc.

Federal Aviation Administration, Office of System Safety

http://nasdac.faa.gov/

This U.S. government site provides access to aviation safety-related databases, safety reports and publications, and information on the Global Analysis Information Network (GAIN) project.

Injury Prevention Research Unit, University of Otago, New Zealand

http://www.otago.ac.nz/Web_menus/Dept_Homepages/IPRU/

The site provides information about the unit, research it undertakes, publications produced.

Investigation of the Safety Implications of Wireless Communications in Vehicles

http://www.nhtsa.dot.gov/people/injury/research/wireless/

This report, available in its entirety, was written by the U.S. National Highway Traffic Safety Administration. The report addresses four specific issues: does cellular telephone use while driving increase crash risk; magnitude of the traffic safety problem as a result; will crashes increase with increased cell phone use; options for safer cell phone use in vehicles.

National Lightning Safety Institute

http://www.lightningsafety.com

This nonprofit organization's lightning safety engineers provide educational and technical services to mitigate lightning hazards. Provides safety checklists and other documents.

National Safety Council

http://www.nsc.org/

Read about the council, its activities and research, publications, etc.

North American Detectors, Inc.

http://www.nadi.com/

A description of the company's carbon monoxide detector is provided. There is also a frequently-asked-questions (FAQ) document about carbon monoxide.

Transportation Safety Board of Canada

http://bst-tsb.gc.ca/

This site contains the full text of TSB's latest occurrence reports, occurrence statistics, significant safety issues, plus information about the board itself.

Programs

Technical Standards and Safety Authority

http://www.tssa.org

TSSA " is an independent, non-government, not-for-profit organization mandated to deliver specific public safety programs and services under Ontario's Safety and Consumer StatutesAdministration Act." The site provides various publications, policies, guidelines, etc.

---

18. Radiation and Laser Safety

Conferences

International Radiation Protection Agency

http://www.irpa.at

Information is provided about this international organization, its congresses, membership, newsletter, etc., plus links to associated societies.

Consulting Services

Bioelectromagnetics Society

http://biomed.ucr.edu/bems.htm

BEMS is a nonprofit organization of biological and physical scientists, physicians, and engineers interested in the interactions of nonionizing radiations with biological systems.

Integrated Environmental Management, Inc.

http://www.iem-inc.com/

IEM delivers "health physics (radiation safety), nuclear engineering, and environmental services to both government and commercial clients." Besides providing information about its services, the company's site provides a section on Radioactivity Basics and a Tool Box.

LaserNet(R)

http://www.iac.net/~rli/index.html

This site, provided by Rockwell Laser Industries, presents information including: RLI online training courses, products, laser safety consulting, and the company's newsletter.

MJW Corporation

http://www.mjwcorp.com/

MJW provides a variety of radiological consulting services and software solutions for health physics and other technical applications. Read about the company's services, download demos of their software, etc.

Nevada Technical Associates

http://www.ntanet.net

The company specializes in training in radiation safety, radiochemistry, and related areas. Read about upcoming courses and the company's related services.

Rebernigg Associates

http://www.sulphurcanyon.com/magiers/default.htm

Read about the company's safety engineering, fire protection, and laser safety services.

Scientech Inc.

http://www.scientech.com

Read about Scientech, a worldwide technical services company, specializing in environmental and safety services to the nuclear industry.

Databases

Dept. of Energy, Office of Human Radiation Experiments Home Page

http://tis.eh.doe.gov/ohre/

"The Office of Human Radiation Experiments, established in March 1994, leads the Department of Energy's efforts to tell the agency's Cold War story of radiation research using human subjects."

Radiation Health Effects Research Resource (RadEFX)

http://radefx.bcm.tmc.edu

Provides links to numerous sites containing ionizing radiation research, case studies, etc.

Directories

Radiation Protection / Health Physics eMail Directory

http://www.cyberfind.com/www/kencoon/rp-hp.html

This is a directory of health physics and radiation specialists at organizations in the U.S. and Canada. Physical address, phone, and email information is provided.

Radiation Protection Health Physics Directory

http://www.cyberfind.com/rp-hp.html

This is a directory of U.S. and Canadian nuclear generating stations and the individuals responsible for radiation protection/health physics.

Educational Resources / Training

LaserNet

http://www.rli.com

Provided by Rockwell Laser Institute, this site provides access to a wide variety of laser safety information: laser tutorials, standards and committees, bioeffects, research labs, and institutes, RLI's newsletter, etc.

LaserNet(R)

http://www.iac.net/~rli/index.html

This site, provided by Rockwell Laser Industries, presents information including RLI online training courses, products, laser safety consulting, and the company's newsletter.

Journals and Newsletters

Health Physics Society

http://www.hps.org/hps/

HPS is a professional organization dedicated to the development, dissemination, and application of both the scientific knowledge of, and the practical means for, radiation protection. Read its newsletter and find out about upcoming meetings.

International Radiation Protection Agency

http://www.irpa.at

Information is provided about this international organization, its congresses, membership, and newsletter, plus links to associate societies.

Journal of Radiological Protection

http://www.iop.org/Journals/jr

This is the official journal of the Society for Radiological Protection, relating to both ionizing and non-ionizing radiations. Information includes tables of contents for current and past issues, descriptions of forthcoming articles, and subscription information.

LaserNet

http://www.rli.com

Provided by Rockwell Laser Institute, this site provides access to a wide variety of laser safety information: laser tutorials, standards and committees, bioeffects, research labs and institutes, and RLI's newsletter.

Radiation Research, Official Journal of the Radiation Research Society

http://www.cjp.com/radres/index1.htm

Abstracts and tables of contents from this journal are provided.

Policy / Procedures Documents

Jefferson Lab EH&S

http://www.cebaf.gov/ehs/

The site provides this nuclear research facility's environmental health and safety manual and emergency management documents.

Michigan State Univ. Office of Radiation, Chemical, and Biological Safety

http://www.orcbs.msu.edu

Various documents are presented, including MSU's chemical hygiene program, their chemical lab safety checklist, radiation safety manual, *Safe Science* newsletter, etc.

Outdoor Workers Exposed to UV Radiation and Seasonal Heat

http://www.usyd.edu.au/su/ohs/outdoor.html

This is the full text of the University of Sydney (Australia) policy document.

San Diego State University, Environmental Health and Safety

http://tns.sdsu.edu/~ehs/

This site provides EHS contact information, some radiation safety information, etc.

Univ. of Illinois (Urbana), Division of Environmental Health and Safety

http://phantom.ehs.uiuc.edu/dehs.html

Provides links to a number of documents and other resources from the division's chemical safety and waste management, occupational safety and health, and radiation safety sections.

University of Illinois at Urbana-Champaign, Radiation Safety Section

http://phantom.ehs.uiuc.edu/~rad/

This site contains the university's Radiation Safety manual, a laser safety tutorial, and assorted links and documents. Information about radioactive materials is also provided, including a radioactive decay calculator.

University of Waterloo Safety Office

http://www.safetyoffice.uwaterloo.ca/

The site provides access to information about the safety office, plus documents such as their *Health and Safety Program Manual*, *Radiation Manual* and *Laser Safety Manual*.

Products

Canberra Industries

http://www.canberra.com

This radiation detection and analysis instrumentation manufacturer provides information about products and services, frequently asked questions (FAQ), and application notes.

Matcor Global Products

http://www.matcor.ca/

Provides information about the company's sensing products, including a UV intensity meter.

SafeSun Personal UV Meter

http://www.SafeSun.com/

Read about this device that "lets you know how much UV you've absorbed throughout the day." The site also provides basic UV information and technical scientific data.

Publications

Baylor College of Medicine, Environmental Safety

http://www.bcm.tmc.edu/envirosafety/

Read biosafety, chemical safety, and radiation safety policy documents/guidelines at Baylor.

Bioelectromagnetics Society

http://biomed.ucr.edu/bems.htm

BEMS is a nonprofit organization of biological and physical scientists, physicians, and engineers interested in the interactions of nonionizing radiations with biological systems.

Department of Energy, Technical Information Service

http://www.tis.eh.doe.gov

This site provides access to a wealth of publications, alerts, fact sheets, and other health and safety related information. It was designed for use by U.S. DOE staff, and as such has an orientation towards electrical generating facilities, but has many useful resources for non-DOE people. More detail about this site is available in the Top 50 Sites section.

EMF Research Activities Completed under the Energy Policy Act of 1992 (1995)

http://www.nap.edu/readingroom/enter2.cgi?NX006605.html

Read a summary of this report from the Committee to Review the Research Activities Completed under the Energy Policy Act of 1992, Board on Radiation Effects Research, Commission on Life Sciences, National Research Council (U.S.A.).

Microwave News

http://www.microwavenews.com/

The site contains selections from, and information about this bimonthly publication on non-ionizing radiation. There are also selections from "key documents" related to electro-magnetic field (EMF) radiation.

National Council on Radiation Protection and Measurements

http://www.ncrp.com/

View information about the organization's mission, membership, and publications. Related links are also accessible.

Nuclear Regulatory Commission

http://www.nrc.gov

Publications, news, and other information produced by the commission are accessible.

Radiation and Health Physics Home Page

http://www.umich.edu/~radinfo/

"This WWW HomePage contains information and links related to radiation. It has been written for three distinct groups: the general public, students and the health physics community at large." This is a well-developed, content-rich site. More detail about this site is available in the Top 50 Sites section.

RSA Publications

http://users.neca.com/rso/pubs.htm

Radiation Safety Associates produce *Radiation Protection Management* (the journal of applied health physics), *RSO Magazine* (devoted to the concerns of radiation safety officers worldwide), and a variety of other health physics publications. Read aboutthe publications and get ordering information.

University of Illinois at Urbana-Champaign, Radiation Safety Section

http://phantom.ehs.uiuc.edu/~rad/

This site contains the university's radiation safety manual, a laser safety tutorial, and assorted links and documents. Information about radioactive materials is also provided, including a radioactive decay calculator.

Programs

Dept. of Energy, Environment Safety and Health, Internat. Health Programs

http://www.eh.doe.gov/ihp/

View information about health effects of radiation and related environmental hazards. Three programs are described: those in Europe, Japan, and the Marshall Islands.

Nuclear Regulatory Commission

http://www.nrc.gov

Publications, news, and other information produced by the commission are accessible.

Radiological and Environmental Sciences Laboratory (RESL) at the Idaho National Engineering Laboratory

http://www.inel.gov/resl/

Read about this U.S. federal laboratory's ongoing programs.

Software

MJW Corporation

http://www.mjwcorp.com/

MJW provides a variety of radiological consulting services and software solutions for health physics and other technical applications. Read about the company's services and download demos of their software.

---

19. Safety and Risk Management

Consulting Services

Behavioral Science Technology, Inc.

http://www.bscitech.com/

Read about the company's services related to behavior-based safety: seminars and research.

Canadian Institute of Stress

http://www.stresscanada.org/

CIS provides stress and risk management services for businesses and individuals. Services include workplace change management services, workplace stress control programs, clinical assessment, counselling and coaching, and speakers for seminars.

Controlled Risk International

http://www5.electriciti.com/risk/

Read about the company's risk management services, courses, OSHA and CAL/OSHA's most frequently cited serious violations.

HRP Consultants

http://www.hrpconsultants.com/

Read about this Ontario, Canada-based consulting company that specializes in "assisting organizations to add value to their environmental, safety, and ISO functions by developing programs and systems that will enhance organization success."

Plant and Machinery Assessing Services

http://www.cyberstrategies.net/pamas/

This Australian consultancy develops safety programmes for the building industry. Services include preparation of safety plans, risk assessment for personal protective equipment, certificates of competency, and training programmes.

Safety Management Services

http://www.cyberg8t.com/~jprice/

This consulting company's site provides access to its Online Safety Program, including an Injury and Illness Prevention Program, Hazardous Communications Program, MSDS and links to other sites.

Ultimate Lifestyle Fitness

http://www.interlynx.net/ultimate/

This Hamilton, Canada-based company designs and delivers workplace wellness programs. Read about its services.

Databases

OHS in the European Union

http://www.occuphealth.fi/e/eu/index.phtml

This site, maintained by the Finnish Institute of Occupational Health, provides many useful resources and links, including the European Health and Safety Database (HASTE).

Projet Regetox—Programme "Protection du travailleur en matière de santé"

http://www.regetox.med.ulg.ac.be/

The object of the study is to develop and install a server equipped with a physico-chemical and toxicological database and interactive risk evaluation software.

RILOSH database

http://www.library.ryerson.ca/molndx

Search this database of over 140,000 records related to labour relations and occupational health and safety. Created by the Ontario Ministry of Labour, it is now maintained by Ryerson Polytechnic University, Toronto (formerly MOLINDEX).

Directories

RiskList

http://pages.prodigy.com/KY/rlowther/risklist.html

This is a directory of Internet resources for risk managers. Topics include risk resources, risk financing, disaster recovery/contingency planning, sources of legal and regulatory risk, breaking risk issues, recommended reading, etc.

Educational Resources / Training

AcuTech Consulting Inc.

http://www.acutech-consulting.com/

Provider of process risk management services to industries handling hazardous materials. Describes the company's consulting services, training,and risk management software.

Applied Risk Management Institute

http://www.erols.com/armi/

ARMI specializes in "helping organizations survive the threat of citations, lawsuits, and worse by teaching their existing staff how to handle wider range of employment-related risks." The site provides a catalog of the institute'scourses.

Behavioral Science Technology, Inc.

http://www.bscitech.com/

Read about the company's services related to behavior-based safety: seminars and research.

Controlled Risk International

http://www5.electriciti.com/risk/

Read about the company's risk management services, courses, OSHA and CAL/OSHA's most frequently cited serious violations.

Plant and Machinery Assessing Services

http://www.cyberstrategies.net/pamas/

This Australian consultancy develops safety programmes for the building industry. Services include preparation of safety plans, risk assessment for personal protective equipment, certificates of competency, and training programmes.

University of Wisconsin-Milwaukee, Environmental Health Safety and Risk Management

http://www.uwm.edu/Dept/EHSRM/

Information about the university's health and safety policies and guidelines is provided, along with links to other organizations.

What's the Risk?

http://www.health.adelaide.edu.au/ComMed/videonet/video.html

This is a training package for hazardous substances risk assessment. Describes training materials produced by the University of Adelaide for WorkCover South Australia.

Work Environment Program, University of Massachusetts Lowell

http://www.uml.edu/Dept/WE

Read about the Work Environment Program (WEP), an academic and research graduate program whose main role is to educate scientists to study and evaluate workplace factors that affect the health of workers.

Policy / Procedures Documents

University of Wisconsin System Administration, Office of Safety and Loss Prevention

http://www.uwsa.edu/oslp/oslp.htm

The page contains links to environmental health and safety, workers compensation, and risk management. Manuals and policy documents are presented.

University of Wisconsin-Milwaukee, Environmental Health Safety and Risk Management

http://www.uwm.edu/Dept/EHSRM/

Information about the University's health and safety policies and guidelines is provided, along with links to other organizations.

Publications

AcuTech Consulting Inc.

http://www.acutech-consulting.com/

Provider of process risk management services to industries handling hazardous materials. Describes the company's consulting services, training, and risk management software.

Corporate Health and Safety, Managing Environmental Issues in the Workplace

http://www.boxelectronics.com/

View a description and ordering information for this industrial hygiene guide.

CR Human Resources Solutions

http://www.sirus.com/users/crhr/

Information is provided about this company's human resources publications and consulting services (U.S.A and Canada). Also "free stuff" available: human resources-related frequently asked questions (FAQ), respiratory protection and lockout/tagout checklists, etc.

Dealing with Workplace Violence: A Guide for Agency Planners

http://www.opm.gov/workplac/

This handbook, developed by the U.S. Office of Personnel Management and the Inter-agency Working Group on Violence in the Workplace, "is intended to assist those who are responsible for establishing workplace violence initiatives at their agencies."

Essentials of Risk Management

http://www.bus.orst.edu/faculty/nielson/rm/toc.htm

This is an online version of parts of the book, *Essentials of Risk Management*, by George L. Head and Stephen Horn, 1991.

Institute for Systems, Informatics and Safety

http://www.jrc.org/isis/index.asp

This Joint Research Centre of the European Commission is involved in the multi-disciplinary analysis of industrial, socio-technical and environmental systems; the innovative application of information and communication technologies, and the science and technology of safety management.Read about its research activities.

Pittsburgh and Spokane Research Ctrs., Mining Health and Safety Research

http://www.usbm.gov

This NIOSH site provides information about its health and safety research, worker health, worker safety, and disaster prevention. Search its database of publications and download MADSS (Mine Accident Decision Support System).

Risk: Health, Safety, and Environment

http://www.fplc.edu/tfield/profRisk.htm

Risk is the official journal of the Risk Assessment and Policy Association. View tables of contents of past and current issues, subscription information, etc.

RiskWorld

http://www.riskworld.com/

This content-rich site provides "news and views on risk management and risk assessment."

Rmis.com (Risk Management Insurance Safety)

http://www.rmis.com

This content-rich site provides access to over 8000 risk management-related resources: databases, publications, templates and checklists, software, etc. Some material is free and some is fee-based.

SAFTEK Information Services

http://www.saftek.com

The site houses a large collection of documents and files related to occupational health and safety and insurance and risk management. This is also the home of the SAFETYBOOKS mailing list.

Target Risk

http://pavlov.psyc.queensu.ca/target/

Target Risk: Dealing with the Danger of Death, Disease, and Damage in Everyday Decidions by Gerald J. S. Wilde. Full text of this risk management book is presented here.

Programs

CR Human Resources Solutions

http://www.sirus.com/users/crhr/

Information is provided about this company's human resources publications and consulting services (U.S.A and Canada). Also "free stuff" available: human resources-related frequently asked questions (FAQ), respiratory protection and lockout/tagout checklists, etc.

Nebraska Dept. of Administrative Services, Risk Management Health Benefit Program

http://www.das.state.ne.us/das_dorm/index.html

View information about this state government program.

Software

AcuTech Consulting Inc.

http://www.acutech-consulting.com/

Provider of process risk management services to industries handling hazardous materials. Site describes the company's consulting services, training and risk management software.

Aurora Data Systems

http://www.alaska.net/~aurora/

Read about the company's database applications: Chemical Information Management System, Equipment Maintenance System, Training Records Management System, etc.

Dakota Decision Support Software

http://www.dakotasoft.com

Read about and view sample screens from Dakota's EH&S auditing software.

Dawnbreaker Consultants

http://204.112.119.2:80/dawnbreaker/

Information about and downloadable evaluation copies of their software are available. There are several programs of interest, including WorkSafe Tools (Occupational Safety and Health Software) and EnviroSafe (Environmental Risk Assessment Software).

Dyadem International

http://www.dyadem.com

Read about and/or download trial versions of this company's "software tools for the Risk Industry."

Environment and Safety Data Exchange (ESDX)

http://www.esdx.org/esdhome.html

Find out about ESDX, an industry association organized to participate in setting standards for more cost-effective management information technology in environmental, health and safety (EHS) activities.

Pittsburgh and Spokane Research Ctrs., Mining Health and Safety Research

http://www.usbm.gov

This NIOSH site provides information about its health and safety research, worker health, worker safety and disaster prevention. Search its database of publications, and download MADSS (Mine Accident Decision Support System).

Reason—Decision Systems Inc.

http://www.rootcause.com/

View a description of the company's REASON software—software that "brings professional-level Root Cause Analysis capabilities for operations problem-solving to the Quality / Safety / Operations professional."

RISK—An OSH Risk Assessment and Control System

http://www.safetyassistant.com/RISK.html

"The RISK program provides a step-by-step approach to risk assessment. Each hazard is assessed individually, providing you with a quantitative value and an action plan for each. Then an action plan is developed for the assessment as a whole. The final printed report becomes your 'worksheet'."

RiskSafe 98

http://www.lican.com/lrs/risksafe_home.html

This site describes and provides access to a downloadable demo of RiskSafe 98, a software package to help you "identify, prioritize, and decide where to spend your resources to manage identified risks." The software is produced and provided by Liberty Risk Services.

Rmis.com (Risk Management Insurance Safety)

http://www.rmis.com

This content-rich site provides access to over 8000 risk management-related resources: databases, publications, templates and checklists, software, etc. Some material is free and some is fee-based.

Star Solutions

http://www.safestar.com

Read about "Safe Star," a software package that manages employee safety information, accident reporting, etc.

WIZARD

http://www.paho.org/english/hepwizar.htm

Read about WIZARD (Workplace Health Information System for Surveillance and Risk Detection,) public domain software for IBM-compatible PCs that helps organize workplace activities for injury and illness surveillance, disability management, and health and safety followup.

---

20. Workers' Compensation, Rehabilitation and Disability

Consulting Services

Amolins and Associates Inc.

http://www.amolins.com

Read about the consulting company's workers compensation auditing services.

Assure Health Management

http://www.assure.ca

This company designs and delivers occupational health and disability management programs to Canadian business and industry. Read about the company and its services.

McCuaig Russell Barristers

http://www.mccuaigrussel.on.ca

This Ottawa, Canada law firm specializes in workers compensation cases, representing management. Read about the firm's services.

Databases

GLADNET

http://www.gladnet.org

GLADNET, the Global Applied Disability Research and Information Network is an initiative of the Vocational Rehabilitation Branch of the International Labour Organization (ILO). As well as having information about the organization itself, the site provides access to the GLADNET Infobase—full text documents and bibliograhic references concerning employment and disability.

Directories

At Work with Julie—Workers' Compensation questions and answers

http://www.abag.ca.gov/govnet/julie/julie.html

This Ann Landers-style site provides workers' compensation information in a question and answer format. The site, and therefore presumably the answers, are California-based. Julie (Carroll) is directory of the Association of Bay Area Governments' Workers Compensation Admin. Program.

Educational Resources / Training

Canadian Council on Rehabilitation and Work

http://www.ccrw.org/

Information includes an employment service for people with disabilities, job accommodation information, descriptions of the council's products and training programs.

Policy / Procedures Documents

Americans with Disabilities Act Document Center

http://janweb.icdi.wvu.edu/kinder/

This site contains copies of the Americans with Disabilities Act (ADA) of 1990, ADA regulations and technical assistance manuals.

Cornell University, School of Industrial and Labor Relations

http://www.ilr.cornell.edu/

The purpose of this site is to provide information about the School of Industrial and Labor Relations at Cornell University and its library, and to disseminate information on all aspects of employer-employee relations and workplace issues.

University of Wisconsin System Administration, Office of Safety and Loss Prevention

http://www.uwsa.edu/oslp/oslp.htm

The page contains links to Environmental Health and Safety, Workers Compensation and Risk Management. Manuals and policy documents are presented.

Publications

Ability Site

http://www.ability.org.uk

This U.K.-based site is presented by and for people with disabilities. It provides access to information about disability resources.

Cornell University, School of Industrial and Labor Relations

http://www.ilr.cornell.edu/

The purpose of this site is to provide information about the School of Industrial and Labor Relations at Cornell University and its library, and to disseminate information on all aspects of employer-employee relations and workplace issues.

Workers Compensation Board of British Columbia

http://www.wcb.bc.ca/

This site provides access to: information about the WCB itself, answers a series of frequently asked questions (FAQ), information about what's new (initiatives, legislation, etc.), and online publications.

Workplace Safety and Insurance Appeals Tribunal (Ontario)

http://www.wsiat.on.ca

The tribunal is the final level of appeal to which workers and employers may bring disputes concerning workers' compensation matters in Ontario. The site provides a description of the appeal process, a list of recent decisions, research publications, etc. (Formerly the Workers' Compensation Appeals Tribunal)

Programs

Assure Health Management

http://www.assure.ca

This company designs and delivers occupational health and disability management programs to Canadian business and industry. Read about the company and its services.

California Department of Industrial Relations (DIR)

http://www.dir.ca.gov/

This site contains information about workers compensation, occupational safety and health, California labor law, etc.

Software

First Report—Workers' Compensation and OSHA Recordkeeping Software

http://www.firstreport.com/

Really Useful Software provides information about this software package. They also provide "commentary and information" about workers' compensation and safety issues.

WIZARD

http://www.paho.org/english/hepwizar.htm

Read about WIZARD (Workplace Health Information System for Surveillance and Risk Detection) public domain software for IBM-compatible PCs that helps organize workplace activities for injury and illness surveillance, disability management, and health and safety followup.

Section C

Web Sites of General Interest

1. Search Engines and Directories

AltaVista

http://www.altavista.com

This search engine provides keyword access to the text of millions of Web pages, plus the text of messages from Usenet Newsgroups.

BigFoot

http://www.bigfoot.com

Look up people's email addresses at this site.

Excite

http://www.excite.com

This search engine provides keyword access to Web pages, plus a general directory of sites with content in a number of general subject areas: Business and Investing, Careers and Education, Computers and Internet, Entertainment, Games, Health, Lifestyle, etc.

HotBot

http://www.hotbot.com

This search engine allows you to search the content of millions of Web pages using a variety of criteria. HotBot claims to refresh its entire database every few days. You can also search on the content of Usenet Newsgroups and select news sites (CNN, various newspapers, etc.)

InfoSeek

http://www.infoseek.com

InfoSeek provides keyword access to the content of Web pages, Usenet Newsgroups and selected news services. A number of other services are provided, including a United Parcel Service (UPS) tracking service.

InReference, Inc.

http://www.reference.com

This site archives messages sent to thousands of Usenet Newsgroups and mailing lists.

Internet Address Finder

http://www.iaf.net

As the name implies, this site's main purpose is to allow you to look up people's email addresses. There is also a "reverse lookup" feature that allows you to enter an email address and find out the name of the addressee.

Lycos

http://www.lycos.com

This site describes itself as "Your Personal Internet Guide." It provides a variety of services: a Web search engine, directories of sites in a variety of general categories, yellow pages, a people finder, road maps, etc.

MetaCrawler

http://www.go2net.com/search.html

This is a "meta search" engine. Type in a word or phrase and MetaCrawler will search for your term on AltaVista, Excite, InfoSeek, Lycos, Thunderstone, WebCrawler, and Yahoo!

Northern Light

http://www.northernlight.com

This search engine provides search results based on the content of Web pages and a large collection of full-text documents available for sale. Search results include dynamically generated "custom search folders"—subsets of the information available as a result of your search.

OpenText

http://index.opentext.com

This site allows keyword search access to Web sites, plus provides access to Usenet Newsgroups, news, and other services.

WebCrawler

http://www.webcrawler.com

This search engine provides Web search capabilities, plus a directory of general subject areas: arts, business, chat, computers, education, entertainment, games, health, Internet.

Yahoo!

http://www.yahoo.com

This is probably the best known general purpose Web site directory on the Internet. Although it provides keyword search capabilities, its strength is as a browsing tool. It is an excellent place to start researching a subject.

Yahoo! People Search (formerly Four11)

http://people.yahoo.com

Allows you to search for individuals' email addresses and for phone numbers in the U.S..

---

2. Reference Information

AltaVista Translations

http://babelfish.altavista.com/cgi-bin/translate?

Translates either text that you type or the content of a Web page that you point to. Translations to and from English, French, German, Portuguese, and Spanish are available.

Area Code Lookup

http://www.555-1212.com/aclookup.html

Look up the area code for any city in the U.S. or Canada, or see what city/state/province a particular area code applies to.

Canada Post, Postal Code Lookup

http://www.mailposte.ca/CPC2/addrm/pclookup/pclookup.html?

Look up the postal code for any address in Canada or find out what streets, cities, etc. are associated with a particular postal code.

Citing Electronic Sources

http://owl.english.purdue.edu/Files/110.html

If you use the Internet and/or other electronic media for research purposes, this site shows you how to cite that information in a bibliography.

Currency Converter

http://www.oanda.com/converter/classic

This site provides current and historical currency conversion for 164 different currencies.

International Calling Codes

http://www.the-acr.com/codes/cntrycd.htm

This site, provided by American Computer Resources Inc., lists telephone country codes for all countries.

Local Times around the World

http://www.hilink.com.au/times/

This site provides current date and time information for all parts of the world.

NCSA Beginner's Guide to HTML

http://www.ncsa.uiuc.edu/General/Internet/WWW/HTMLPrimer.html

As the name implies, this site provides basic information for using Hypertext Markup Language (HTML) to create Web pages.

OneLook Dictionaries

http://www.onelook.com/

Type a word in the fill-in form. Your word will be searched for in over 150 online dictionaries and glossaries.

SiteBuilder Network Authoring

http://www.microsoft.com/workshop/default.asp

This Microsoft site provides an author's guide and HTML reference manual. It begins with the basic elements of creating Web pages and goes on to cover more complex issues and technologies.

Travlang's Translating Dictionaries

http://dictionaries.travlang.com/

This site provides access to a number of online translating dictionaries: English to French, German, Spanish, Dutch, Portuguese, Danish, Finnish, and a number of others.

U.S. Postal Service, ZIP Code Lookup

http://www.usps.gov/ncsc/

Look up the ZIP Code for a U.S. address or see what ZIP codes are associated with what cities and states.

Web Pages That Suck

http://www.webpagesthatsuck.com

This site provides an illustrated tour of badly designed Web sites. Learn what not to do when designing your own site.

WhoWhere

http://www.whowhere.lycos.com/

This directory allows you to look up individuals' email addresses, search for U.S. postal addresses, and phone numbers.

Yale C/AIM Web Style Guide

http://info.med.yale.edu/caim/manual/index.html

This is a manual to help you design useful and usable Web sites. To quote from the introduction, "The advice here is aimed at the practical concerns of bending and adapting a relatively primitive authoring and layout tool(HTML) to purposes it was never really intended to serve (graphic page design)."

---

3. Sources of Computer Software

Info-Mac HyperArchive

http://hyperarchive.lcs.mit.edu/HyperArchive.html

This site contains a wide variety of Macintosh freeware and shareware.

Tucows

http://www.tucows.com

This is an archive of shareware and freeware for use with the Internet, including Web browsers, email software, news readers, utilities of various kinds, add-on programs, etc. Software for Windows 3.1, Windows 95/NT, and Apple Macintosh is available.

---

4. Internet Culture

MediaCulture Review

http://www.mediademocracy.org/MediaCultureReview/

This is a compendium of the best features, commentary, and criticism from the alternative press and elsewhere on media, technology, and culture.

Red Rock Eater News Service

http://commons.somewhere.com/rre/index.html

The RRE News Service is a mailing list, most of the messages of which concern the social and political aspects of computing and networking. Various "articles" from the list are available, as well as a searchable archive.

Scout Toolkit

http://wwwscout.cs.wisc.edu/scout/toolkit/

This site provides a number of useful tools and discussions related to the Internet in general. There are links to quick reference guides for searching, Web tools, Internet publications, etc.

_____

—————————Section D—————————

Geographic Listing of Web Sites
(*excluding U.S. sites*)

Argentina

Boletin Argentino de Ergonomia

http://www.geocities.com/CapeCanaveral/6616/

News about Argentine ergonomists and their laboratories. (Spanish)

Australia

1997 Australian Vice—Chancellors' Committee (AVCC) OH&S Conference

http://www.adelaide.edu.au/HR/OH&S/avccohs/conference.htm

Read about and download the proceedings from this conference.

Adaptive Model of Thermal Comfort

http://atmos.es.mq.edu.au/~rdedear/ashrae_rp884_home.html

Macquarie University's ASHRAE RP-884 Project. This is a world database of thermal comfort field experiments.

Australian and New Zealand Society of Occupational Medicine

http://www.anzsom.org.au/

"ANZSOM is a professional and social organisation which provides a focal point for the advancement of knowledge in those registered medical practitioners who are actively involved in or are interested in Occupational Medicine."

Australian Institute of Health and Welfare, National Injury Surveillance Unit

http://www.nisu.flinders.edu.au/welcome.html

The National Injury Surveillance Unit (NISU) of the Australian Institute of Health and Welfare provides access to epidemiological and statistical studies conducted by the agency.

Australian Institute of Occupational Hygienists (AIOH)

http://www.curtin.edu.au/org/aioh/

This is the official home page of the AIOH. It contains information about the institute's publications, membership, office holders, conferences, etc.

CareFlight

http://www.usyd.edu.au/su/radiology/caref/index.htm

Based in and servicing Sydney, Australia, the role of CareFlight is to extend a teaching hospital standard of medicine to rescue and accident sites, as well as to smaller hospitals, in cooperation with emergency authorities.

Cyberg 1996

http://www.curtin.edu.au/conference/cyberg/

Proceedings of the first Ergonomics conference in CyberSpace, presented by the Ergonomics Society of Australia and the International Ergonomics Association. Preliminary information about Cyberg 1999 is also available.

Emergency Management Australia

http://www.ema.gov.au/

EMA's mission is to promote and support comprehensive, integrated, and effective emergency management in Australia and its region of interest. Information includes publications and manuals, training information, and links to related Australian sites.

Ergonomics in Australia

http://www.ergonomics.com.au/index.html

"On this page you will find interesting free information about ergonomics, safety and risk management. Resources for your ergonomics projects, services, and links to extraordinarily helpful ergonomics all over the world." Information is also provided about the services of this ergonomics consultancy.

Firenet Information Network

http://www.csu.edu.au/firenet/

The information at this site "concerns all aspects of fire science and management—including fire behaviour, fire weather, fire prevention, mitigation and suppression, plant and animal responses to fire and all aspects of fire effects."

Griffith University, Faculty of Health and Behavioural Sciences

http://www.ua.gu.edu.au/hbk/hbs_ind.htm

Information about courses provided by this Queensland Australia University is available. The School of Occupational Health and Safety is within this faculty.

Griffith University, Faculty of Environmental Sciences

http://www.ens.gu.edu.au

View information about the Faculty itself, plus links to other Australian and international environmental sites.

HAZCON Pty. Ltd.

http://www.dcscomp.com.au/hazcon/

Information about this Victoria, Australia-based Environmental and Occupational Health and Safety Consulting company and its services.

Human Heat Balance

http://atmos.es.mq.edu.au/~rdedear/pmv/

This JavaScript program allows you to enter both environmental and personal parameters to calculate human reaction to heat. Results generated include skin temperature, metabolic rate, core temperature, total evaporative heat loss at skin surface, etc.

National Occupational Health and Safety Commission (WorkSafe Australia)

http://www.worksafe.gov.au

Read about this national OH&S agency, view statistical information, Australian regulatory requirements, fact sheets, and alerts.

OSHNET—Occupational Safety and Health Network of Western Australia

http://sage.wt.com.au/safetyline/oshnet/osh_ind.htm

"Western Australian occupational safety and health service and product providers, who have a common interest in export markets, have formed a network called 'OSHNET'."

Outdoor Workers Exposed to UV Radiation and Seasonal Heat

http://www.usyd.edu.au/su/ohs/outdoor.html

This is the full text of the University of Sydney (Australia) policy document.

Performance Safety Home Page

http://www.uow.edu.au/crearts/theater/safehome.html

The Performance Safety Home Page is an expanding site dedicated to peforming arts safety. It includes a range of information including Australian OH&S legislation, fire safety, and hazardous materials.

Plant and Machinery Assessing Services

http://www.cyberstrategies.net/pamas/

This Australian consultancy develops safety programmes for the building industry. Services include preparation of safety plans, risk assessment for personal protective equipment, certificates of competency, and training programmes.

Prevention and Control of Legionnaires' Disease

http://www.wt.com.au/safetyline/codes/legion.htm

This detailed code of practice is presented in its entirety by WorkSafe Western Australia.

Safety Assistant—Software for Workplace Health and Safety

http://www.safetyassistant.com

View descriptions and sample screens from this Australian company's various OSH software packages.

Safety Network

http://www.safetynews.com/index.html

This Australian site describes itself as "the online hub for people interested in Occupational Health and Safety issues." It provides access to the online edition of *Australian Safety News*, the National Safety Council of Australia, and other safety resources.

SafetyCare Group of Companies

http://www.safetycare.com.au

SafetyCare is a worldwide supplier of occupational health and safety video programs, courses and manuals. View their online video catalog and their list of offices in various countries.

SafetyLine

http://www.safetyline.wa.gov.au

SafetyLine is an online information service provided by WorkSafe Western Australia. This is a very content-rich site, with safety and health solutions in a variety of subject areas, training materials, Australian federal and state legislation, etc. More detail about this site is available in the Top 50 Sites section.

SafetyLine Institute

http://www.safetyline.wa.gov.au/institute/

"The SafetyLine Institute is an online education and training facility established by WorkSafe Western Australia to provide high quality education and training materials in occupational safety and health." Available courses include OSH Management, OSH Information Sources, Safe Systems of Work, and OSH Promotion.

University of Adelaide, Occupational Health and Safety Unit

http://www.adelaide.edu.au/HR/OH&S/

Detailed information is provided about the unit. Read the university's guidelines and hazard alerts.

University of New South Wales, Department of Safety Science

http://argus.appsci.unsw.edu.au/

Read about the department's courses and research.

University of Sydney, Occupational Health and Safety

http://www.usyd.edu.au/su/planning/ohs/index.html

View the university's health and safety policy and staff guide, information about the Risk Management Office and accident investigation. A series of policy documents are also presented: asbestos safety, children on campus, confined spaces, manual handling, and disposal of sharps.

University of Technology, Sydney—Environment Health and Safety

http://www.hru.uts.edu.au/ehs/index.htm

View the university's policies and procedures for finding and fixing hazards, making your workplace safe, accidents reacting and reporting, disposal of hazardous waste, and emergencies.

Violence Network

http://www.space.net.au/~talktome/

This site contains a series of documents, news releases, articles, and links related to workplace violence

---

What's the Risk?

http://www.health.adelaide.edu.au/ComMed/videonet/video.html

This is a training package for hazardous substances risk assessment. Describes training materials produced by the University of Adelaide for WorkCover South Australia.

Austria

International Radiation Protection Agency

http://www.irpa.at

Information is provided about this international organization, its congresses, membership, and newsletter, plus links to associate societies.

University of Vienna (Austria), Department of Occupational Medicine

http://www.univie.ac.at/Innere-Med-4/Arbeitsmedizin/

Information is provided about the department, its research, and personnel. Most information is in German. Summary information is in English.

Belgium

Belgian Biosafety Server

http://biosafety.ihe.be

This server, located at the Institute of Hygiene and Epidemiology (IHE) in Brussels provides detailed information about biosafety, particularly in the European Union.

Comité National d'Action pour la Sécurité et l'Hygiène dans la Construction

http://www.cobonet.be/navb-cnac.be/

National Action Committee for Safety and Health in Construction. Read about its services, publications, and training.

European Biological Safety Association

http://biosafety.ihe.be/EBSA/HomeEBSA.html

EBSA was formed in 1996 "to promote biosafety as a scientific discipline and serve the growing needs of the biosafety professionals throughout Europe."

IDEWE Occupational Health and Safety Services on the Internet

http://www.idewe.be/

IDEWE is a nonprofit occupational health service in Flanders, Belgium. Read about the organization's services.

Institut pour la Prévention, la Protection et le Bienêtre au Travail (PREVENT)

http://www.prevent.be

Read about the work of this Belgian health and safety agency. Most information is presented in French and Dutch.

MediBel-Net—the Belgian Information Kiosk for Health and Environment

http://www.health.fgov.be/

View information about MediBel-Net, the responsibilities and work of the Belgian Ministry of Social Affairs, Health and Environment. Most information is in French and Dutch.

OMCO Air Treatment

http://www.omco.be/

Read about this Belgian company's ventilation products. Information is in French and Dutch.

Projet Regetox—Programme "Protection du travailleur en matière de santé"

http://www.regetox.med.ulg.ac.be/

The object of the study is to develop and install a server equipped with a physico-chemical and toxicological database and interactive risk evaluation software.

Université catholique de Louvan, Unité de Toxicologie et de Medecine du Travail

http://www.md.ucl.ac.be/entites/esp/toxi/

The site contains a description of the unit, its members, educational programs, and a list of scientific publications.

Brazil

RACCO Safety

http://www.raccosafety.com.br/

Describes this Brazilian company's safety, industrial hygiene, and occupational health products and services. Most information is in Portuguese, with a general description in English.

Canada

4th ICOH International Conference on Occupational Health for Health Care Workers

http://www.santepub-mtl.qc.ca/icoh1999/icoh1999.html

Read about this conference in Montreal Canada, September 28 to October 1, 1999.

ABSA Canada

http://www.nanook.com/absa_canada/

ABSA Canada is a Canadian affiliate of the American Biological Safety Association. Its objective is to establish a Canadian network of individuals interested in a wide variety of biological safety issues.

Alberta Construction Safety Association

http://www.compusmart.ab.ca/acsa/

The site provides information about this industry association, its training courses, a description of resources available to members, products, etc.

Alberta Health

http://www.health.gov.ab.ca

The site provides information about this provincial government department. It includes public health information and details about emergency services.

Alberta Labour

http://www.gov.ab.ca/lab/index1.html

This Alberta government department's site provides access to the full text of acts and regulations, accident and injury statistics, and publications.

Alberta Safety Net

http://abirc.com/safetynet/home.htm

This site is intended to be a virtual mall for safety related information, providing information about safety consultants and training, safety equipment suppliers.

Amolins and Associates Inc.

http://www.amolins.com

Read about the consulting company's workers compensation auditing services.

Asbestos Institute

http://www.asbestos-institute.ca

Access the Montreal-based Institute's Biomedical Data Bank, view reports and press releases, and get information on their printed publications.

Association for Canadian Registered Safety Professionals (ACRSP)

http://www.acrsp.ca/

ACRSP's site provides information about association membership, professional designation, and the association's code of ethics.

Association of Ontario Health Centres

http://www.aohc.org/

The Association of Ontario Health Centres (AOHC) is the nonprofit organization that represents community health centers (CHCs) and some health service organizations (HSOs) in the province of Ontario, Canada. The site contains information about AOHC, its position papers, and health promotion programs.

Association of Public Safety Communications Officials (APCO)

http://www.apco.ca/

Information about this nonprofit association is provided: membership information, conferences, training. This is the Web site of the Canadian chapter of APCO International.

Association paritaire pour la santé et la sécurité du travail, secteur affaires municipales

http://www.apsam.com/

Contains a description of this Quebec health and safety agency, its mission, and services. Bulletins, data sheets, and guidelines are also available. All information is in French.

Association québécoise pour l'hygiène, la santé et la sécurité au travail (AQHSST)

http://www.aqhsst.qc.ca/

Information about the association, membership, its annual conference. The site also provides links to other sites and information about a forthcoming mailing list. All information on the site is in French.

Assure Health Management

http://www.assure.ca

This company designs and delivers occupational health and disability management programs to Canadian business and industry. Read about the company and its services.

BLMC Computer Based Training Programs

http://www.blmc.com/

Information is provided about this Canadian company's computer based health and safety training programs.

Campus Safety Health and Environmental Management Association (CSHEMA)

http://www.ualberta.ca/~rrichard/cshema.html

The site includes information on the CSHEMA organization, its newsletter, training and policy information. More detail about this site is available in the Top 50 Sites section.

Canada Safety Council (CSC)

http://www.safety-council.org

CSC is an independent, national, nonprofit membership safety organization. The Web site offers information about the organization, its training programs, and publications.

Canadian Auto Workers

http://www.caw.ca/caw/

Information about the union, its work, and studies undertaken. Its health, safety and environment newsletter and health and safety fact sheets are also available.

Canadian Bioethics Report

http://www.cma.ca/cbr/

This quarterly electronic journal of the Canadian Medical Association (CMA) Department of Ethics and Legal Affairs is complete with some full papers and abstracts of others.

Canadian Centre for Occupational Health and Safety (CCOHS)

http://www.ccohs.ca

CCOHS's site provides access to Canadian environmental and OSH legislation, plus major OSH databases (*e.g.*, NIOSHTIC, RTECS, HSELine) on a subscription basis. Other content includes information and hazard alerts and a large directory of health and safety Internet resources. More detail about this site is available in the Top 50 Sites section.

Canadian Chemical Producers' Association (CCPA)

http://www.ccpa.ca

Information about CCPA and its responsible care program is provided.

Canadian Council on Rehabilitation and Work

http://www.ccrw.org/

Information includes an employment service for people with disabilities, job accommodation information, descriptions of the council's products and training programs, etc.

Canadian Environmental Law Association

http://www.web.net/cela/

The Canadian Environmental Law Association (CELA) is a nonprofit, public interest organization established in 1970 to use existing laws to protect the environment and to advocate environmental law reforms. Provides information about CELA, a list of publications, subscription information for its newsletter, and descriptions of current cases and causes.

Canadian General Standards Board

http://www.pwgsc.gc.ca/cgsb/

View CGSB's online catalogue and read about the standards development process.

Canadian Health Network

http://www.canadian-health-network.ca/

CHN is currently in the developmental stages of a national health information service. The information contained within the framework of the CHN will draw from Canadian federal, provincial, and territorial governments, nongovernment organizations, universities, community-based and private sector organizations.

Canadian Injured Workers Alliance

http://www.ciwa.ca/

"The Canadian Injured Workers Alliance is a national organization, supporting and strengthening local and provincial IWG's across Canada. We provide resources, training and a forum for information exchange for these groups, as well as for trade unions."

Canadian Institute of Stress

http://www.stresscanada.org/

CIS provides stress and risk management services for businesses and individuals. Services include workplace change management services, workplace stress control programs, clinical assessment, counselling and coaching, and speakers for seminars.

Canadian Medical Association, Occupational Medicine, Clinical Practice Guidelines

http://www.cma.ca/cpgs/occup.htm

Links are provided to a series of documents describing specific guidelines.

Canadian Network of Toxicology Centres (CNTC)

http://www.uoguelph.ca/cntc/

Information is available about the organization, online newsletters and documents, plus links to other toxicology-related sites.

Canadian Reflection

http://www.reflection.gc.ca

This site describes a study launched by the Federal Labour Minister to examine how the Canadian workplace is changing.

Canadian Registration Board of Occupational Hygienists

http://www.crboh.ca/

The CRBOH is a national, nonprofit organization that sets standards of professional competence for occupational hygienists and occupational hygiene technologists in Canada. The site provides information about ROH and ROHT certification, exams, and maintenance.

Canadian Society of Safety Engineering (CSSE)

http://www.csse.org

Information is provided about the society, its programs, publications, and courses.

Canadian Society of Safety Engineering, Edmonton Chapter

http://www.freenet.edmonton.ab.ca/csse/

The site provides information about this CSSE chapter, membership, and meetings.

Canadian Standards Association (CSA)

http://www.csa.ca

CSA's Web site includes information about its products and services, how to buy standards, information about standards development, and its newsletter *Harmonization News*.

CCINFOWeb

http://ccinfoweb.ccohs.ca

This is the home page for the commercial OHS database service on the Web, provided by the Canadian Centre for Occupational Health and Safety (CCOHS). Databases include MSDS, CHEMINFO, RTECS, NIOSHTIC, and HSELine, plus the full text of all Cdn. environmental and OSH legislation. Other databases will be added on an ongoing basis.

CCOHS's Health and Safety Internet Directory

http://www.ccohs.ca/resources/

This site has links to about 1500 resources—Web sites, mailing lists, and Usenet Newsgroups.

Centre de Toxicologie du Québec

http://www.ctq.qc.ca

The CTQ's mission is to provide quality services in the area of human toxicology, primarily for the population of Quebec. Read about the organization and its services.

Christie Communications

http://www.mrg.ab.ca/christie/

Safety training courses and a list of safety-related Internet resources.

Christie Communications—WHMIS Course

http://www.mrg.ab.ca/course/whindex.htm

This inexpensive Web-based course provides a general introduction to WHMIS (Workplace Hazardous Materials Information System).

City of Toronto

http://www.city.toronto.on.ca

Access public health documents and information about emergency services.

Civil Air Search and Rescue Association

http://www.casara.ca

Read about CASARA, "a Canada-wide volunteer aviation association dedicated to the promotion of Aviation Safety, and to the provision of air search support services to the National Search and Rescue Program."

Clinic of Injury and Disease Response

http://web.idirect.com/~cidr/

CIDR is a medical clinic specializing in the assessment, treatment, and research of work related musculoskeletal disorders (WMSD or RSI) and other soft tissue injuries. The site provides general RSI information and a bibliography of RSI-related materials.

Commission de la santé et de la sécurité du travail du Québec (CSST)
http://www.csst.qc.ca/

Describes the role and services of the commission and the laws it administers. Other information includes CSST news releases, summaries of their magazine *Prévention au travail* and a publications list. Most information is in French and some is also in English.

Confédération des syndicats nationaux, santé sécurité
http://www.csn.qc.ca/Pageshtml2/SSOutils.html

Confederation of National Unions, Health and Safety. The site contains descriptions of publications and other health and safety tools, plus an annotated directory of more than 250 Web sites. All information is in French.

Construction Safety Association of Ontario
http://www.csao.org/

As well as providing information about the association itself, the site provides a number of basic safety awareness documents. The full text of their magazine *Construction Safety* and their newsletter are also available.

Conveyerail Systems Ltd.
http://www.conveyerail.com/

View descriptive information about VacuMove, a vacuum lifting system.

Dawnbreaker Consultants
http://204.112.119.2:80/dawnbreaker/

Information about and downloadable evaluation copies of their software are available. There are several programs of interest, including WorkSafe Tools (Occupational Safety and Health Software) and EnviroSafe (Environmental Risk Assessment Software).

Disaster Recovery Information Exchange (DRIE)
http://www.utoronto.ca/security/drie.htm

Content on this Toronto, Canada-based organization's site includes membership information, disaster reference materials, and upcoming conferences.

Disaster Recovery Institute Canada
http://www.dri.ca

A nonprofit corporation established in 1996 to create a base of common disaster recovery planning knowledge through education, assistance, and development of a resource base; certify qualified individuals; and promote the credibility of certified professionals.

Domestic Substances List/ Non-Domestic Substances List (DSL/ NDSL)

http://www2.ec.gc.ca/cceb1/cas_e.html

This site provides access to the Canadian DSL/ NDSL. Searching is by CAS number only.

Education Safety Association of Ontario

http://www.esao.on.ca/

A nonprofit Ontario "Safe Workplace Association." Site has links to documents, facts and figures, information about the organization itself, and a health and safety discussion group for the educational sector.

Effective Prevention

http://www.interlog.com/~effprev/

Information is provided about Effective Prevention, a skin cream that "repels bodily fluids and many biological hazards in the medical and food services profession."

EHS Diagnostics Inc.

http://www.ehsdiagnostics.com

EHS Diagnostics Inc. (EDI) is a Canadian environment, health and safety consulting firm based in Calgary, Alberta, Canada, which provides both consulting services and health and safety software development. Read about the company's services.

Electrical Utilities Safety Association of Ontario

http://www.eusa.on.ca/

Read about this safe workplace association, its consulting services, training programs, and accident prevention solutions.

Emergency Preparedness Information Exchange (EPIX)

http://hoshi.cic.sfu.ca:80/~anderson/index.html

Content includes links and documents about various emergency topics, emergency related organizations, upcoming conferences, and alerts.

Emergency Services Registry and Search Site

http://www.district.north-van.bc.ca/admin/depart/fire/ffsearch/mainmenu.cfm

This is a directory of emergency services professionals.

Environment Canada

http://www.ec.gc.ca

Information about this Canadian federal government department, its initiatives, structure, and legislation.

Eric R. Rumack

http://www.total.net/~erumack/homepage.html

Read about the services provided by this Ontario, Canada-based occupational health physician and disability management consultant.

Farm Safety Association of Ontario

http://www.fsai.on.ca

The site provides information about the association itself, plus safety information for agriculture/ horticulture and landscapers.

First Aid 1st

http://www.seadrive.com/first.aid.1st/

This Alberta-based company provides first aid supplies and services to business, sports groups, nonprofit organizations, and individuals. View product descriptions, locate distributors.

Front-Line Safety

http://www.frontlinesafety.com/

This Nova Scotia-based company provides health and safety training programs, safety equipment rentals, consulting/safety audits. They also provide an online WHMIS course.

Halifax Work Hardening and Sports Injury Clinic

http://www3.ns.sympatico.ca/tom.stanley/index.htm

Read about the work of this Canadian rehabilitation centre.

Hazard Alert Training Supplies Canada Inc. (HATSCAN)

http://www.hatscan.com/

HATSCAN is an occupational health, safety and environment training and consulting firm, providing consulting and training in WHMIS, TDG, OH & S, due diligence, WCB, environmental compliance, hazardous waste, and more. Find out about the company's courses, publications, and computer assisted training.

HazardNet

http://hoshi.cic.sfu.ca/hazard/

HazardNet is a prototype natural and technological hazard information sharing network under development as a collaborative demonstration project of the International Decade for Natural Disaster Reduction (IDNDR).

Health Canada, Environmental Health Program

http://www.hc-sc.gc.ca/ehp/

The Environmental Health Program consists of four bureaus: Chemical Hazards, Product Safety, Tobacco Control, and Radiation Protection. The site provides access to numerous technical reports and publications and general interest fact sheets.

Health Canada, Health Information Network

http://www.hc-sc.gc.ca/

This Canadian federal government department presents its health alerts, public health documents, and news releases, plus links to its various directorates—Laboratory Centre for Disease Control, Environmental Health Directorate, etc. More detail about this site is available in the Top 50 Sites section.

Health Canada, Laboratory Centre for Disease Control

http://www.hc-sc.gc.ca/hpb/lcdc/

The site contains information about the centre itself, public health documents, travel health advisories, and disease prevention guidelines. More detail about this site is available in the Top 50 Sites section.

Health Canada, Laboratory Centre for Disease Control, Laboratory Biosafety Guidelines

http://hwcweb.hwc.ca/hpb/lcdc/bmb/biosafty/

This site provides the full text of LCDC's guidelines document. More detail about this site is available in the Top 50 Sites section.

Health Care Health and Safety Association

http://www.hchsa.on.ca

HCHSA is a designated Safe Workplace Association in Ontario, Canada. The site provides information about the association, its mandate, and news.

Health in Action

http://www.health-in-action.org/

Health in Action is the Alberta Information Clearinghouse on Prevention and Promotion. Find out about the organization's programs and access its databases.

Healthcare EnviroNet

http://healthcare-environet.com/

The Healthcare Enviro-Net is a network of environmental and healthcare organizations committed to the promotion of sustainable development in Canadian healthcare. Its goal is to provide healthcare professionals with easy access to reliable online environmental and health information that will enable them to make informed decisions.

HRP Consultants

http://www.hrpconsultants.com/

Read about this Ontario, Canada-based consulting company that specializes in "assisting organizations to add value to their environmental, safety and ISO functions by developing programs and systems that will enhance organization success."

Human Factors Association of Canada

http://www.hfac-ace.ca/

Read about the association itself, ergonomics conferences in Canada, programs and courses in ergonomics, and view a directory of consultants.

Human Resources Development Canada—Labour Program

http://labour-travail.hrdc-drhc.gc.ca/eng/

"The mission of the Labour Program of Human Resources Development Canada (HRDC) is to promote a fair, safe, healthy, stable, cooperative and productive work environment that contributes to the social and economic well-being of all Canadians."

Human Resources Development Canada, Labour Operations, Occupational Safety and Health

http://info.load-otea.hrdc-drhc.gc.ca/~oshweb/homeen.shtml

Information is provided to allow Canadian federally regulated organizations to meet the requirements under the Canada Labour Code Part II. There is an overview of the code and information about setting up an OSH program in your workplace. Information of general interest includes occupational injury statistics.

Industrial Accident Prevention Association (IAPA)

http://www.iapa.on.ca

Information is provided about this Ontario, Canada safety association, its products, services, and training programs.

Institute for Work and Health

http://www.iwh.on.ca

Information about this Toronto, Canada-based organization, its publications, programs, special events, and newsletter.

International Society of Indoor Air Quality And Climate

http://www.cyberus.ca/~dsw/toc_a.html

Read about the organization and its membership. Other information includes publications, upcoming conferences, and IAQ-related Internet links.

International Society of Indoor Air Quality and Climate (ISIAQ)

http://www.cyberus.ca/%7Edsw/index.html

ISIAQ "is an international, independent, multidisciplinary, scientific, nonprofit organization whose purpose is to support the creation of healthy, comfortable and productivity-encouraging indoor environments." Read about the organization, membership, publications, conferences, and symposia.

Job Accommodation Network Canada

http://janweb.icdi.wvu.edu/english/homecan.htm

JAN is an international consulting service that provides information about job accommodations and the employability of people with disabilities. The site provides information about the organization and its services and the job accommodation process.

Justice Canada

http://canada.justice.gc.ca/Loireg/index_en.html

This site provides access to the full text of many Canadian federal acts and regulations.

Justice Institute of British Columbia

http://www.jibc.bc.ca/

"The Justice Institute of British Columbia develops and delivers training in all areas of justice, public safety, and human services." Read about its various programs.

KnowledgeWare Communications Corp.

http://www.kccsoft.com/

This N. Vancouver, BC company provides several software packages: Simply Safety! Safety Management, WHMIS CBT Suite, and Trax and Fax MSDS Software. View information and download demonstrations of these software packages.

Lenco Enterprises

http://www.lara.on.ca/~lenco/

The site describes this Hamilton, Canada-based consulting company. Lenco gives seminars and training on electrical safety and lockout/tagout.

Major Industrial Accidents Council of Canada (MIACC)

http://hoshi.cic.sfu.ca/~miacc/

View information about the council, major industrial accidents, related Canadian legislation, meetings, and workshops.

Manitoba Prehospital Professions Association

http://www.escape.ca/~mppa/

Information about the association, upcoming events, and links to both government and EMS sites is provided.

Manitoba Workplace Safety and Health Division

http://www.gov.mb.ca/labour/safety/index.html

Information about the various branches of this provincial government division (Workplace Safety and Health, Mining Inspection and Occupational Health), plus access to the full text of a number of its WorkSafe publications. More detail about this site is available in the Top 50 Sites section.

Matcor Global Products

http://www.matcor.ca/

Information is provided about this company's sensing products, including a UV intensity meter.

Material Safety Data Sheets (MSDS) for Infectious Substances

http://www.hc-sc.gc.ca/hpb/lcdc/biosafty/msds/index.html

Health Canada, Health Protection Branch, Laboratory Centre for Disease Control, presents these MSDSs "for personnel working in the life sciences as quick safety reference material relating to infectious microorganisms." Data sheets are available in English and French.

Mathews, Dinsdale & Clark

http://www.mdclabourlaw.com/

This is a Toronto, Canada-based law practice "restricting its practice to labour and employment law." The site provides a number of interpretive documents about Ontario law with quizzes to test your knowledge.

McCuaig Russell Barristers

http://www.mccuaigrussel.on.ca

This Ottawa, Canada law firm specializes in workers compensation cases, representing management. Read about the firm's services.

McGill University, Environmental Safety Office

http://www.mcgill.ca/eso/

Provides information about staff and services provided by the office, links to McGill's Occupational Health Programme, and various McGill policy documents, including its Laboratory Safety Manual.

McGill University, Occupational Health Sciences

http://www.mcgill.ca/occh/

The site provides information about their academic programs, activities/news/seminars, research groups/labs, and publications.

McMaster Univ., Faculty of Engineering and Science, Safety Handbook

http://www.eng.mcmaster.ca/safety/

This is the full text of the Safety Handbook from this Hamilton Canada university.

McMaster Univ., Occupational and Environmental Health Laboratory

http://www-fhs.mcmaster.ca/oehl/index.html

This AIHA accredited laboratory in Hamilton, Canada provides occupational hygiene services to industry, labour and the general public. Read about the lab's research projects and services.

MediTrav Web

http://www.meditrav.com

The site for French-speaking occupational physicians contains numerous documents, information about training, legislation, and software.

Mines and Aggregates Safety and Health Association

http://www.masha.on.ca/

MASHA is a designated Safe Workplace Association in Ontario, Canada. The site provides information about the association, its publication catalogue, and a calendar of events.

MSDS2

http://www.whmis.com

Read about and "test drive" BC Hydro's material safety data sheet management system.

Municipal Health and Safety Association of Ontario

http://www.mhsao.com/

Read about the association, its mission, publications, and courses.

National Clearinghouse on Tobacco and Health (Canada)

http://www.cctc.ca/ncth/

This site provides access to documents about the health effects of tobacco smoke (smoking and second-hand smoke) and smoke-free environments. Links to related sites.

National HVAC

http://www.nationalhvac.com/

The content of the site describes this Canadian company's services, including: indoor environment consulting, building ventilation assessments, field investigations, air testing, and remediation management services.

National Institute of Disability Management and Research

http://www.nidmar.ca/

The Institute is a labour-management initiative established by by the Disabled Workers Foundation of Canada. Its mandate is to promote integration of disabled people into the workplace. Read about their training programs, publications, videos, and other services.

Nepean Fire Department

http://www.city.nepean.on.ca/firedept/

Read about this Ontario, Canada fire department, its news releases, and fire prevention information.

New Brunswick Workplace Health, Safety and Compensation Commission

http://www.gov.nb.ca/whscc/

This site provides information about the commission and its programs, calculating WCB claims in New Brunswick, and industry assessment rates.

New Substances Notification Regulations

http://www2.ec.gc.ca/cceb1/eng/nsnr_cp.htm

"Environment Canada and Health Canada have prepared this brochure to alert persons who manufacture or import chemical substances and polymers of their obligation to submit a notification package before they import or manufacture a substance that is considered new to Canada." The brochureis presented in its entirety.

Newfoundland Environment and Labour

http://www.gov.nf.ca/env/default.asp

Contains information about this provincial government department and press releases.

North American Detectors, Inc.

http://www.nadi.com/

A description of the company's carbon monoxide detector is provided. There is also a frequently-asked-questions (FAQ) document about carbon monoxide.

Northern Industrial Supply Company

http://www.nisco.net/

This Ontario, Canada company sells industrial heating, ventilation and air conditioning equipment. Read about the products they provide.

Nova Scotia Department of Labour

http://www.gov.ns.ca/labr/

Information is provided about the department, the minister, departmental publications.

Nova Scotia Environmental Industry Association

http://www.nseia.ns.ca/

Information is provided about the association, membership, newsletter, and codes of practice. Also, Nova Scotia environmental legislation, permit and licensing information.

Nova Scotia Workers Compensation Board

http://www.gov.ns.ca/bacs/acns/paal/ndxwork.htm

Information is provided about registration with the WCB.

Occupational Disease Panel Electronic Library

http://www.ccohs.ca/odp/

The Occupational Disease Panel (ODP) was created under the Workers' Compensation Act of Ontario. In its 10 years of operation, ending in 1997, it produced numerous publications on issues relating to workplace/occupational diseases and their compensation. These publications are available in their entirety on this site.

Occupational Hygiene Association of Ontario (OHAO)

http://www.ohao.org/

Read about this provincial association, membership, upcoming events, and the association's newsletter.

Office of the Employee Advisor, Ontario

http://www.gov.on.ca/lab/oea/

The Office of the Employer Adviser (OEA), an independent agency of the Ontario Ministry of Labour, is in business to provide employers with advisory services, education, and information on Workplace Safety Insurance. Read about OEA's services.

Office of the Fire Marshal (Ontario Canada)

http://www.gov.on.ca/OFM/

The site provides information about the office, public fire safety information, fire statistics, legislation (including the Ontario Fire Code), guidelines, and technical papers.

OH&S Consulting Services

http://www.hronline.com/dr/

Dilys Robertson is a Toronto-based occupational health and safety consultant and writer. Read about her consulting services, which include training design and delivery and research work with joint health and safety committees.

OHS Canada

http://www.ohscanada.com/

The site provides information about *OHS Canada* magazine and selected content from the current and previous issues.

Ontario Forestry Safe Workplace Association

http://www.ofswa.on.ca/

"This site showcases the many products and services provided to the Ontario forest industry by OFSWA."

Ontario Ministry of Labour

http://www.gov.on.ca/LAB/main.htm

The site contains information about the ministry, provincial government OSH discussion papers, and a list of ministry publications.

Ontario Service Safety Alliance

http://www.ossa.com

"The OSSA was developed to meet the needs of the service industry by providing education and training to assist in preventing injuries and illnesses." View information about the organization, its educational programmes, research, and studies.

Ontario Statutes and Regulations

http://204.191.119.105/

This site presents the full text of Ontario statutes and regulations. At the time of writing, it was current to January 1, 1998.

OSH Answers

http://www.ccohs.ca/oshanswers/

This is a collection of frequently asked health and safety questions, presented by the Canadian Centre for Occupational Health and Safety (CCOHS).

Pest Management Regulatory Agency (PMRA)

http://www.hc-sc.gc.ca/pmra-arla/

This agency, part of Health Canada, has as its mandate "to protect human health safety and the environment by minimizing risks associated with pesticides while enabling access to pest management tools, namely pest control products and pest management strategies."

Petroleum Industry Training Service

http://www.pits.ca/

Read about this Calgary, Alberta-based organization's training services for the petroleum industry.

Prairie Agricultural Machinery Institute, "Be Seen, Be Safe!"

http://www.pami.ca/beseennf/beseennf.htm

This site has been adapted from the guidebook of the same name, originally produced by the Prairie Agricultural Machinery Institute (PAMI) under the direction of Manitoba Highways and Transportation.

Preventive Action Safety Services Ltd.

http://www.autobahn.mb.ca/~passltd/

Read about this Winnipeg Canada based health and safety company and its services.

Pro Grip Safety Sole

http://www.uleth.ca/~gadd/index.htm

Information about Pro-Grip nonslip safety soles.

Production Video L. M. Inc.

http://www.generation.NET/~pvlm/

This commercial site provides detailed information about Production Video's occupational health and safety training videos.

Proformix

http://www.proformix.com

The company provides ergonomic keyboard platforms and related products, plus a suite of ergonomic management software.

Queen's University, Environmental Health and Safety

http://www.safety.queensu.ca:443/safety/safety.htm

The site provides access to a number of policy documents and guidelines, plus links to MSDS sites, and other sites of interest.

Real Stories—Community Health in Action

http://www.opc.on.ca/realstories/

The site contains a number of "authentic stories of real people who benefited by their involvement at their local Community Health Centre (CHC)" in Ontario.

Resource Environmental Associates

http://www.rea4ehs.com/

Information is provided about this Toronto, Canada-based environment, health and safety consulting company, its services, and publications.

Restoration Environmental Contractors

http://Home.InfoRamp.Net/~restcon/

This Canadian company provides services such as asbestos abatement, lead abatement, fire restoration, mold and fungus removal, and plant cleanups. The Web site provides documents outlining step-by-step actions to be taken for most of these types of cleanups/restorations.

RILOSH database

http://www.library.ryerson.ca/molndx

Search this database of over 140,000 records related to labour relations and occupational health and safety. Created by the Ontario Ministry of Labour, it is now maintained by Ryerson Polytechnic University, Toronto (formerly MOLINDEX).

Safe Communities Foundation

http://www.safecommunities.ca

"The Safe Communities Foundation is a unique partnership between the private and public sectors that is dedicated to making Canada the safest place to work, live and play in the world." Includes safety tips, lists of safe communities, and the foundation's newsletter.

Safe Guard Network

http://hoshi.cic.sfu.ca/safeguard/

SAFE GUARD is a national public recognition program aimed at increasing public awareness of emergency preparedness in Canada. There is information about Safe Guard itself, its publications, and public service announcements.

Safe-tec Canada

http://www.safetec.com

This is an online catalog of the company's infection control and personal safety products.

Safety Check Consulting

http://www.lights.com/safetycheck/

Information about this Saskatchewan-based safety consulting company's services: safety audits, training programs, emergency preparedness, accident/incident investigation.

Safety Smart! Magazine

http://safetysmart.com

The site provides information about this magazine (selected articles available) and the company's other safety publications, posters, and clip art.

Safety Superstore

http://www.safetysuperstore.com

The site contains this company's catalog of safety products. This Canadian company will ship anywhere in the world.

Safety Training and Consulting Services

http://www.stacs.com

Provides information about this Canadian consulting company. Its services include training, consulting, and recruiting.

SAREC

http://www.sarec.ca

Read about the expertise and services of this occupational hygiene consulting company.

SAREC Bookstore

http://www.sarec.ca/books/

SAREC catalogues occupational health and safety books and provides online ordering capabilities in association with Amazon.com.

Saskatchewan Department of Labour

http://www.gov.sk.ca/govt/labour/

Describes the department's mandate and provides information about the Minister.

Shiftwork like Clockwork

http://shiftwork.sdhu.com/

Read about this workplace wellness program, produced by the Sudbury (Canada) District Health Unit. The program is made available on a cost-recovery basis.

Simply Safety Store

http://www.island.net/~dnagle/

Offers training, information, and products for safety in the workplace and at home. Read about the company's offerings.

Southern Alberta Occupational Medicine

http://www.med.ucalgary.ca/oemweb/

This Web site provides a variety of learning resources in occupational medicine to assist undergraduate students, medical residents, and practicing physicians in Alberta.

Standards Council of Canada

http://www.scc.ca

Access the database of Canadian standards, find out about ISO 9000 and ISO 14000, read a concept paper on occupational health and safety systems standards, view the full contents of *Consensus, Canada's News Magazine of Standardization*.

Statutes and Regulations of Ontario

http://204.191.119.105/

Provides access to the full text of acts and regulations of the government of Ontario.

Strategis

http://strategis.ic.gc.ca

This Industry Canada site provides information about many aspects of Canadian industry: industry profiles, technology, etc.

Sulowski Fall Protection Inc.

http://ourworld.compuserve.com/homepages/AndrewSulowski/sulowski.htm

This consulting company's site provides information about its services, fall protection courses, publications, and upcoming events.

Surety Manufacturing & Testing Ltd.

http://www.Suretyman.com/

This Edmonton-based company manufactures and sells equipment for fall protection and confined space entry. View the company's catalogue and read about training videos.

Surviving the Workplace Jungle

http://vvv.cqm/m2/swj/

The site presents detailed information about this workplace violence video and workbook.

Target Risk

http://pavlov.psyc.queensu.ca/target/

Target Risk: Dealing with the Danger of Death, Disease, and Damage in Everyday Decisions by Gerald J. S. Wilde. The full text of this risk management book is presented here.

Technical Standards and Safety Authority

http://www.tssa.org

TSSA "is an independent, nongovernment, not-for-profit organization mandated to deliver specific public safety programs and services under Ontario's Safety and Consumer StatutesAdministration Act." Site provides various publications, policies, and guidelines.

Transport Canada

http://www.tc.gc.ca

The site contains current transportation of dangerous goods (TDG) regulations, draft regulations, and other transportation safety information—aviation safety, marine, railway, and road.

Transportation Safety Association of Ontario

http://www.tsao.org

This safe workplace association's site provides access to publications, information about training programmes, and the association's services.

Transportation Safety Board of Canada

http://bst-tsb.gc.ca/

This site contains the full text of TSB's latest occurrence reports, occurrence statistics, significant safety issues, and information about the board itself.

Ultimate Lifestyle Fitness

http://www.interlynx.net/ultimate/

This Hamilton Canada-based company designs and delivers workplace wellness programs. Read about its services.

University and College Emergency Planning Mailing List

http://www.uwo.ca/emerg/list.html

This page describes the email-based discussion group, emerg-univ.

University of Alberta, Medical Laboratory Science Program

http://www.ualberta.ca/~medlabsc/

Read about their undergraduate and graduate programs and Internet resources available to students.

University of British Columbia, Occupational Hygiene Program

http://www.interchg.ubc.ca/occhyg/

The site describes UBC's Master's program in Occupational Hygiene.

University of Calgary, Safety Office

http://www.acs.ucalgary.ca/~ucsafety/

View policy and procedure documents and training course information.

University of Toronto, Office of Environmental Health and Safety

http://www.utoronto.ca/safety/ehshome.htm

The site provides access to U of T's Health and Safety manual, emergency procedures documents, and MSDS sites.

University of Waterloo Safety Office

http://www.safetyoffice.uwaterloo.ca/

The site provides access to information about the safety office, and documents such as their *Health and Safety Program Manual*, *Radiation Manual,* and *Laser Safety Manual*.

University of Western Ontario Emergency Planning

http://www.uwo.ca/emerg/index.html

Read about the university's emergency plans: University Policy on Emergency Planning and Response and the Master Emergency Plan.

Water Technology International

http://www.wti.cciw.ca

WTI is an environmental technology and services company. Read about their products and services (hazardous waste, site remediation, etc.). Of particular interest is their 40-hour health and safety course, which satisfies general Canadian (federal and provincial) and U.S. (29 CFR1910.120, "HAZWOPER") health and safety training requirements for contaminated site workers.

West Arm Consulting Services

http://www.westarm.bc.ca/

This company specializes in WWW page development for EMS. Their page has extensive links to various EMS and police sites.

Westray Mine Public Inquiry

http://www.gov.ns.ca/legi/inquiry/westray/

The site presents the findings of a public inquiry into the 9 May 1992 underground explosion at the Westray coal mine in Nova Scotia that killed 26 miners.

WHMIS 96

http://www.bbsi.net/50para/p_web.html-ssi

Read about this CD-ROM based WHMIS (Canadian right-to-know legislation) training package, produced by C.D. 50e PARALLELE INC.

WHMIS Software Training

http://www.niagara.com/blmc/

Contains a brief description of BLMC's WHMIS training software, with contact information.

Windsor Occupational Health Information Service (WOHIS)

http://www.mnsi.net/~wohis/

WOHIS is an occupational health and safety information/referral service located in Windsor, Canada. Its mandate is "to promote community awareness of occupational health and safety issues in order to prevent illness and injury." Read about its services and view publications.

Workers Compensation Board of British Columbia

http://www.wcb.bc.ca/

This site provides access to information about the WCB itself, answers a series of frequently asked questions (FAQ), information about what's new (initiatives, legislation, etc.), and online publications.

Workers Compensation Board of Manitoba

http://www.wcb.mb.ca/

The site provides information about the WCB, claims, employer and employee responsibilities, etc. The WCB Act, Policy Manual, and fact sheets are provided.

Workers Health and Safety Centre

http://www.whsc.on.ca/

The centre is a worker-driven health and safety delivery organization in Ontario, Canada. The site provides information about the centre and its services, including the full text of publications and links to labor and health and safety resources.

Workers' Compensation Board of Nova Scotia

http://www.wcb.ns.ca/

The site provides information about the WCB, claims, employer and employee responsibilities, etc. There is also a "library," including the Workers' Compensation Act, regulations, a WCB policy manual, and fact sheets.

Workers' Compensation Board of Alberta

http://www.wcb.ab.ca/

The site provides information about the WCB, claims, employer and employee responsibilities, etc. The WCB Publications section contains the full text of the WCB Policies and Information Manual, the Workers' Compensation Act and Regulation, various booklets, informationand reports, and forms.

Workplace Safety and Insurance Appeals Tribunal (Ontario)

http://www.wsiat.on.ca

The tribunal is the final level of appeal to which workers and employers may bring disputes concerning workers' compensation matters in Ontario. The site provides a description of the appeal process, a list of recent decisions, and research publications. (Formerly the Workers' Compensation Appeals Tribunal)

Workplace Safety and Insurance Board, Ontario, Canada

http://www.wsib.on.ca

This content-rich site contains information about the board itself, information for employers and workers, news releases, facts and figures, and the Ontario Workers' Compensation Act. This board has replaced theWorkers Compensation Board of Ontario.

Worksite News

http://www.networkx.com/worksitenews/

Selections from and subscription information about this Western Canadian occupational health, safety, and environment news magazine.

Worplace Health, Safety and Compensation Commission (Newfoundland)

http://www.whscc.nf.ca/

The site provides information about the commission, claims, and employer and employee responsibilities. The Workers Compensation Act and Regulations, Policy Manual, and other publications are available on the site.

Young Worker Awareness, Workplace Health and Safety Agency (Ontario)

http://www.whsa.on.ca

The site contains information for young workers, their parents, teachers, principals, employers, and others. Information categories include what are the risks, the law (Ontario), your rights, what if I get hurt, resources, and true health and safety stories. More detail about this site is available in the Top 50 Sites section.

China

ChinaHawk Enterprises Ltd.

http://home.hkstar.com/~chinahwk/home.html

Information about this Beijing-based company's disposable safety products for the health care, industrial, and food services industries. Offices are in Beijing and Hong Kong.

Hong Kong Education and Manpower Bureau

http://www.info.gov.hk/emb/welcome.htm

This government agency's site provides information about occupational health and safety policies and labor relations.

Hong Kong Occupational Safety and Health Association

http://www.hk.super.net/~hkosha/

Information is provided about the association, Hong Kong safety and health legislation and standards, and the association's publications, along with links to other sites.

Hong Kong University of Science and Technology, Safety and Environmental Protection Office

http://www.ab.ust.hk/sepo/

The site contains the university's health and safety manual and various guidance documents.

Occupational Safety and Health Council (Hong Kong)

http://www.oshc.org.hk/

This well-developed site provides information in a variety of areas: consultancy services, safety and health laws, education and training, approved personal protective equipment, and OSHC publications.

Czech Republic

Ceske pracovni zdravi a bezpecnost

http://www.markl.cz

Czech occupational health and safety. The site contains a description of this consulting company's products, a directory of OSH organizations in the Czech republic, and Czech OSH legislation. All information is available in Czech, and some in English.

Synthesia

http://www.synthesia.cz/

The site provides information in Czech and English about this Czech Republic chemical company.

Estonia

Estonian Newsletter on Occupational Health and Safety

http://www3.occuphealth.fi/eng/info/estonia/

The full text of this newsletter is available online. It is published 3 to 4 times per year, by the Estonian Centre of Occupational Health, Estonian Institute of Experimental and Clinical Medicine, and the Finnish Institute of Occupational Health. It is published in English, Estonian, and Russian.

European Union

AINnet

http://www.ain.es/basedat.htm

The site contains a Spanish and European Union legislative database. All information is in Spanish.

CELEX

http://europa.eu.int/celex/

"CELEX is a comprehensive and authoritative information source on European Community law. It offers multilingual, full text coverage of a wide range of legal acts including the founding treaties, binding and nonbinding legislation, opinions and resolutions issued by the EU Institutions and consultative bodies, and the case law of the European Court of Justice."

EC Chemical Regulation Reporter

http://www.incadinc.com/edr/ecchemicalregulationreporter.html

This site provides basic information about European Community chemical regulations and subscription information about "EC Chemical Regulation," a monthly notification of EC actions affecting the chemical industry.

Environmental Chemicals Data and Information Network (ECDIN)

http://ecdin.etomep.net/

"ECDIN is a factual databank, created under the Environmental Research Programme of the Joint Research Centre (JRC) of the Commission of European Communities at the Ispra Establishment." "ECDIN deals with the whole spectrum of parameters and properties that might help the user to evaluate real or potential risk in the use of a chemical and its economical and ecological impact."

Europa—Governments online

http://europa.eu.int/en/gonline.html

This is a directory to government pages for European Union member countries.

European Agency for Safety and Health at Work

http://www.eu-osha.es/

This site (will) contain(s) European information in the following areas: legislation and standards, research, practice, strategies and programmes, statistics, information, and news. More detail about this site is available in the Top 50 Sites section.

European Chemicals Bureau

http://ecb.ei.jrc.it/

"The principal task of the ECB is to carry out and coordinate the scientific/technical work needed for the implementation of EU-legislation (directives and regulations) in the area of chemical control." Read about ECB's projects and activities and publications.

European Environment Agency

http://www.eea.dk/

The site provides a description of the agency and numerous documents and document descriptions related to the environment in Europe.

European Federation, Biotechnology Working Party on Safety in Biotechnology

http://www.boku.ac.at/iam/efb/efb_wp.htm

Information available includes a list of publications from and membership information for the Working Party, biosafety links and a bibliography on laboratory acquired infections.

Health and Safety Promotion in the European Union

http://www.hsa.ie/hspro/index1.html

HSPro—EU will set up a remotely accessible telematics system providing a one-stop-shop for occupational health and safety information. Categories of information include news and events, guidance, OSH information, OSH databases, bibliographies, library catalogs, and publications. More detail about this site is available in the Top 50 Sites section.

Institute for Systems, Informatics and Safety

http://www.jrc.org/isis/index.asp

This Joint Research Centre of the European Commission is involved in the multidisciplinary analysis of industrial, socio-technical, and environmental systems; the innovative application of information and communication technologies; and the science and technology of safety management. Read about its research activities.

International Centre for Genetic Engineering and Biotechnology, Biosafety Web Pages

http://www.icgeb.trieste.it/biosafety/

ICGEB's site provides the full text of numerous biosafety documents, a bibliographic database of biosafety papers, European and U.S. biosafety regulations, a biosafety mailing list, and links to other biosafety resources. More detail is available in the Top 50 Sites section.

OHS in the European Union

http://www.occuphealth.fi/e/eu/index.phtml

This site, maintained by the Finnish Institute of Occupational Health, provides many useful resources and links, including the European Health and Safety Database (HASTE).

Finland

Finnish Institute of Occupational Health

http://www.occuphealth.fi

Information is provided about the Institute, its purpose, and its work. Lists of institute publications and journals are provided. There is also a list of OH&S conferences.

OSHWeb

http://oshweb.me.tut.fi/cgi-bin/oshweb.pl

This is Teuvo Uusitalo's popular directory of health and safety resources on the Web.

Scandinavian Journal of Work, Environment and Health

http://www.occuphealth.fi/eng/dept/sjweh/

The site provides information about upcoming articles in the journal and about subscribing to the paper version. The journal appears bimonthly.

France

Institut national de recherche et de sécurité (INRS)

http://www.inrs.fr/

The role of the INRS in France "is to contribute technically, using appropriate means, to the prevention of occupational accidents and diseases to ensure the protection of human health and safetyat work." The site provides information about INRS's research, consulting, and publications.

JOB-MDT

http://www.meditrav.com/emploi.htm

This page describes the mailing list, JOB-MDT, a system for posting job offers and requests for occupational health professionals. All information is in French.

LegiFrance

http://www.legifrance.gouv.fr

This site contains the entire legislative code of France. All information is in French.

Germany

Waldmann Lichttechnik

http://www.waldmann.de/

View information about this German company's range of workplace lighting products.

Iceland

Disaster Connection

http://www.itn.is/~gro/disaster/

This is an extensive list of Internet disaster-related resources and documents.

India

EHS Management Technologies Private Limited, India

http://www.corporatepark.com/ehsindia/

Read about this consulting company's environmental, health and safety, and management services.

International

African Newsletter on Occupational Health and Safety

http://www.occuphealth.fi/eng/info/anl/

The full text of current and previous issues are available at the site. The newsletter is published by the Finnish Institute of Occupational Health, in association with the ILO - CIS International Occupational Health and Safety Information Centre.

Asian-Pacific Network on Occupational Safety and Health Information

http://www.ilo.org/public/english/270asie/asiaosh/

Presented by the ILO/FINNIDA Asian-Pacific Regional Programme on Occupational Safety and Health, this site was established with a view to sharing OSH information in and about the Asian-Pacific region.

Asian-Pacific Newsletter on Occupational Health and Safety

gopher://gopher.nectec.or.th/11/bureaux/ASIA-OSH/newsletter

The full text of the current and all past issues of the newsletter is available online. The newsletter is produced by the Asian-Pacific Regional Network on Occupational Safety and Health, a project of the International Labour Organization.

Baltic Sea Network on Occupational Health and Safety

http://www3.occuphealth.fi/eng/project/baltic/

The purpose of this site, largely as a result of World Health Organization initiatives, is the "establishment of a telematic information network on occupational health and safety in the countries around the Baltic Sea." It (will) contain(s) OHS contact information on these countries, national OHS information, and information about ongoing projects and events in the Baltic region.

Biosafety Information Network and Advisory Service (BINAS)

http://binas.unido.org/binas/binas.html

BINAS, a service of the United Nations Industrial Development Organization, monitors global developments in regulatory issues in biotechnology. Site highlights include links to regulations in many countries, UNIDO publications, the BINAS newsletter, and OECD Environment Monographs.

Central and Eastern European Occupational Safety and Health Net (CEE OSHNet)

http://siri.uvm.edu/ceeosh/

CEE OSHNet is an "unofficial" home page of the American Industrial Hygiene Association's International Affairs Subcommittee for Central and Eastern Europe (CEE). Ultimately, the CEE OSHNet will contain 'links to general and specific country demographic information, specific safety and health information, and resources available on the Internet.

GLADNET

http://www.gladnet.org

GLADNET, the Global Applied Disability Research and Information Network is an initiative of the Vocational Rehabilitation Branch of the International Labour Organization (ILO). As well as having information about the organization itself, the site provides access to the GLADNET Infobase—full text documents and bibliograhic references concerning employment and disability.

International Agency for Research on Cancer, Overall Evaluations of Carcinogenicity to Humans

http://www.iarc.fr/monoeval/crthall.htm

As evaluated in IARC Monographs Volumes 1-69 (a total of 836 agents, mixtures, and exposures), this list contains all hazards evaluated to date, according to the type of hazard posed and the type of the exposure.

International Agency for Research on Cancer (IARC)

http://www.iarc.fr

IARC's mission is "to coordinate and conduct research on the causes of human cancer and to develop scientific strategies for cancer control. The Agency is involved in both epidemiological and laboratory research, and disseminates scientific information through meetings, publications, courses and fellowships." The full text of some IARC publications is available on the site.

International OH&S conferences

http://www.occuphealth.fi/e/congress.phtml

This site lists worldwide conferences related to occupational health and safety

---

International Register on Biosafety

http://irptc.unep.ch/biodiv/

This site, provided by the United Nations Environment Programme, presents a number of biosafety resources. Highlights include a series of international directories, guidelines and fact sheets, and descriptions of international biosafety activities.

North American Emergency Response Guidebook (1996)

http://www.tc.gc.ca/canutec/english/guide/menug_e.htm

This is an online version of the *Guidebook*, developed jointly by Transport Canada, U.S. Department of Transportation, and the Secretariat of Communications and Transportation of Mexico. More detail about this site is available in the Top 50 Sites section.

Policies for Occupational Health and Safety Management: Systems and Workplace Change

http://www.irob.unsw.edu.au/conference.htm

A report of the proceedings and abstracts of this September 1998 conference in Amsterdam are available in HTML and PDF formats.

SilverPlatter

http://www.silverplatter.com

The site provides information about databases commercially available from SilverPlatter, both on CD-ROM and via the Internet.

UNEP Chemicals (International Registry of Potentially Toxic Chemicals—IRPTC)

http://irptc.unep.ch/irptc/

UNEP Chemicals is the center for all chemicals-related activities of the United Nations Environment Programme. Highlights include an inventory of information sources on chemicals, the UNEP Chemicals database, and information about UNEP programs.

Weekly Epidemiological Record

http://www.who.int/wer/

WER provides weekly information about outbreaks of diseases of public health importance. This World Health Organization publication is available in English and French.

Ireland

BLIND-DEV Home Page

http://www.cs.tcd.ie/blind-dev/

View a description of, and archives for, the BLIND-DEV mailing list—a discussion group for issues concerning the development of computer products and adaptive equipment for blind and visually impaired computer users.

Health and Safety Authority

http://www.hsa.ie/osh/welcome.htm

Information is provided about the Health and Safety Authority (Ireland), a list of publications, a list of acts and regulations, and information about the authority's programs.

Italy

Euroware Associates Software

http://www.euroware.com/index.html

Information is provided about the company's MSDS and chemical-related software packages.

Nikos International

http://www.antinfortunistica.it

The site describe's the company's industrial cleaning products and personal protective equipment. Most information is in Italian.

Japan

Global Health Disaster Network

http://hypnos.m.ehime-u.ac.jp/GHDNet/index.html

GHDNet is maintained by Department of Emergency Medicine, Ehime University, Japan. Contains documents, bibliographies, conference information, and links to related sites.

Global Information Network for Chemicals (GINC)

http://www.nihs.go.jp/GINC

GINC is a world wide information network for safe use of chemicals. Read about the GINC project and access its integrated chemical databases.

International Center for Disaster-Mitigation Engineering (INCEDE)

http://incede.iis.u-tokyo.ac.jp/Incede.html

This site contains documents on disasters, including floods, earthquakes, etc. Most information is related to Asia.

National Institute of Health Sciences—Japan

http://www.nihs.go.jp

Information is provided about the various divisions within NIHS, along with links to NIHS-sponsored sites.

Research Institute for Paradigm on Occupational Safety

http://www.cc.rim.or.jp/~tosikazu/index.htm

Provides accident injury statistics for Japan. The institute is studying fundamental theories on the occurrence of accidents and attempting to explain accident statistics.

Netherlands

Elsevier Science

http://www.elsevier.nl

View information about the company's publications. Subjects include chemistry and chemical engineering, clinical medicine, and environmental science and technology.

Hyginist

http://home.wxs.nl/~ihpc/

Produced by Scheffers Industrial Hygiene Publishing and Consultancy in the Netherlands, HYGINIST is an "industrial hygiene statistical tool. It evaluates the exposure data collected with the standard exposure assessment strategies."

International Journal of Industrial Ergonomics

http://www.elsevier.nl/locate/ergon

View the tables of contents of printed issues, information about special issues, an author index, and subject index for the journal.

New Zealand

Entropy Software, OOS Software

http://www.albatross.co.nz/~miker/main.htm

Describes this New Zealand company's OOS software, a program that tells you when to take a break from your computer to help prevent/alleviate OOS (occupational overuse syndrome or repetitive strain injuries). Download a copy of the software from this site.

Injury Prevention Research Unit, University of Otago, New Zealand

http://www.otago.ac.nz/Web_menus/Dept_Homepages/IPRU/

The site provides information about the unit, research it undertakes, and publications produced.

Massey University (New Zealand), Dept. of Human Resource Management

http://www.massey.ac.nz/~wwhrm/teachosh.htm

This site contains a description of Massey's Postgraduate Diploma in Occupational Safety and Health (Dip OSH).

New Zealand's Health and Safety Net

http://www.OSH.dol.govt.nz/

This is the home page of the Occupational Safety and Health Service of the Department of Labour, New Zealand. The site provides information about the organization, work hazards, health and safety law, and training programs.

Occupational Overuse Syndrome/RSI resources

http://www.mcs.vuw.ac.nz/comp/General/OOS/

The site contains articles and bulletins from New Zealand Occupational Safety and Health, the typing injuries FAQ, and links to other RSI information.

Safeguard Publications

http://www.OSH.dol.govt.nz/order/safeguard/index.html

View a description of and ordering information for *Safeguard* magazine, *Safeguard Update* and *Safeguard Buyers Guide*, a trio of New Zealand health and safety publications from Colour Workshop.

Shiftwork Services

http://host.mwk.co.nz/shiftwork/

Shiftwork Services is a division of the Auckland (New Zealand) Sleep Management Centre. The site provides access to the full text of the organization's newsletter, sleep tips, and guidelines for managers.

University of Canterbury (New Zealand), Health and Safety

http://www.mech.canterbury.ac.nz/dinfo/helth.htm

Read about the university's health and safety department.

University of Otago (New Zealand), Health and Safety

http://wsweb.otago.ac.nz/wsweb/health.htm

Information about health and safety at the university is provided.

University of Waikato (New Zealand), Health and Safety

http://www.waikato.ac.nz/fmd/hsc/

A number of documents are available, including the University's Occupational Health and Safety Policy and information about Occupational Overuse Syndrome.

Norway

Primary Care Internet Guide, Occupational and Environmental Medicine

http://www.uib.no/isf/guide/occu.htm

This is a directory of mailing lists, newsgroups, and sites related to occupational and environmental medicine. The directory is international, with some emphasis on Norwegian material.

Peru

Ministry of Health, Peru

http://www.digesa.sld.pe

This site provides information about the Peruvian Ministry of Health, along with links to the Directorate of Environmental Health, General Office of Epidemiology, and Peruvian Association of Public Health. All information is in Spanish.

Poland

Central Institute for Labour Protection (Poland)

http://www.ciop.waw.pl

Information is provided about the institute and its work, as well as subscription information for the institute's journal, *International Journal of Occupational Safety and Ergonomics*.

Singapore

Poisons Information Database

http://vhp.nus.sg/PID/

The site provides information about poisonous plants and animals, a directory of antivenoms, a directory of toxinologists, and a directory of poison control centers around the world.

South Africa

Association of Societies for Occupational Safety and Health (Southern Africa)

http://www.asosh.org/

The initial focus of this site will be on OHS in South Africa, and it will be expanded to provide information on OHS in the Southern African Development Community (SADC) region. Find out about the member associations, their resources, training programmes, and legislation. More detail about this site is available in the Top 50 Sites section.

Chempute Software

http://www.chempute.com

This South African company produces and provides a variety of software packages for chemical process industries. Plant safety and quality control programs are available. Download demos and tutorials.

Southern Africa Occupational Health Web

http://www.und.ac.za/und/med/comhlth/occ_hlth.html

Maintained by the Occupational Health Programme, Dept. of Community Health, University of Natal, the site describes the university's Occupational Health Programme, there is a discussion paper on women in the workplace, and there are links to various occupational health sites.

Spain

Asociacion para la Prevencion de Accidentes

http://www.safetyonline.net/apa/home.htm

Information is provided about this Spanish association, its services (training, databases, publications, etc.), subscriptions to APA's journal *Prevencion*, a series of safety posters and conference information. All information is in Spanish.

Sweden

Arbetslivsinstitutet

http://www.niwl.se/niwl.htm

National Institute of Working Life (Arbetslivsinstitutet) is Sweden's Research and Development center for occupational health and safety, working life and the labor market. Access information about the insitute, its library catalog, and the full text of its publications.

Department of Radiation Physics, Lund University

http://www.fysik.lu.se/~radiofys/rfihomee.htm

Read about the radiation physics research being performed at this Swedish university. Some abstracts and research papers are available in English.

FEB—The Swedish Association for the ElectroSensitive

http://www.feb.se

This site contains various documents, a conference list, and links to other sites primarily related to Electromagnetic Fields (EMFs) and "Electrical Hypersensitivity." Information is provided in English and Swedish.

Rådet för arbetslivsforskning

http://www.ralf.se

The Swedish Council for Work Life Research is a government agency which plans and funds Swedish research on subjects related to the work environment and working life. The site provides contact information, descriptions of its programmes, a newsletter, and its publications. Information is in Swedish and English.

Sick Building Syndrome

http://www.occmed.uu.se/indoor.html

This document provides a description of sick building syndrome, its causes and diagnosis. It is provided by the Department of Occupational and Environmental Medicine, University Hospital in Uppsala, Sweden.

Swedish Occupational Safety and Health Administration

http://www.arbsky.se/arbskeng.htm

View Swedish work environment legislation, selected documents produced by the organization, and information about the organization itself. Content is in Swedish and English.

Swedish Society of Radiation Physics

http://www.fysik.lu.se/~radiofys/sfrengho.htm

Read about the society itself, upcoming meetings and conferences, and links to radiation sites. Content is in Swedish and English.

Vetenskap och Erfarenhet i TrafiksakerhetsArbetet (VETA)

http://www.veta.se/

VETA (Science and Experience in Traffic Safety Work) is a nonprofit organization for traffic safety research. Summary information is provided about its studies and publications,along with links to other sites.

Volvo Group—Environment

http://www.volvo.se/environment/index.html

Volvo's Environmental policies, programs, and environmental report are provided. Volvo's "blacklist" of chemicals is included in the 1995 environmental report.

Switzerland

Global Strategy on Occupational Health for All

http://www.ccohs.ca/who/contents.htm

This document describes recommendations arising from a meeting of World Health Organization Collaborating Centres.

ILOLEX—the ILO's Database on International Labour Standards

http://ilolex.ilo.ch:1567/public/english/50normes/infleg/iloeng/index.htm

This is a searchable database of the full text of the International Labour Organization's international labor standards. More detail is available in the Top 50 Sites section.

International Labour Organization

http://www.ilo.org/

The site provides information about the ILO, its various international programs, and publications. Includes links to ILO departmental home pages and an ILO directory.

International Occupational Safety and Health Information Centre—International Labour Office

http://www.ilo.org/public/english/90travai/cis/

Information is available about the organization's collaborating centers, publications and directories, and international programs. More detail is available in the Top 50 Sites section.

International Organization for Standardization (ISO)

http://www.iso.ch

View information about this standards-setting organization, the standard-setting process, technical committees, and the ISO standards catalog.

World Health Organization

http://www.who.int

WHO's programs, its regional offices around the world, and international health information reports and publications. More detail is available in the Top 50 Sites section.

Taiwan

Institute of Occupational Safety and Health, Taiwan

http://192.192.46.66/INDEXE0.HTM

The missions of this research institute include "application of scientific technology, surveys and analyses of various risk factors in the working environment, as well as development of countermeasures." The site describes the institute and the research areas of its various divisions.

United Kingdom

Ability Site

http://www.ability.org.uk

This U.K.-based site is presented by and for people with disabilities. It provides access to information about disability resources.

Annals of Occupational Hygiene

http://www.elsevier.nl:80/inca/publications/store/2/0/1/

Get tables of contents, ordering information, etc., for this British Occupational Hygiene Society publication. A free table of contents service via email is available.

British Medical Journal

http://www.bmj.com/bmj/

Abstracts of papers from current and past issues of the journal are available.

British Occupational Hygiene Society (BOHS)

http://www.bohs.org/

This page describes the society, its mission, members, publications, conferences, and meetings. There are also links to related sites.

British Safety Council

http://www.britishsafetycouncil.co.uk

Provides information about the council itself, its mission, history, and membership.

British Standards Institution

http://www.bsi.org.uk/

View BSI's catalog of standards, information about certification and testing, and descriptions of BSI's electronic products.

British Toxicology Society

http://www.bts.org

The site provides information about the society, its publications, and upcoming events.

Chartered Institute of Environmental Health (U.K.)

http://www.cieh.org.uk

CIEH is a nongovernmental organisation dedicated to the promotion of environmental health and the dissemination of knowledge about environmental health issues for the benefit of the public. It is also responsible for the training and professional development of environmental health officers and provides educational services to the profession

---

Chemical Toxicology—an introduction to chemical hazards in the workplace

http://www.med.ed.ac.uk/hew/chemical/toxicol.html

An introductory teaching module by Dr. Alison Jones of the Royal Infirmary of Edinburgh.

Construction Industry Research and Information Association

http://www.ciria.org.uk/

CIRIA is a nonprofit U.K. organization that provides best practice guidance to professionals. Read about the organization, its research, and publications.

Department of Health (U.K.)

http://www.doh.gov.uk/dhhome.htm

The site provides information about the department itself, news releases, departmental programs, and publications (some full text).

Directory of Sites in Occupational and Environmental Health

http://www.med.ed.ac.uk/hew/links/default.htm

This site, maintained by Dr. Raymond Agius at the University of Edinburgh, provides access to a variety of occupational and environmental health sites. The emphasis is on European information.

Ergonomics Society

http://www.ergonomics.org.uk/

The U.K.-based Ergonomics Society Web site contains information about the society and its members, lists of primarily U.K. ergonomics consultants, training courses, and jobs.

European Recycling and the Environment

http://www.tecweb.com/recycle/eurorec.html

This site is a comprehensive source for recycling in the U.K. specifically and Europe generally. It contains a library of recycling articles and a link to *Recycling World* magazine.

Faculty of Occupational Medicine, Royal College of Physicians (U.K.)

http://www.facoccmed.ac.uk/

The faculty is "a professional and academic body empowered to develop and maintain high standards of training, competence and professional integrity in occupational medicine." Read about its various educational programmes, publications, and newsletter.

FAQ: Sources of EMC and Safety Compliance Information

http://world.std.com/~techbook/compliance_faq.html

This is the frequently-asked-questions document (FAQ) for the sci.engr.electrical.compliance newsgroup.

Fire Safety Engineering Group

http://fseg.gre.ac.uk/

This Greenwich U.K. consultancy group describes itself as "one of the largest research groups in the world dedicated to the development and application of mathematical modelling tools suitable for the simulation of fire related phenomena." Read about the company's services.

H and H Scientific Consultants Ltd.

http://dspace.dial.pipex.com/hhsc/

Read about the various health and safety publications and databases distributed by this U.K. company.

Hascom

http://www.hascom.com

Find out about this U.K.-based health and safety consultancy. Information is provided about the company itself, its services, links to selected sites, and an online newsletter. The newsletter is particularly useful to those involved in health and safety in the U.K..

Health and Safety Division

http://www.dedni.gov.uk/hsd/

This is the Health and Safety Division of the Northern Ireland Department of Economic Development. Information is provided about the division itself and its responsibilities.

Health and Safety Executive (U.K.)

http://www.open.gov.uk/hse/hsehome.htm

Information is provided about this U.K. regulatory agency, its publications and videos, research, etc.

Health and the Environment: Frequently Asked Questions

http://www.med.ed.ac.uk/hew/faqhe.html

This is a series of questions and answers maintained by Dr. Raymond Agius at the University of Edinburgh, Environmental Health. Excellent resource!

HEBSWeb

http://www.hebs.scot.nhs.uk/

This is the site of the Health Education Board for Scotland. Bibliographies and assorted documents on a variety of health topics are provided.

Indoor Air 99

http://www.ia99.org/

The site provides detailed information about the Indoor Air 99 conference in Edinburgh, Scotland, August 8-13, 1999.

Institute of Occupational Hygienists

http://www.ed.ac.uk/~alfred/ioh.html

Read about this British Institute's membership, code of ethics, and officers.

Institute of Safety in Technology and Research

http://www.bham.ac.uk/istr/

"The Institute was founded as the Institute of University Safety Officers (IUSO) by a group of U.K. university safety officers who were seeking to establish a body to represent their individual interests on a professional basis and in their specialist area of research and high technology."

Institution of Occupational Safety and Health (IOSH)

http://www.iosh.co.uk/

The site contains information on all the major areas of activity that this U.K.-based institution is involved in: membership and career information documents, U.K. conferences, etc.

International Institute of Risk and Safety Management

http://www.britishsafetycouncil.co.uk/iirsm/Default.htm

Information is provided about the institute itself and membership options.

International Occupational Hygiene Association

http://www.ed.ac.uk/~robin/ioha.html

The IOHA is, "an international voice of the occupational hygiene profession through its status as a nongovernmental organization (NGO) of the World Health Organization."

Journal of Radiological Protection

http://www.iop.org/Journals/jr

This is the official journal of the Society for Radiological Protection, relating to both ionizing and non-ionizing radiations. Information includes tables of contents for current and past issues, descriptions of forthcoming articles, and subscription information.

Map 80 Systems Ltd.

http://www.map80.co.uk/

Read about this U.K. company's labelling software for chemical, pharmaceutical, industrial, and retail settings.

National Access and Rescue Centre

http://www.narc.co.uk/

This U.K. organization "is a training, development and equipment supply centre for rope access, line rescue, confined space entry, mast climbing, fall prevention and all work involving high places or other difficult access." Read about the organization's training programmes and services.

National Association of Healthcare Safety and Risk Practitioners

http://www.nahsrp.org.uk/

NAHSRP is an association for safety and risk professionals who work in the healthcare sector in the United Kingdom. Its aim is to promote best practice in the provision of safer health care. Read about the organization and membership, view back issues of its newsletter, and participate in its message forum.

National Radiological Protection Board

http://www.nrpb.org.uk

This U.K. agency was set up "to advance the acquisition of knowledge about the protection of mankind from radiation hazards" and "to provide information and advice to persons (including Government Departments) with responsibilities in the United Kingdom."

New United Kingdom Official Publications Online

http://www.soton.ac.uk/~nukop/index.html

NUKOP provides bibliographic and ordering information for thousands of U.K. government publications in a variety of subject areas, including health and safety.

Occupational Asthma

http://www.med.ed.ac.uk/hew/work/ocasthma.html

This is a detailed paper on the subject, including aspects of primary, secondary, and tertiary prevention.

Occupational Health and Safety Information Group

http://panizzi.shef.ac.uk/oshig/oshig.html

This professional association, set up in 1990, "caters for the needs of health and safety information specialists." Read about membership in the association, meetings, and its newsletter.

Occupational Safety and Health Information Group

http://panizzi.shef.ac.uk/oshig/oshig.html

"The purpose of the group is to provide a forum for the exchange of views, ideas and expertise on matters of health, safety and environmental information" to practitioners in the U.K. Read about its aims and objectives, membership, and current executive.

OHS&E—The Journal of Occupational Health, Safety, and Environment

http://www.ohse.co.uk

To quote the editors of this U.K. magazine, "Besides containing details of the magazine and our previous and forthcoming features, the site has news, features, legislation information, and an Open Forum in which occupational health, safety and environment matters can be aired and responded to by anyone in the world."

Professional Development Unit, uksafety.net

http://www.uksafety.net/

This U.K. safety and technical training consultancy's site provides a database of U.K. consultants; a bulletin board that includes "hot gossip," frequently asked questions, news of upcoming events, and a "trading post"; and a catalogue of the company's training programmes.

Royal Environmental Health Institute of Scotland

http://www.ed.ac.uk/~tbell/rehis.html

"The main aims of the Institute are to promote the advancement of all aspects of health and hygiene, to stimulate interest in public health, and to disseminate knowledge on health matters to the benefit of the community." Read about the institute's activities, including the courses it provides.

Royal Society of Chemistry

http://www.rsc.org/

"The Royal Society of Chemistry is the Learned Society for chemistry and the Professional Body for chemists in the U.K. with 46,000 members worldwide." Read about membership, the society's information resources, and library.

SilverPlatter Health and Safety World

http://www.silverplatter.com/oshinfo.htm

These pages, edited by Sheila Pantry, provide links to health and safety resources on the Internet, an editorial, a diary of upcoming events, and a large directory of links to health and safety sites.

Society of Chemical Industry

http://sci.mond.org

Information is provided about this U.K.-based learned society, its membership, meetings, publications, etc. There are details about their health and safety group.

Society of Occupational Medicine

http://www.ed.ac.uk/~rma/som/

This U.K. Society aims to stimulate interest, research, and education in occupational medicine. Read about its organization, activities, and publications.

U.K. Maximum Exposure Limits

http://physchem.ox.ac.uk/MSDS/mels.html

This site provides maximum exposure limits in the U.K. for a number of chemicals, based on Schedule 1 of the COSHH regulations.

University of Birmingham (U.K.), Institute of Occupational Health

http://www.bham.ac.uk/IOH/

Read about the institute's postgraduate programs, upcoming local, national, and international conferences and meetings.

University of Edinburgh, Health and Safety Services

http://www.admin.ed.ac.uk/safety/

This site contains a mission statement, contact information about the various health and safety services at the university, and "health and safety news flashes."

University of Edinburgh, Health Environment and Work

http://www.med.ed.ac.uk/hew/

The site provides a long list of "tutorials" (educational resources) on a variety of occupational and environmental health topics, as well as information about Edinburgh's academic programs. There is also an excellent directory of primarily European Internet resources. More detail about this site is available in the Top 50 Sites section.

---

8 Discussion Directory

The Internet is an important medium for general discussion of safety and health issues. These discussion groups operate either by email or through newsgroups. This directory of discussion groups on the Internet is designed to be wide-ranging and includes many discussions related to fields associated with health and safety professionals as well as discussions directly dealing with these issues.

Using Email lists

Mailing lists are email-based discussion groups. To participate in a mailing list, you must first subscribe to it. This is generally free. Different mailing lists have different subscription procedures. In general, if the subscription address begins with "listserv" or "listproc," the body of your message should be

subscribe listname <your name>

where <your name> is your regular name, not your email address. The email address is gotten from the "From:" line of your email note.

If the subscription address begins with "majordomo," the body of your message should be

subscribe listname

Your name and email address will come from the header of your email message.

If neither of these apply, send a message to the subscription address. The body of the message should be the single word "help" (without the quotes). This is likely to elict some form of useful response, either from the computer program or the person who manages the list.

Biosafety

BIOSAFTY

A list for biosafety discussions.

> Subscribe by sending email to listserv@mitvma.mit.edu
> Send messages to the list at biosafty@mitvma.mit.edu

For more details, contact rfink@mitvma.mit.edu

NURSENET

NURSENET is an open unmoderated list for discussions about a variety of nursing issues.

> Subscribe by sending email to listserv@listserv.utoronto.ca
> Send messages to the list at nursenet@listserv.utoronto.ca

For more details, contact Judy Norris: judy.norris@ualberta.ca

Chemical Safety

CHEMED-L

Chemistry education discussion list.

> Subscribe by sending email to listproc@atlantis.cc.uwf.edu
> Send messages to the list at chemed-l@atlantis.cc.uwf.edu

CHEMLAB_L

The CHEMLAB_L list is for people interested in chemistry laboratories (both academic and research), students' experiments (high school, college, and university), classroom demonstrations and shows for the public of chemical processes, chemistry stockroom management, lab safety, and small-scale chemical waste handling procedures. To subscribe, put the word SUBSCRIBE in the subject line of your message.

Subscribe by sending email to chemlab_l@vax1.bemidji.msus.edu
Send messages to the list at chemlab_l@vax1.bemidji.msus.edu

For more details, contact Dr. Gerald Morine: ghmo@vax1.bemidji.msus.edu

CHMINF-L

Chemical information sources discussion list.

Subscribe by sending email to listserv@iubvm.ucs.indiana.edu
Send messages to the list at cheminf-l@iubvm.ucs.indiana.edu

For more details, contact Gary Wiggins: wiggins@indiana.edu

CMTS-L

CMTS-L is a a forum for the exchange of ideas and information among organizations and corporations engaged in the establishment of chemical management and tracking systems or among those concerned with improving chemical management processes in general.

Subscribe by sending email to listserv@ls.emax.com
Send messages to the list at cmts-l@ls.emax.com

CONS-EQST-PESTICIDE-FORUM

The purpose of the CONS-EQST-PESTICIDE-FORUM list is to facilitate discussion and exchange of information among Sierra Club members on issues associated with the manufacture, use, and licensing of pesticides. This forum is open to all current Sierra Club members.

Subscribe by sending email to listserv@lists.sierraclub.org
Send messages to the list at cons-eqst-pesticide-forum@lists.sierraclub.org

For more details, contact mm6@doc.mssm.edu

CSBNEWS

This is a distribution list for news releases from the U.S. Chemical Safety and Hazard Investigation Board.

Subscribe by sending email to majordomo@csb.gov
Send messages to the list at csbnews@csb.gov

For more details, contact the owner: csbnews@csb.gov

DOE-ATCS

U.S. Department of Energy, Action Team for Chemical Safety.

Subscribe by sending email to listserv@bbs.pnl.gov
Send messages to the list at doe-atcs@bbs.pnl.gov

DOECHEMSAFETY

U.S. Department of Energy Chemical Safety.

Subscribe by sending email to listserv@listserv.pnl.gov
Send messages to the list at doechemsafety@listserv.pnl.gov

DRIFTERS

Drifters is an open forum for educators, trainers, regulators, and industry, including applicators, to share information, data, and training activities on managing pesticide drift.

Subscribe by sending email to majordomo@reeusda.gov
Send messages to the list at drifters@reeusda.gov

For more details, contact John W. Impson: jimpson@reesusda.gov

EPA-TOX

EPA-TOX is a one-directional mailing list for the distribution of select U.S. *Federal Register* documents automatically on the day of publication. It contains the Office of Pollution Prevention and Toxic Substances documents excluding Community Right-to-Know Toxic Release Inventory documents.

Subscribe by sending email to listserver@unixmail.rtpnc.epa.gov
Send messages to the list at epa-tox@unixmail.rtpnc.epa.gov

For more details, contact John Richards: richards.john@epamail.epa.gov

EPA-TRI

EPA-TRI is a one-directional mailing list for the distribution of select U.S. *Federal Register* documents automatically on the day of publication. It contains Community-Right-To-Know Toxic Release Inventory documents.

Subscribe by sending email to listserver@unixmail.rtpnc.epa.gov
Send messages to the list at epa-tri@unixmail.rtpnc.epa.gov

For more details, contact John Richards: richards.john@epamail.epa.gov

FTSS-INFO

Monthly change notifications for the MSDS database from the Canadian Centre for Occupational Health and Safety (CCOHS).

Subscribe by sending email to majordomo@ccohs.ca
Send messages to the list at ftss-info@ccohs.ca

For more details, contact Chris Moore: chrism@ccohs.ca

HAZMAT-L

Transportation, storage, and reporting of hazardous materials, plus general hazmat discussions.

Subscribe by sending email to listserv@coloradocollege.edu
Send messages to the list at hazmat-l@coloradocollege.edu

For more details, contact John DeLaHunt: jdelahunt@cc.colorado.edu

HMEP

U.S. Federal Emergency Management Agency, Hazardous Materials Emergency Planning.

Subscribe by sending email to majordomo@fema.gov
Send messages to the list at hmep@fema.gov

For more details, contact eipa@fema.gov

LABSAFETY-L

LABSAFETY-L is a discussion list for members of Laboratory Safety Workshop and National Association of Chemical Hygiene officers interested in lab health and safety issues.

Subscribe by sending email to listserv@siu.edu
Send messages to the list at labsafety-l@siu.edu

For more details, contact James A. Kaufman: labsafe@aol.com

LEPC

Hazardous materials emergency response planning.

Subscribe by sending email to listserv@list.uvm.edu
Send messages to the list at lepc@list.uvm.edu

For more details, contact Ralph Stuart: rstuart@esf.uvm.edu

MCS-IMMUNE-NEURO

MCS/Chemical Injury Support. Information exchange, support group, and management strategies for people adversely effected by chemicals. Owner Ginny Kloth is the volunteer Director of Electronic Communications for the Chemical Injury Information Network (CIIN).

Subscribe by sending email to listserv@maelstrom.stjohns.edu
Send messages to the list at mcs-immune-neuro@maelstrom.stjohns.edu

For more details, contact bijou@blrg.tds.net (Ginny Kloth)

MSDS-INFO

Monthly change notifications for the MSDS database from the Canadian Centre for Occupational Health and Safety (CCOHS).

Subscribe by sending email to majordomo@ccohs.ca
Send messages to the list at msds-info@ccohs.ca

For more details, contact Chris Moore: chrism@ccohs.ca

NAOSMM

This list is a vehicle whereby members of the National Association of Scientific Materials Managers (NAOSMM) can maintain contact.

Subscribe by sending email to listserv@listserv.rice.edu
Send messages to the list at naosmm@listserv.rice.edu

For more details, contact Tom Blakeney: blakeney@ruf.rice.edu
John Elliott: jelliott@ruf.rice.edu

NAPIAP

National Agricultural Pesticide Impact Assessment Program.

Subscribe by sending email to majordomo@reeusda.gov
Send messages to the list at napiap@reeusda.gov

For more details, contact Dennis Kopp: dkopp@reeusda.gov

NIOSH-CASC

NIOSH - Chemical Analytical Services.

> Subscribe by sending email to listserv@listserv.cdc.gov
> Send messages to the list at niosh-casc@listserv.cdc.gov

For more details, contact ddd1@cdc.gov

NIOSH-CASC-PROJ

NIOSH - Chemical Analytical Services, Special Projects.

> Subscribe by sending email to listserv@listserv.cdc.gov
> Send messages to the list at niosh-casc-proj-request@listserv.cdc.gov

For more details, contact ddd1@cdc.gov

PANUPS

PANUPS—the Pesticide Action Network Updates Service—is a weekly news service featuring updates about pesticides and sustainable agriculture. PANUPS also includes the Resource Pointer, which summarizes and gives ordering information for recent publications.

> Subscribe by sending email to majordomo@igc.org
> Send messages to the list at panups@igc.org

For more details, contact panna@igc.org

PESTCOM

Chemical pesticide residues

> Subscribe by sending email to listproc@colostate.edu
> Send messages to the list at pestcom@colostate.edu

PESTCON

PESTCON (Integrated Pest Management at Colleges and Universities) is a networking forum which brings together the many resources of knowledge about insects, mammals, birds, and other life forms which impact the health and safety of, or conflict with the daily mission of, human occupancy at learning institutions, private, public, state, and federal government buildings and grounds, and institutional farms.

Subscribe by sending email to majordomo@list.uiowa.edu
Send messages to the list at pestcon@list.uiowa.edu

For more details, contact Dave Jackson: dave-jackson@uiowa.edu
Suresh Prabhakaran: sprabhak@pps1-po.phyp.uiowa

PRINTING

Printing industry discussions.

Subscribe by sending email to majordomo@nexus.netconcepts.com
Send messages to the list at printing@printer.net

TRANSDERM

Exchange and dissemination of ideas, information and data—including un-published material—on all aspects of skin penetration by chemicals. Seminars, conferences, and research posts can be publicized. It should interest workers in dermatology, pharmaceuticals, cosmetics, and risk assessment.

Subscribe by sending email to mailbase@mailbase.ac.uk
Send messages to the list at transderm@mailbase.ac.uk

For more details, contact John Pugh: pugh@cardiff.ac.uk

UNIFY-TRAINING

This list enables members of the Superfund Unified Training Team and stake-holders to share information on training development and delivery.

Subscribe by sending email to listserver@valley.rtpnc.epa.gov
Send messages to the list at unify-training@valley.rtpnc.epa.gov

For more details, contact Gary Turner: turner.gary@epamail.epa.gov

Construction Safety

CIBW99-L

International construction safety and health.

Subscribe by sending email to listproc@hawaii.edu
Send messages to the list at cibw99-l@hawaii.edu

COMFORTZONE

The purpose of COMFORTZONE is to open up communications among members of the international thermal comfort community.

Subscribe by sending email to majordomo@penman.es.mq.edu.au
Send messages to the list at comfortzone@penman.es.mq.edu.au

For more details, contact Dr. Richard de Dear: rdedear@laurel.ocs.mq.edu.au

CONST

Construction management discussion list.

Subscribe by sending email to listproc@lists.colorado.edu
Send messages to the list at const@lists.colorado.edu

IAQ LIST

The IAQ List discusses issues and concerns regarding indoor air quality. A variety of things can pollute and contaminate the indoor air quality of a home, office, or building—from molds and bioaerosols to combustion by-products like carbon monoxide, soot, PAHs, and VOCs. Many can cause serious health concerns and, in some cases, extensive property damage. This list will address the identification and mitigation of contaminants.

Subscribe by sending email to http://www.onelist.com
Send messages to the list at iaq@onelist.com

For more details, contact C. Flanders, IAQ List Manager: rkfabf@aol.com

TOTAL-QUALITY-CONSTRUCTION

A forum for the exchange of ideas and information by researchers and practitioners involved in the pursuit of excellence in the construction sector.

Subscribe by sending email to mailbase@mailbase.ac.uk
Send messages to the list at total-quality-construction@mailbase.ac.uk

For more details, contact Tony Moody: tmoody@pmc1.pmc.port.ac.uk
Christopher Preece: c.n.preece@leeds.ac.uk

Disaster/Emergency Response & Management/Fire

A-ENEWS-L

Oxford Medical Publications: Accident and Emergency News

Subscribe by sending email to listserv@webber.oup.co.uk
Send messages to the list at a-enews-l@webber.oup.co.uk

For more details, contact Clare Marl: marlc@oup.co.uk

ACAD-AE-MED

The list will be of relevance to all trainees, including undergraduates and postgraduate practitioners, in accident and emergency medicine. It will seek to promote and foster academic research in the world of accident and emergency medicine.

Subscribe by sending email to mailbase@mailbase.ac.uk

Send messages to the list at acad-ae-med@mailbase.ac.uk

CAEMS

California Emergency Medical Services Mail List. CAEMS is a mailing list dedicated to information sharing, problem solving, and any other issues related to Emergency Medical Services in the State of California. List subscribers include people at all levels of EMS.

Subscribe by sending email to listproc@ucdavis.edu

Send messages to the list at caems@ucdavis.edu

For more details, contact Mel Ochs: melochs@aol.com

R. Steven Tharratt: rstharratt@ucdavis.edu

CCEP NEWS E-ZINE

Weekly online newsletter from the Canadian Centre for Emergency Preparedness.

Subscribe by sending email to ccep-news@new-focus.org

Send messages to the list at ccep-news@new-focus.org

For more details, contact Michael Bittle: mbittle@new-focus.org

DPRA

Disaster Prevention and Recovery Alliance discussion list.

Subscribe by sending email to majordomo@new-focus.org

Send messages to the list at dpra@new-focus.org

For more details, contact Michael Bittle: mbittle@new-focus.org

EM-NSG-L

The Emergency Nursing List. Practice and issues of emergency nursing, care of emergency patients, and the discussion of shared concerns with other emergency professionals.

Subscribe by sending email to listserv@itssrv1.ucsf.edu
Send messages to the list at em-nsg-l@itssrv1.ucsf.edu

For more details, contact ttrimble@hooked.net

EMED-L

EMED-L is a list for hospital based emergency medicine practitioners. It was created for the purpose of the free exchange of issues pertaining to the care of patients in emergency medicine as well as topics and controversies related to the practice of emergency medicine.

Subscribe by sending email to listserv@itssrv1.ucsf.edu
Send messages to the list at emed-l@itssrv1.ucsf.edu

For more details, contact David English: english@itsa.ucsf.edu

EMERG-L

Emergency Services Discussion List.

Subscribe by sending email to listserv@vm.marist.edu
Send messages to the list at emerg-l@vm.marist.edu

EMERG-UNIV

The University and College Emergency Planning Mailing List.

Subscribe by sending email to majordomo@sfu.ca
Send messages to the list at emerg-univ@sfu.ca

For more details, contact Peter S. Anderson: anderson@sfu.ca

EMERGENCY-MANAGEMENT

The EMERGENCY-MANAGEMENT list is a resource for sharing of news, ideas, and points-of-view about mitigation of hazards, preparedness for, response to, and recovery from major emergencies and disasters. Primary list users are emergency management coordinators of municipal and county governments. The EMERGENCY-MANAGEMENT list is an information service of Public Safety America.

Subscribe by sending email to listserv@listserv.aol.com
Send messages to the list at emergency-management@listserv.aol.com

For more details, contact Doug Crichlow: PSATDoug@aol.com

EMERGENCY-TELECOMS

The emergency-telecoms mailing list has been established for the exchange of information among the members of the Working Group on Emergency Telecommunications (WGET). Such information can relate to all subjects covered by the Terms of Reference of the WGET and the related Project on Emergency Telecommunications of the United Nations Department of Humanitarian Affairs (DHA).

Subscribe by sending email to mailserv@itu.int
Send messages to the list at emergency-telecoms@itu.int

For more details, contact Hans Zimmermann: hans.zimmermann@itu.ch

EMS-EDU-L

EMS-EDU-L is the Emergency Medical Services Educators List. The list is moderated, which means that messages will not be distributed until they have been reviewed and approved by the list owner(s).

Subscribe by sending email to listserv@informatics.sunysb.edu
Send messages to the list at ems-edu-l@informatics.sunysb.edu

EMSLEAD

International EMS Leadership Discussion Group.

Subscribe by sending email to listserv@hermes.circ.gwu.edu
Send messages to the list at emslead@hermes.circ.gwu.edu

For more details, contact Brian Maguire: maguire@gwis2.circ.gwu.edu

EMSNY-L

Emergency Medical Services issues for NY State providers.

Subscribe by sending email to listserv@health.state.ny.us
Send messages to the list at emsny-l@health.state.ny.us

FIRE-ALARM

Sponsored by the Automatic Fire Alarm Association, this list is for technical discussions on all aspects of fire alarm systems including smoke detection, flame detection, and anything else having to do with fire alarm systems. Subscription is open, but postings are restricted to subscribers. This is an unmoderated list with no digest.

Subscribe by sending email to majordomo@halcyon.com
Send messages to the list at fire-alarm@halcyon.com

For more details, contact Ed Robisheaux: eroby@halcyon.com

FIRE-L

FIRE-L is the place to discuss any firefighting related topic—from new apparatus to the "big one" that happened near you. Feel free to discuss anything related to firefighting.

Subscribe by sending email to listproc@cornell.edu
Send messages to the list at fire-l@cornell.edu

For more details, contact Gil Emery: gemery@tiac.net

FIRE-LIST

The Fire-List is devoted to the fire protection community. Its mission is to facilitate communication among professionals in the fire protection community: fire equipment dealers, fire protection engineers, fire inspectors, insurance personnel, and others with an interest in the proper selection and maintenance of fire suppression equipment. The list is sponsored by the National Association of Fire Equipment Distributors (NAFED).

Subscribe by sending email to majordomo@halcyon.com
Send messages to the list at fire-list@halcyon.com

For more details, contact Ed Robisheaux: eroby@halcyon.com

FIRENET

Fire fighting and emergency discussions.

Subscribe by sending email to majordomo@online.anu.edu.au
Send messages to the list at firenet@online.anu.edu.au

For more details, contact chris.trevitt@anu.edu.au

HAZARDS

This is the *Disaster Research* electronic newsletter. DR comes out approximately twice monthly.

Subscribe by sending email to listproc@lists.colorado.edu
Send messages to the list at hazards@lists.colorado.edu

HAZMIT

Hazard Mitigation Discussion. HAZMIT is a moderated discussion list for the global hazard mitigation community. The list is in English and covers both natural and technological disasters.

Subscribe by sending email to hazmit-request@mitigation.com
Send messages to the list at hazmit@mitigation.com

For more details, contact Christian Stalberg: listmaster@mitigation.com

HMEP

U.S. Federal Emergency Management Agency, Hazardous Materials Emergency Planning.

Subscribe by sending email to majordomo@fema.gov
Send messages to the list at hmep@fema.gov

For more details, contact eipa@fema.gov

IAEM-Y2K

From the International Association of Emergency Managers, this list is for individuals working in emergency management to share thoughts and concerns about Y2K (year 2000).

Subscribe by sending email to majordomo@new-focus.org
Send messages to the list at iaem-y2k@new-focus.org

For more details, contact Michael Bittle: mbittle@new-focus.org

LEPC

Hazardous materials emergency response planning.

Subscribe by sending email to listserv@list.uvm.edu
Send messages to the list at lepc@list.uvm.edu

For more details, contact Ralph Stuart: rstuart@esf.uvm.edu

NATURAL-HAZARDS-DISASTERS

This is a discussion group in the field of natural hazards and disasters (earthquakes, volcanoes, floods, hurricanes, tornadoes, tsunamis, droughts, lightning, hailstorms, landslides, etc.). It includes scholars from different disciplines.

Subscribe by sending email to mailbase@mailbase.ac.uk
Send messages to the list at natural-hazards-disasters@mailbase.ac.uk

For more details, contact Josie Difrancesco: josie1@mdx.ac.uk
Kathy Ingrey: k.ingrey@mdx.ac.uk

NCEMSF-L

The National Collegiate EMS Foundation (NCEMSF) list was established to help campus EMS groups communicate. NCEMSF is an "open" list, meaning that people can post whatever they want. Members are encouraged to participate in discussions. Anyone who has an active interest in campus emergency medicine may join the list.

Subscribe by sending email to listproc@hubcap.clemson.edu
Send messages to the list at ncemsf-l@hubcap.clemson.edu

For more details, contact Scott C. Savett: webmaster@ncemsf.org

NEWS

This list is used to distribute news releases and advisories from the U.S. Federal Emergency Management Agency (FEMA).

Subscribe by sending email to majordomo@fema.gov
Send messages to the list at news@fema.gov

For more details, contact eipa@fema.gov

PUBLICSAFETYNEWS

News and information for public safety and emergency services professionals.

Subscribe by sending email to listserv@listserv.aol.com
Send messages to the list at publicsafetynews@listserv.aol.com

For more details, contact D. L. Crichlow: PSATDoug@aol.com

SES

Australian State Emergency Services (SES) discussions. To subscribe, send email to ses@gec.com.au with the word SUBSCRIBE as the subject line.

Subscribe by sending email to ses@gec.com.au
Send messages to the list at ses@gec.com.au

For more details, contact David Muller: davidm@zip.com.au

SITREP

This list is used to distribute major incident abridged situation reports from the U.S. Federal Emergency Management Agency.

Subscribe by sending email to majordomo@fema.gov

Send messages to the list at sitrep@fema.gov

For more details, contact eipa@fema.gov

URG-L

URG-L is a French language mailing list for international discussions among emergency physicians. URG-L is the official list of AMUQ, l'association des médecins d'urgence du Québec (Quebec Emergency Physicians Association), but is open to anyone with an interest in emergency medicine or related disciplines.

Subscribe by sending email to listserv@home.ease.lsoft.com

Send messages to the list at urg-l@home.ease.lsoft.com

For more details, contact Alain Vadeboncoeur: alainvad@MBR.centra.ca

Y2K-NEWS

This is a one-way broadcast newsletter of current news releases on Y2K issues, particularly those pertaining to emergency management and public safety. To submit items for broadcasting, send to mekabay@compuserve.com.

Subscribe by sending email to majordomo@new-focus.org

Send messages to the list at y2k-news@new-focus.org

For more details, contact Michael Bittle: mbittle@new-focus.org

Ergonomics/Human Factors

ENGPSY

Discussion and notices list for the British Psychological Society Engineering Psychology Special Interest Group. All aspects of human factors and ergonomics research covered. Also for notices of meetings and enquiries.

Subscribe by sending email to mailbase@mailbase.ac.uk

Send messages to the list at engpsy@mailbase.ac.uk

For more details, contact Mark Young: myoung@soton.ac.uk

ERGOMED

Ergonomics and human factors in medicine.

> Subscribe by sending email to listproc@ucdavis.edu
> Send messages to the list at ergomed@ucdavis.edu

ERGONOMICS

This list aims to promote the discussion of ideas and principles relating to ergonomics and human factors. The list is aimed primarily at practicing ergonomists, but anyone from other fields with an interest in ergonomics is welcome to join.

> Subscribe by sending email to mailbase@mailbase.ac.uk
> Send messages to the list at ergonomics@mailbase.ac.uk

For more details, contact Jayne and Tim: avru.erg@derby.ac.uk

ERGOWEB-L

This list is a place where registered ErgoWeb(R) users can exchange information and ideas and discuss current topics related to ergonomics. Registration is free. The list is moderated. Register through the ErgoWeb web site at http://www.ergoweb.com.

> Subscribe by sending email to http://www.ergoweb.com
> Send messages to the list at ergoweb-l@ergoweb.com

HFS-L

Human Factors and Ergonomics Society, Virginia Tech Chapter.

> Subscribe by sending email to listserv@listserv.vt.edu
> Send messages to the list at hfs-l@listserv.vt.edu

For more details, contact John Kelso: kelso@vt.edu

PCHEALTH

PCHealth is intended to promote sharing of information, experiences, concerns, and advice about the health effects of working with and near computers.

> Subscribe by sending email to listserv@listserv.aol.com
> Send messages to the list at pchealth@listserv.aol.com

For more details, contact Ballew Kinnaman: kinnaman@Emissary.Net

RSI-EAST

St. John's University Repetitive Strain Injury List.

Subscribe by sending email to listserv@maelstrom.stjohns.edu

Send messages to the list at rsi-east@maelstrom.stjohns.edu

For more details, contact   drz@dorsai.org

or   rik@world.std.com

SOREHAND

A mailing list for the discussion of carpal tunnel syndrome (CTS), tendonitis, repetitive strain injuries (RSI), etc.

Subscribe by sending email to listserv@itssrv1.ucsf.edu

Send messages to the list at sorehand@itssrv1.ucsf.edu

For more details, contact Deanna McHugh: deannam@itsa.ucsf.edu

Industrial Safety

BUSHEA-L

Health related information for business and industry.

Subscribe by sending email to listserv@siu.edu

Send messages to the list at bushea-l@siu.edu

For more details, contact Mark Kittleson: kittle@siu.edu

CONSPACE-LIST

This is a forum for discussions related to confined space. It is intended to serve as a forum for discussion by the Confined Space Technical Committee of the American Industrial Hygiene Association, and for general exchanges on the subject.

Subscribe by sending email to majordomo@lists.aiha.org

Send messages to the list at conspace-list@lists.aiha.org

For more details, contact John Meagher: jmeagher@aiha.org

INDUSTRIAL-RELATIONS-RESEARCH

A forum for academic discussion of industrial relations broadly conceived. It covers current research, methods, results, and theories about employment relations, collective relationships, trade unions, HRM, and employment law.

Subscribe by sending email to mailbase@mailbase.ac.uk
Send messages to the list at industrial-relations-research@mailbase.ac.uk

For more details, contact Robert Blackburn: r.a.blackburn@kingston.ac.uk

ISO14000

This list is designed for the discussion of the ISO 14000 certification guidelines for environmental and related industries. It is currently unmoderated.

Subscribe by sending email to majordomo@quality.org
Send messages to the list at iso14000@quality.org

For more details, contact Gene Tatsch: cet@rti.org

MHSSN-L

The Maquiladora Health and Safety Support Network is a volunteer network of occupational health and safety professionals who have placed their names on a resource list to provide information, technical assistance, and onsite instruction regarding workplace hazards in the over 3,800 *maquiladora* (foreign-owned assembly plants) along the U.S.-Mexico border.

Subscribe by sending email to majordomo@igc.org
Send messages to the list at mhssn-l@igc.org

PRINTING

Printing industry discussions.

Subscribe by sending email to majordomo@nexus.netconcepts.com
Send messages to the list at printing@printer.net

QUALITY

Total Quality Management (TQM) in manufacturing and service industries.

Subscribe by sending email to listserv@pucc.princeton.edu
Send messages to the list at quality@pucc.princeton.edu

QUEST

The emerging national and international requirement to combine quality, environment, and safety structures and systems assessments is creating the need for the development of a uniform and integrated management discipline. The purpose of this discussion is to define and develop a management system which integrates these three disciplines into an effective operational management system. Discussion will include the interrelation of support and responsibility and the identification of benefits, resources, and assessment objectives.

> Subscribe by sending email to listserv@listserv.nodak.edu
>
> Send messages to the list at quest@listserv.nodak.edu

For more details, contact jennejohn@uwstout.edu

SAFETYPAPER

This group is for individuals with an interest in occupational health and safety as it relates to the pulp and paper and corrugated paper industries. To subscribe, send a message to pandre@pphsa.on.ca with SUBSCRIBE SAFETYPAPER in the subject line.

> Subscribe by sending email to pandre@pphsa.on.ca
>
> Send messages to the list at safetypaper@egroups.com

For more details, contact Paul Andre: pandre@pphsa.on.ca

SAFTNEWS

SAFTNEWS is a free biweekly newsletter specifically designed for U.S. and Canadian businesses interested in staying up-to-date with the latest in safety and industrial news. The focus is on new and upcoming regulations and how they will affect your workplace.

> Subscribe by sending email to listserv@labsafety.com
>
> Send messages to the list at saftnews@labsafety.com

SHIFTWORK

Shiftwork is a moderated forum that provides a place for shiftworkers and their family members, shiftwork managers, researchers, and other experts in the shiftworking field to discuss the many different aspects of working in round-the-clock operations.

Subscribe by sending email to majordomo@world.std.com
Send messages to the list at shiftwork@world.std.com

For more details, contact Keith C. Bagley: kcb@circadian.com

STEEL-TALK

Open discussion of steel industry issues.

Subscribe by sending email to majordomo@igc.org
Send messages to the list at steel-talk@igc.org

For more details, contact Ted Kuster: tkuster@igc.apc.org

TOTAL-QUALITY-ISOSTDS

A forum for the exchange of ideas and information by researchers and practitioners involved in the certification, development, and management of environments to ISO Standards.

Subscribe by sending email to mailbase@mailbase.ac.uk
Send messages to the list at total-quality-isostds@mailbase.ac.uk

For more details, contact

Jim McMenamin: jmcmenam@pmc1.pmc.port.ac.uk
Tony Moody: tmoody@pmc1.pmc.port.ac.uk
Christopher Seow: mcyi6cks@fs1.sm.umist.ac.uk

Industrial/Occupational Hygiene

CONSPACE-LIST

This is a forum for discussions related to confined space. It is intended to serve as a forum for discussion by the Confined Space Technical Committee of the American Industrial Hygiene Association, and for general exchanges on the subject.

Subscribe by sending email to majordomo@lists.aiha.org
Send messages to the list at conspace-list@lists.aiha.org

For more details, contact John Meagher: jmeagher@aiha.org

GLOBALOCCHYG-LIST

The Global Occupational Hygiene list was formed in order to discuss initiatives to advance the profession of industrial/occupational hygiene and prevent occupational injury and diseases globally. While serving as the official list of the International Affairs Committee of the American Industrial Hygiene Association, participation from other International Occupational Hygiene Association (IOHA) members is being actively solicited.

Subscribe by sending email to majordomo@intr.net
Send messages to the list at globalocchyg-list@intr.net

For more details, contact Jim Platner: jplatner@usa.net

IH-LIST

IH-list is a list for the general use of the industrial hygienists and interested professionals who wish to discuss topics of general industrial hygiene or ask questions of the listserv community. This list is open to all subscribers and is unmonitored.

Subscribe by sending email to majordomo@lists.aiha.org
Send messages to the list at ih-list@lists.aiha.org

For more details, contact Ed Bartosh: ebartosh@bartosh.com

INDUSTHYG

INDUSTHYG is an electronic mailing list designed to widely distribute industrial hygiene information to the industrial hygiene community. As a member of INDUSTHYG, you receive the *Army Industrial Hygiene Newsletter* approximately three times per year, Monthly Regulatory Summaries, job announcements in the field of industrial hygiene, and other items of interest to the industrial hygiene community.

Subscribe by sending email to bwolbert@aeha1.apgea.army.mil
Send messages to the list at bwolbert@aeha1.apgea.army.mil

For more details, contact Brenda Wolbert: bwolbert@aeha1.apgea.army.mil

OCCHYGPRO

Developed by the Canadian Registration Board of Occupational Hygienists (CRBOH) in collaboration with the Canadian Centre for Occupational Health and Safety (CCOHS), the intent of the list is to provide a high-quality, low-volume forum for exchange of information and ideas, where substantive, cutting-edge matters of relevance to the hygiene profession can be addressed effectively by eliciting comment and/or consensus among other experienced and insightful professionals, in conformity to our codes of ethics.

Subscribe by sending email to majordomo@ccohs.ca
Send messages to the list at occhygpro@ccohs.ca

For more details, contact Ugis Bickis: uib@phoenix-ohc.on.ca

UKOH

This list is for the discussion of matters relating to occupational hygiene in the U.K.

Subscribe by sending email to http://www.findmail.com
Send messages to the list at ukoh@findmail.com

For more details, contact david@bloor.demon.co.uk

Laboratory Safety

CAARL-L

Forum for the Colorado Association of Academic and Research Laboratories, an environmental regulation reinvention focus group in Colorado.

Subscribe by sending email to listserv@coloradocollege.edu
Send messages to the list at caarl-l@coloradocollege.edu

For more details, contact John DeLaHunt: jdelahunt@coloradocollege.edu

CHEMLAB_L

The CHEMLAB_L list is for people interested in chemistry laboratories (both academic and research), students' experiments (high school, college, and university), classroom demonstrations and shows for the public of chemical processes, chemistry stockroom management, lab safety, and small-scale chemical waste handling procedures. To subscribe, put the word SUBSCRIBE in the subject line of your message.

Subscribe by sending email to chemlab_l@vax1.bemidji.msus.edu
Send messages to the list at chemlab_l@vax1.bemidji.msus.edu

For more details, contact Dr. Gerald Morine: ghmo@vax1.bemidji.msus.edu

LAB-XL

Performance-oriented environmental regulation of laboratories.

Subscribe by sending email to listserv@list.uvm.edu
Send messages to the list at lab-xl@list.uvm.edu

For more details, contact Ralph Stuart: rstuart@esf.uvm.edu

LABSAFETY-L

LABSAFETY-L is a discussion list for members of Laboratory Safety Workshop and National Association of Chemical Hygiene officers interested in lab health and safety issues.

Subscribe by sending email to listserv@siu.edu
Send messages to the list at labsafety-l@siu.edu

For more details, contact James A. Kaufman: labsafe@aol.com

LAB_SAFETY

Laboratory safety discussion list.

Subscribe by sending email to listserv@listserv.vt.edu
Send messages to the list at lab_safety@listserv.vt.edu

For more details, contact Donald Conner: dcon@vt.edu

Natural Hazards

NATURAL-HAZARDS-DISASTERS

This is a discussion group in the field of natural hazards and disasters (earthquakes, volcanoes, floods, hurricanes, tornadoes, tsunamis, droughts, lightning, hailstorms, landslides, etc.). It includes scholars from different disciplines.

Subscribe by sending email to mailbase@mailbase.ac.uk
Send messages to the list at natural-hazards-disasters@mailbase.ac.uk

For more details, contact Josie Difrancesco: josie1@mdx.ac.uk
Kathy Ingrey: k.ingrey@mdx.ac.uk

Noise/Hearing

AUDIOL-L

This is an open forum for audiologists and others who have an interest in all aspects of clinical audiology, audition, and professional issues in the field of audiology.

Subscribe by sending email to listproc2@ecnet.net
Send messages to the list at audiol-l@ecnet.net

BEYOND-HEARING

This list is intended to provide a communication vehicle for people who have a hearing loss and who seek to overcome the barriers of hearing loss between themselves, other people, and the environment.

Subscribe by sending email to majordomo@acpub.duke.edu
Send messages to the list at beyond-hearing@acpub.duke.edu

For more details, contact Miriam (Mimi) Clifford: dmimi@acpub.duke.edu

NOISE_AND_HEARING

Basic and applied questions and discussions of noise-induced hearing loss.

Subscribe by sending email to listserv@listserv.cdc.gov
Send messages to the list at noise_and_hearing@listserv.cdc.gov

For more details, contact Rick Davis: rrd1@cdc.gov

Occupational Medicine

AEROSO-L

A discussion group for researchers involved with the determination of health effects associated with exposure to various concentrations and classes of particulate matter (PM) in ambient air (*e.g.*, toxicologists and epidemiologists). List members discuss topics such as generation and characterization methods of aerosols' biological effect parameters (*e.g.*, biochemical and immunological) and respiratory tract dosimetry models.

Subscribe by sending email to listserv@nic.surfnet.nl
Send messages to the list at aeroso-l@nic.surfnet.nl

For more details, contact FR.Cassee@rivm.NL

ALLERGY

The Allergy mailing list discusses human allergies of all types: how allergies impact our health and lifestyles, treatments for allergies from the consumer perspective and experience, self-help prevention of allergy symptoms, allergy self-care, allergy support systems, and basic facts about these topics.

Subscribe by sending email to listserv@listserv.tamu.edu
Send messages to the list at allergy@listserv.tamu.edu

For more details, contact Ballew Kinnaman: kinnaman@immune.com

CO-CURE

Co-Cure was established in 1996 with the goal of furthering cooperative efforts towards finding the cure for Chronic Fatigue Syndrome (CFS) and the related Fibromyalgia (FM) through the open distribution and exchange of information between the medical/clinical, political, and patient groups, as well as other organizations/institutions.

Subscribe by sending email to listserv@listserv.nodak.edu
Send messages to the list at co-cure@listserv.nodak.edu

For more details, contact Donna Tabish: djtabish@islandnet.com

HEALTH-WORK

North American Health Care Workers Network.

Subscribe by sending email to majordomo@icg.org
Send messages to the list at health-work@icg.org

HEPC

U.K.-based Hepatitis C discussion group.

Subscribe by sending email to hepc-request@lists.vossnet.co.uk
Send messages to the list at hepc@lists.vossnet.co.uk

For more details, contact Daniel Dimitriou: crina@vossnet.co.uk

HIV-CONF

HIV-CONF is maintained by the U.S. Centers for Disease Control and Prevention, and moderated by the Division of HIV/AIDS Prevention, Technical Information and Communications Branch (DHAP-TICB). The purpose of the HIV-CONF mailing list is to notify the public when HIV/AIDS conferences and trainings are announced and where to obtain more information. Usually, the frequency of notification is not more than once every month.

Subscribe by sending email to listserv@listserv.cdc.gov

Send messages to the list at hiv-conf@listserv.cdc.gov

HIV-HASR

HIV-HASR is maintained by the U.S. Centers for Disease Control and Prevention and moderated by the Division of HIV/AIDS Prevention, Technical Information and Communications Branch (DHAP-TICB). The purpose of the HIV-HASR mailing list is to notify the public when the HIV/AIDS Surveillance Report has been released and where the report can be accessed. Usually, the frequency of notification is not more than twice every year.

Subscribe by sending email to listserv@listserv.cdc.gov

Send messages to the list at hiv-hasr@listserv.cdc.gov

HIV-MMWR

HIV-MMWR is maintained by the U.S. Centers for Disease Control and Prevention and moderated by the Division of HIV/AIDS Prevention, Technical Information and Communications Branch (DHAP-TICB). The purpose of the HIV-MMWR mailing list is to notify the public as soon as HIV/AIDS related issues of the Morbidity and Mortality Weekly Report (MMWR) have been released and where the reports can be accessed. Usually, the frequency of notification is not more than twice every month.

Subscribe by sending email to listserv@listserv.cdc.gov

Send messages to the list at hiv-mmwr@listserv.cdc.gov

HIV-PUBS

HIV-PUBS is maintained by the U.S. Centers for Disease Control and Prevention, and moderated by the Division of HIV/AIDS Prevention, Technical Information and Communications Branch (DHAP-TICB). The purpose of the HIV-PUBS mailing list is to notify the public when HIV/AIDS Publications have been released and where the reports can be accessed. Usually, the frequency of notification is not more than twice every month.

Subscribe by sending email to listserv@listserv.cdc.gov

Send messages to the list at hiv-pubs@listserv.cdc.gov

HIV-STAT

HIV-STAT is maintained by the U.S. Centers for Disease Control and Prevention, and moderated by the Division of HIV/AIDS Prevention, Technical Information and Communications Branch (DHAP-TICB). The purpose of the HIV-STAT mailing list is to notify the public about updates to HIV/AIDS Basic Statistics reports and where they can be accessed. Usually, the frequency of notification is not more than twice every year.

Subscribe by sending email to listserv@listserv.cdc.gov
Send messages to the list at hiv-stat@listserv.cdc.gov

IMMUNE

We discuss immune system disorders such as Chronic Fatigue Syndrome, Multiple Chemical Sensitivity, Fibromyalgia, Lupus, Candida, Asthma, etc. It is a support group for people with any of these conditions and also for those who know and/or work with them.

Subscribe by sending email to immune@best.com
Send messages to the list at immune@best.com

JOB-MDT

This list is a French language distribution channel for job offers for and requests for employment by occupational health professionals (occupational health physicians, ergonomists, industrial hygienists, occupational health nurses, etc.). To subscribe, send a blank message to liste@meditrav.com with the word SUBSCRIBE in the subject line.

Subscribe by sending email to liste@meditrav.com
Send messages to the list at job-mdt@meditrav.com

For more details, contact admJOB@meditrav.com

MCS-IMMUNE-NEURO

MCS/Chemical Injury Support. Information exchange, support group, and management strategies for people adversely effected by chemicals. Owner Ginny Kloth is the volunteer Director of Electronic Communications for the Chemical Injury Information Network (CIIN).

Subscribe by sending email to listserv@maelstrom.stjohns.edu
Send messages to the list at mcs-immune-neuro@maelstrom.stjohns.edu

For more details, contact bijou@blrg.tds.net (Ginny Kloth)

MMWR-ASC

U.S. Centers for Disease Control and Prevention's Morbidity and Mortality Weekly Report (MMWR), ASCII version.

> Subscribe by sending email to listserv@listserv.cdc.gov
> Send messages to the list at mmwr-asc@listserv.cdc.gov

For more details, contact mmwrq@epo.em.cdc.gov

MMWR-TOC

This is a distribution list for Tables of Contents of the Morbidity and Mortality Weekly Report (MMWR), from the U.S. Centers for Disease Control and Prevention.

> Subscribe by sending email to listserv@listserv.cdc.gov
> Send messages to the list at mmwr-toc@listserv.cdc.gov

For more details, contact mmwrq@epo.em.cdc.gov

NURSENET

NURSENET is an open unmoderated list for discussions about a variety of nursing issues.

> Subscribe by sending email to listserv@listserv.utoronto.ca
> Send messages to the list at nursenet@listserv.utoronto.ca

For more details, contact Judy Norris: judy.norris@ualberta.ca

OCC-ENV-MED-L

Occupational and environmental medicine represents a growing clinical and public health discipline seeking to evaluate and prevent the diseases and health effects that may be related to exposures at work and from other environments (*e.g.*, pollution). This list provides a moderated forum for announcements, dissemination of text files, and academic discussion and allows the presentation of clinical vignettes, synopses of new regulatory issues, and reports of interesting items from publications elsewhere (both the medical and the nonmedical journals).

> Subscribe by sending email to listserv@listserv.duhc.duke.edu
> Send messages to the list at occ-env-med-l@listserv.duhc.duke.edu

For more details, contact
> Dr. Gary Greenberg: green011@mc.duke.edu
> (or): AOEC (aoec@DGS.dgsys.com)

OCCENVMED

This list will be of interest to all practitioners of occupational and environmental medicine and occupational health. Its aim is to promote discussion about current issues and to foster a global approach to research and teaching. Occenvmed is a United Kingdom-based list. Nonetheless, membership is encouraged from other European member states, as well as from other parts of the world.

Subscribe by sending email to mailbase@mailbase.ac.uk
Send messages to the list at occenvmed@mailbase.ac.uk

OEM-ANNOUNCE

This is a moderated list for the distribution of notices about upcoming events and items of interest to occupational and environmental medicine professionals.

Subscribe by sending email to listserv@listserv.duhc.duke.edu
Send messages to the list at oem-announce@listserv.duhc.duke.edu

For more details, contact Dr. Gary Greenberg: gary.greenberg@duke.edu

OHN-LIST

Ohn-list is an open, unmoderated, international mailing list for occupational health nurses and allied professionals.

Subscribe by sending email to majordomo@oise.utoronto.ca
Send messages to the list at ohn-list@oise.utoronto.ca

For more details, contact Jane Lemke: klemke@hookup.net
Suzanne Arnold: smarnold@oise.utoronto.ca

PAIN-L

The Pain Forum covers all aspects of chronic pain: physical and political, health and health-politics, for physicians and patients and all interested parties. Everyone is confronted with pain at one time or another in their lives. This is an educational forum.

Subscribe by sending email to listserv@maelstrom.stjohns.edu
Send messages to the list at pain-l@maelstrom.stjohns.edu

For more details, contact Hank Roth: odin@netline.net

RUBBER

The rubber and latex allergy mailing list discusses all aspects of this allergy from research and diagnosis to treatment, support, and economic ramifications.

Subscribe by sending email to listserv@listserv.tamu.edu
Send messages to the list at rubber@listserv.tamu.edu

For more details, contact rubber-request@listserv.tamu.edu

TOTAL-QUALITY-HEALTHCARE

A forum to present information, comment, and debate on quality management issues, trends, and developments which affect the healthcare industry.

Subscribe by sending email to mailbase@mailbase.ac.uk
Send messages to the list at total-quality-healthcare@mailbase.ac.uk

For more details, contact Christopher Seow: mcyi6cks@fs1.sm.umist.ac.uk

Office Safety

AFSCME-OSH

This list was designed for American Federation of State, County and Municipal Employees (AFSCME) members and other interested persons to exchange information about health and safety conditions in their workplaces, and solutions and strategies that they have found to be successful or unsuccessful. It can be a place to discuss political action ideas and strategies concerning health and safety issues—either on the local or national level. It can also be used to distribute announcements and alerts.

Subscribe by sending email to majordomo@igc.org
Send messages to the list at afscme-osh@igc.org

For more details, contact Jordan Barab: jbarab@afscme.org

COMFORTZONE

The purpose of COMFORTZONE is to open up communications among members of the international thermal comfort community.

Subscribe by sending email to majordomo@penman.es.mq.edu.au
Send messages to the list at comfortzone@penman.es.mq.edu.au

For more details, contact Dr. Richard de Dear: rdedear@laurel.ocs.mq.edu.au

CTDNEWS

Cumulative Trauma Disorder (CTD) News Update List

Subscribe by sending email to listserv@lrp5.lrp.com

Send messages to the list at ctdnews@lrp5.lrp.com

IAQ LIST

The IAQ List discusses issues and concerns regarding indoor air quality. A variety of things can pollute and contaminate the indoor air quality of a home, office, or building—from molds and bioaerosols to combustion by-products like carbon monoxide, soot, PAHs, and VOCs. Many can cause serious health concerns and, in some cases, extensive property damage. This list will address the identification and mitigation of contaminants.

Subscribe by sending email to http://www.onelist.com

Send messages to the list at iaq@onelist.com

For more details, contact C. Flanders, IAQ List Manager: rkfabf@aol.com

PCHEALTH

PCHealth is intended to promote sharing of information, experiences, concerns, and advice about the health effects of working with and near computers.

Subscribe by sending email to listserv@listserv.aol.com

Send messages to the list at pchealth@listserv.aol.com

For more details, contact Ballew Kinnaman: kinnaman@Emissary.Net

RSI-EAST

St. John's University Repetitive Strain Injury List.

Subscribe by sending email to listserv@maelstrom.stjohns.edu

Send messages to the list at rsi-east@maelstrom.stjohns.edu

For more details, contact drz@dorsai.org

(or) rik@world.std.com

SOREHAND

A mailing list for the discussion of carpal tunnel syndrome (CTS), tendonitis, repetitive strain injuries (RSI), etc.

Subscribe by sending email to listserv@itssrv1.ucsf.edu
Send messages to the list at sorehand@itssrv1.ucsf.edu

For more details, contact Deanna McHugh: deannam@itsa.ucsf.edu

Professional Associations/Societies

CONS-EQST-PESTICIDE-FORUM

The purpose of the CONS-EQST-PESTICIDE-FORUM list is to facilitate discussion and exchange of information among Sierra Club members on issues associated with the manufacture, use, and licensing of pesticides. This forum is open to all current Sierra Club members.

Subscribe by sending email to listserv@lists.sierraclub.org
Send messages to the list at cons-eqst-pesticide-forum@lists.sierraclub.org

For more details, contact mm6@doc.mssm.edu

EHS-BC

Environmental Health and Safety - Boston Consortium

Subscribe by sending email to listserv@mitvma.mit.edu
Send messages to the list at ehs-bc@mitvma.mit.edu

For more details, contact Greenley@MIT.Edu
(or) Zbitnoff@Babson.Edu

ENGPSY

Discussion and notices list for the British Psychological Society, Engineering Psychology Special Interest Group. All aspects of human factors and ergonomics research covered. Also for notices of meetings and inquiries.

Subscribe by sending email to mailbase@mailbase.ac.uk
Send messages to the list at engpsy@mailbase.ac.uk

For more details, contact Mark Young: myoung@soton.ac.uk

FORUM

National Environmental Training Association Trainer's Forum

Subscribe by sending email to majordomo@envirotraining.org
Send messages to the list at forum@envirotraining.org

For more details, contact Charles (Rick) L. Richardson:

rick@envirotraining.org

HFS-L

Human Factors and Ergonomics Society, Virginia Tech Chapter.

Subscribe by sending email to listserv@listserv.vt.edu
Send messages to the list at hfs-l@listserv.vt.edu

For more details, contact John Kelso: kelso@vt.edu

NAOSMM

This list is a vehicle whereby members of the National Association of Scientific Materials Managers (NAOSMM) can maintain contact.

Subscribe by sending email to listserv@listserv.rice.edu
Send messages to the list at naosmm@listserv.rice.edu

For more details, contact Tom Blakeney: blakeney@ruf.rice.edu
John Elliott: jelliott@ruf.rice.edu

Public Health

ALLERGY

The Allergy mailing list discusses human allergies of all types: how allergies impact our health and lifestyles, treatments for allergies from the consumer perspective and experience, self-help prevention of allergy symptoms, allergy self-care, allergy support systems, and basic facts about these topics.

Subscribe by sending email to listserv@listserv.tamu.edu
Send messages to the list at allergy@listserv.tamu.edu

For more details, contact Ballew Kinnaman: kinnaman@immune.com

CO-CURE

Co-Cure was established in 1996 with the goal of furthering cooperative efforts towards finding the cure for Chronic Fatigue Syndrome (CFS) and the related Fibromyalgia (FM) through the open distribution and exchange of information between the medical/clinical, political, and patient groups, as well as other organizations/institutions.

Subscribe by sending email to listserv@listserv.nodak.edu
Send messages to the list at co-cure@listserv.nodak.edu

For more details, contact Donna Tabish: djtabish@islandnet.com

EID-ASCII

U.S. Centers For Disease Control and Prevention's EMERGING INFEC-
TIOUS DISEASES (EID). ASCII format.

Subscribe by sending email to listserv@listserv.cdc.gov
Send messages to the list at eid-ascii@listserv.cdc.gov

For more details, contact eidlistmgr@cdc.gov

EID-PDF

This is a distribution list for EMERGING INFECTIOUS DISEASES (EID),
from the U.S. Centers for Disease Control and Prevention in PDF (Adobe
Acrobat) format.

Subscribe by sending email to listserv@listserv.cdc.gov
Send messages to the list at eid-pdf@listserv.cdc.gov

EID-TOC

This is a distribution list for the Tables of Contents of EMERGING INFEC-
TIOUS DISEASES (EID), from the U.S. Centers for Disease Control and
Prevention.

Subscribe by sending email to listserv@listserv.cdc.gov
Send messages to the list at eid-toc@listserv.cdc.gov

For more details, contact eidlistmgr@cdc.gov

HEALTH-WORK

North American Health Care Workers Network.

Subscribe by sending email to majordomo@icg.org
Send messages to the list at health-work@icg.org

HEPC

U.K.-based Hepatitis C discussion group.

Subscribe by sending email to hepc-request@lists.vossnet.co.uk
Send messages to the list at hepc@lists.vossnet.co.uk

For more details, contact Daniel Dimitriou: crina@vossnet.co.uk

HEPCSC

Hepatitis C support and information mailing list for Canadians.

Subscribe by sending email to majordomo@mail.island.net

Send messages to the list at hepcbc@mail.island.net

For more details, contact hepcbc@iforward.com

HIV-CONF

HIV-CONF is maintained by the U.S. Centers for Disease Control and Prevention and moderated by the Division of HIV/AIDS Prevention, Technical Information and Communications Branch (DHAP-TICB). The purpose of the HIV-CONF mailing list is to notify the public when HIV/AIDS conferences and trainings are announced and where to obtain more information. Usually, the frequency of notification is not more than once every month.

Subscribe by sending email to listserv@listserv.cdc.gov

Send messages to the list at hiv-conf@listserv.cdc.gov

HIV-HASR

HIV-HASR is maintained by the U.S. Centers for Disease Control and Prevention and moderated by the Division of HIV/AIDS Prevention, Technical Information and Communications Branch (DHAP-TICB). The purpose of the HIV-HASR mailing list is to notify the public when the HIV/AIDS Surveillance Report has been released and where the report can be accessed. Usually, the frequency of notification is not more than twice every year.

Subscribe by sending email to listserv@listserv.cdc.gov

Send messages to the list at hiv-hasr@listserv.cdc.gov

HIV-MMWR

HIV-MMWR is maintained by the U.S. Centers for Disease Control and Prevention and moderated by the Division of HIV/AIDS Prevention, Technical Information and Communications Branch (DHAP-TICB). The purpose of the HIV-MMWR mailing list is to notify the public as soon as HIV/AIDS related issues of the Morbidity and Mortality Weekly Report (MMWR) have been released and where the reports can be accessed. Usually, the frequency of notification is not more than twice every month.

Subscribe by sending email to listserv@listserv.cdc.gov

Send messages to the list at hiv-mmwr@listserv.cdc.gov

HIV-PUBS

HIV-PUBS is maintained by the U.S. Centers for Disease Control and Prevention and moderated by the Division of HIV/AIDS Prevention, Technical Information and Communications Branch (DHAP-TICB). The purpose of the HIV-PUBS mailing list is to notify the public when HIV/AIDS publications have been released and where the reports can be accessed. Usually, the frequency of notification is not more than twice every month.

Subscribe by sending email to listserv@listserv.cdc.gov

Send messages to the list at hiv-pubs@listserv.cdc.gov

HIV-STAT

HIV-STAT is maintained by the U.S. Centers for Disease Control and Prevention and moderated by the Division of HIV/AIDS Prevention, Technical Information and Communications Branch (DHAP-TICB). The purpose of the HIV-STAT mailing list is to notify the public about updates to HIV/AIDS Basic Statistics reports and where they can be accessed. Usually, the frequency of notification is not more than twice every year.

Subscribe by sending email to listserv@listserv.cdc.gov

Send messages to the list at hiv-stat@listserv.cdc.gov

IMMUNE

We discuss immune system disorders such as Chronic Fatigue Syndrome, Multiple Chemical Sensitivity, Fibromyalgia, Lupus, Candida, Asthma, etc. It is a support group for people with any of these conditions and those who know and/or work with them.

Subscribe by sending email to immune@best.com

Send messages to the list at immune@best.com

LEAD-POISON-EDU

Lead Poisoning Education from the U.S. Centers for Disease Control and Prevention.

Subscribe by sending email to listserv@listserv.cdc.gov

Send messages to the list at lead-poison-edu@listserv.cdc.gov

For more details, contact axn0@cdc.gov & amd2@cdc.gov

LEAD-POISON-SURV

Lead Poisoning Surveillance, from the U.S. Centers for Disease Control and Prevention.

Subscribe by sending email to listserv@listserv.cdc.gov
Send messages to the list at lead-poison-surv@listserv.cdc.gov

For more details, contact axn0@cdc.gov & amd2@cdc.gov

MMWR-ASC

U.S. Centers for Disease Control and Prevention's Morbidity and Mortality Weekly Report (MMWR), ASCII version.

Subscribe by sending email to listserv@listserv.cdc.gov
Send messages to the list at mmwr-asc@listserv.cdc.gov

For more details, contact mmwrq@epo.em.cdc.gov

MMWR-TOC

This is a distribution list for Tables of Contents of the Morbidity and Mortality Weekly Report (MMWR), from the U.S. Centers for Disease Control and Prevention.

Subscribe by sending email to listserv@listserv.cdc.gov
Send messages to the list at mmwr-toc@listserv.cdc.gov

For more details, contact mmwrq@epo.em.cdc.gov

OILREFINE-ACT

Citizens and workers for cleaner, safer oil refineries.

Subscribe by sending email to majordomo@icg.org
Send messages to the list at oilrefine-act@icg.org

PATIENTSAFETY-L

Patientsafety-l is an unmoderated mailing list "devoted to thoughtful conversation toward the development of a safer health care system. Patientsafety-l is primarily oriented to the interests of health professionals, systems and human factors specialists, and others involved in exploring the professional, consumer, and systems issues related to patient safety."

Subscribe by sending email to listserv@list.dis.net
Send messages to the list at patientsafety-l@list.dis.net

For more details, contact Lorri Zipperer: Lorri_Zipperer@ama-assn.org

PUBLIC-HEALTH

This list provides a discussion forum and information resource for those working in epidemiology and public health. It aims to facilitate information sharing (*e.g.*, workshops, seminars, conferences, and new research) and promote links, collaborative working, joint problem-solving, and mutual support.

Subscribe by sending email to mailbase@mailbase.ac.uk
Send messages to the list at public-health@mailbase.ac.uk

For more details, contact Richard Edwards: p.r.edwards@newcastle.ac.uk

URBAN-ENVIRONMENTAL-HEALTH

A list for information exchange and dissemination on the health impacts of urban environment related diseases and hazards. This includes health issues related to and/or affected by urban water supply, wastewater, solid waste, air pollution, food hygiene, and tropical diseases such as urban malaria.

Subscribe by sending email to mailbase@mailbase.ac.uk
Send messages to the list at urban-environmental-health@mailbase.ac.uk

Public Safety

CARGO-L

Cargo-L is meant for all parties involved in international transactions or that have to move cargo to and/or from anywhere.

Subscribe by sending email to MAISER@distart.ing.unibo.it
Send messages to the list at cargo-l@distart.ing.unibo.it

For more details, contact pingu@tin.it

DOT

DOT is intended for governmental transportation agencies at all levels: federal, state/provincial, and local. Topics such as highway design, light rail, HOV, CADD/GIS, computers, planning, right-of-way, environmental, legislation, project management, and more may be discussed. The list is intended for those who work FOR or WITH transportation governmental agencies worldwide.

Subscribe by sending email to listserv@listserv.nodak.edu
Send messages to the list at dot@listserv.nodak.edu

For more details, contact Kevin Andres: keandres@mail.patriot.net

DPRA

Disaster Prevention and Recovery Alliance discussion list.

Subscribe by sending email to majordomo@new-focus.org
Send messages to the list at dpra@new-focus.org

For more details, contact Michael Bittle: mbittle@new-focus.org

EMERGENCY-MANAGEMENT

The EMERGENCY-MANAGEMENT list is a resource for sharing of news, ideas and point-of-view about mitigation of hazards, preparedness for, response to, and recovery from major emergencies and disasters. Primary list users are emergency management coordinators of municipal and county governments. The EMERGENCY-MANAGEMENT list is an information service of Public Safety America.

Subscribe by sending email to listserv@listserv.aol.com
Send messages to the list at emergency-management@listserv.aol.com

For more details, contact Doug Crichlow: PSATDoug@aol.com

EMFLDS-L

Electromagnetics in medicine, science, and comunications. This is a list of those interested in electromagnetic fields as they pertain to our new "Electromagnetic Society." Vast scientific literature has been published in recent years on the biological interactions of EM fields with biological systems. It is hoped that EMFLDS-L allows and encourages scientific dialogue between scientists, engineers, clinicians, or anyone interested in furthering knowledge on EM fields, known biological interactions, measurement techniques, or epidemiological studies.

Subscribe by sending email to listserv@listserv.acsu.buffalo.edu

Send messages to the list at emflds-l@listserv.acsu.buffalo.edu

For more details, contact David Rodman: rodman@acsu.buffalo.edu

IAEM-Y2K

From the International Association of Emergency Managers, this list is for individuals working in emergency management to share thoughts and concerns about Y2K (year 2000).

Subscribe by sending email to majordomo@new-focus.org

Send messages to the list at iaem-y2k@new-focus.org

For more details, contact Michael Bittle: mbittle@new-focus.org

IPSP

Listserv for the Institute for Public Safety Partnerships.

Subscribe by sending email to listserv@listserv.uic.edu

Send messages to the list at ipsp@listserv.uic.edu

For more details, contact jchip@uic.edu

(or) marmst2@uic.edu

(or) jengold@uic.edu

PUBLICSAFETYNEWS

News and information for public safety and emergency services professionals.

Subscribe by sending email to listserv@listserv.aol.com

Send messages to the list at publicsafetynews@listserv.aol.com

For more details, contact D. L. Crichlow: PSATDoug@aol.com

ROAD-TRANSPORT-TECHNOLOGY

The list is for the use of academics and others interested in technical, operational, or regulatory aspects of heavy vehicles, road damage, bridges, vehicle/road interaction, weigh-in-motion, heavy vehicle safety, or other related subjects.

Subscribe by sending email to mailbase@mailbase.ac.uk

Send messages to the list at road-transport-technology@mailbase.ac.uk

SNYPSD-L

State University of New York public safety directors list.

Subscribe by sending email to listserv@ls.sysadm.suny.edu
Send messages to the list at snypsd-l@ls.sysadm.suny.edu

For more details, contact Bruce McBride: mcbridrb@sysadm.suny.edu

USAFETY

Detroit area urban safety discussions.

Subscribe by sending email to listserv@lists.wayne.edu
Send messages to the list at usafety@lists.wayne.edu

For more details, contact David Martin: ac2938@wayne.edu

Y2K-NEWS

This is a one-way broadcast newsletter of current news releases on Y2K issues, particularly those pertaining to emergency management and public safety. To submit items for broadcasting, send to mekabay@compuserve.com.

Subscribe by sending email to majordomo@new-focus.org
Send messages to the list at y2k-news@new-focus.org

For more details, contact Michael Bittle: mbittle@new-focus.org

Product Safety

CPSCINFO-L

Consumer Product Safety News.

Subscribe by sending email to listproc@cpsc.gov
Send messages to the list at cpscinfo-l@cpsc.gov

For more details, contact mcohn@cpsc.gov

EMC-PSTC

The EMC-PSTC is an informal group of people interested in product safety regulations and standards worldwide, networked electronically by mailing list. Its purpose is to provide a forum for the sharing of public, but possibly obscure, product safety or regulatory compliance information or related information with limited natural distribution.

Subscribe by sending email to majordomo@ieee.org
Send messages to the list at emc-pstc@ieee.org

For more details, contact Roger Volgstadt: Volgstadt_Roger@Tandem.com

Radiation Safety

MEDPHYS

Medical Physics mailing list.

Subscribe by sending email to listserv@lists.wayne.edu
Send messages to the list at medphys@lists.wayne.edu

For more details, contact Raj K. Mitra: rajm@mcn.net

RADSAFE

The Radiation Safety Distribution list is an electronic mailing list for health physicists, medical physicists, radiological engineers, and others who have a professional interest in matters related to radiation protection.

Subscribe by sending email to listserv@romulus.ehs.uiuc.edu
Send messages to the list at radsafe@romulus.ehs.uiuc.edu

For more details, contact Melissa Woo: melissa@romulus.ehs.uiuc.edu

RADSAFE-EU

RADSAFE-EU is the European equivalent to the international maillist RADSAFE. Purpose of the list: discussion of radiation safety issues of general interest to radiation and health physics professionals in the European Union, *e.g.*, (but not restricted to) harmonization of radiation protection regulations and procedures

Subscribe by sending email to majordomo@fz-juelich.de
Send messages to the list at radsafe-eu@fz-juelich.de

Regulatory Information

ACTUALITES-CCHST

This is a broadcast list for French-language news releases and product announce-ments from the Canadian Centre for Occupational Health and Safety (CCOHS).

Subscribe by sending email to majordomo@ccohs.ca
Send messages to the list at actualites-cchst@ccohs.ca

For more details, contact Chris Moore: chrism@ccohs.ca

CAARL-L

Forum for the Colorado Association of Academic and Research Laboratories, an environmental regulation reinvention focus group in Colorado.

Subscribe by sending email to listserv@coloradocollege.edu

Send messages to the list at caarl-l@coloradocollege.edu

For more details, contact John DeLaHunt: jdelahunt@coloradocollege.edu

EMC-PSTC

The EMC-PSTC is an informal group of people interested in product safety regulations and standards worldwide, networked electronically by mailing list. Its purpose is to provide a forum for the sharing of public, but possibly obscure, product safety or regulatory compliance information or related information with limited natural distribution.

Subscribe by sending email to majordomo@ieee.org

Send messages to the list at emc-pstc@ieee.org

For more details, contact Roger Volgstadt: Volgstadt_Roger@Tandem.com

EPA-GENERAL

EPA-GENERAL is a one-directional mailing list established to distribute select U.S. *Federal Register* documents automatically on the day of publication. It contains general EPA nonprogram-specific documents, presidential documents related to environmental issues, and other agency environmental documents other than environmental impact and endangered species actions.

Subscribe by sending email to listserver@unixmail.rtpnc.epa.gov

Send messages to the list at epa-general@unixmail.rtpnc.epa.gov

For more details, contact John Richards: richards.john@epamail.epa.gov

EPA-IMPACT

EPA-IMPACT is a one-directional mailing list established to distribute select U.S. *Federal Register* documents automatically on the day of publication. It contains environmental impact statements published in the *Federal Register*.

Subscribe by sending email to listserver@unixmail.rtpnc.epa.gov

Send messages to the list at epa-impact@unixmail.rtpnc.epa.gov

For more details, contact John Richards: richards.john@epamail.epa.gov

EPA-MEETINGS

EPA-MEETINGS is a one-directional mailing list established to distribute select U.S. *Federal Register* documents automatically on the day of publication. It contains all meeting notices. Program-specific meeting notices are duplicated under the appropriate area.

> Subscribe by sending email to listserver@unixmail.rtpnc.epa.gov
> Send messages to the list at epa-meetings@unixmail.rtpnc.epa.gov

For more details, contact John Richards: richards.john@epamail.epa.gov

EPA-PEST

EPA-PEST is a one-directional mailing list for the distribution of select U.S. *Federal Register* documents automatically on the day of publication. It contains Office of Pesticide Programs documents.

> Subscribe by sending email to listserver@unixmail.rtpnc.epa.gov
> Send messages to the list at epa-pest@unixmail.rtpnc.epa.gov

For more details, contact John Richards: richards.john@epamail.epa.gov

EPA-SAB

EPA-SAB is a one-directional mailing list for the distribution of select U.S. *Federal Register* documents automatically on the day of publication. It contains material relating to the Science Advisory Board.

> Subscribe by sending email to listserver@unixmail.rtpnc.epa.gov
> Send messages to the list at epa-sab@unixmail.rtpnc.epa.gov

For more details, contact John Richards: richards.john@epamail.epa.gov

EPA-TOX

EPA-TOX is a one-directional mailing list for the distribution of select U.S. *Federal Register* documents automatically on the day of publication. It contains the Office of Pollution Prevention and Toxic Substances documents excluding Community-Right-To-Know (Toxic Release Inventory) documents.

> Subscribe by sending email to listserver@unixmail.rtpnc.epa.gov
> Send messages to the list at epa-tox@unixmail.rtpnc.epa.gov

For more details, contact John Richards: richards.john@epamail.epa.gov

EPA-TRI

EPA-TRI is a one-directional mailing list for the distribution of select U.S. *Federal Register* documents automatically on the day of publication. It contains Community-Right-To-Know Toxic Release Inventory documents.

> Subscribe by sending email to listserver@unixmail.rtpnc.epa.gov
> Send messages to the list at epa-tri@unixmail.rtpnc.epa.gov

For more details, contact John Richards: richards.john@epamail.epa.gov

EPA-WASTE

EPA-WASTE is a one-directional mailing list for the distribution of select U.S. *Federal Register* documents automatically on the day of publication. It contains Hazardous and Solid Waste documents.

> Subscribe by sending email to listserver@unixmail.rtpnc.epa.gov
> Send messages to the list at epa-waste@unixmail.rtpnc.epa.gov

For more details, contact John Richards: richards.john@epamail.epa.gov

EPA-WATER

EPA-WATER has been established to distribute select *Federal Register* documents automatically on the day of publication.

> Subscribe by sending email to listserver@unixmail.rtpnc.epa.gov
> Send messages to the list at epa-water@unixmail.rtpnc.epa.gov

EPAFR-CONTENTS

EPAFR-CONTENTS is a one-directional mailing list for the distribution of select U.S. *Federal Register* documents automatically on the day of publication. It contains the full-text *Federal Register* table of contents with page number citations.

> Subscribe by sending email to listserver@unixmail.rtpnc.epa.gov
> Send messages to the list at epafr-contents@unixmail.rtpnc.epa.gov

For more details, contact John Richards: richards.john@epamail.epa.gov

EXPERT-L

The Expert List is an Internet mailing list for the various technical professionals that engage in expert witness activities. If you perform expert witness services and want to start offering your services as an expert witness, or you are in the legal profession and use expert witnesses, this mailing list is for you.

Subscribe by sending email to majordomo@lists.harvard.net
Send messages to the list at expert-l@lists.harvard.net

For more details, contact lern@legalresearch.ultranet.com

HAZMAT-L

Transportation, storage, and reporting of hazardous materials, along with general hazmat discussions.

Subscribe by sending email to listserv@coloradocollege.edu
Send messages to the list at hazmat-l@coloradocollege.edu

For more details, contact John DeLaHunt: jdelahunt@cc.colorado.edu

INDUSTRIAL-RELATIONS-RESEARCH

A forum for academic discussion on industrial relations broadly conceived. It covers current research, methods, results, and theories on employment relations, collective relationships, trade unions, HRM, and employment law.

Subscribe by sending email to mailbase@mailbase.ac.uk
Send messages to the list at industrial-relations-research@mailbase.ac.uk

For more details, contact Robert Blackburn: r.a.blackburn@kingston.ac.uk

LAB-XL

Performance-oriented environmental regulation of laboratories.

Subscribe by sending email to listserv@list.uvm.edu
Send messages to the list at lab-xl@list.uvm.edu

For more details, contact Ralph Stuart: rstuart@esf.uvm.edu

NTPMAIL

A number of NTP reports are available from the NTP web site. These offerings are continually updated as new information/documents are made available. We will notify you by email of such happenings as upcoming NTP events and new publications. (U.S. National Toxicology Program)

Subscribe by sending email to rowley@niehs.nih.gov
Send messages to the list at ntpmail@niehs.nih.gov

RADSAFE-EU

RADSAFE-EU is the European equivalent to the international maillist RADSAFE. Purpose of the list: discussion of radiation safety issues of general interest to radiation and health physics professionals in the European Union, *e.g.*, (but not restricted to) harmonization of radiation protection regulations and procedures

Subscribe by sending email to majordomo@fz-juelich.de

Send messages to the list at radsafe-eu@fz-juelich.de

SAFTNEWS

SAFTNEWS is a free biweekly newsletter specifically designed for U.S. and Canadian businesses interested in staying up-to-date with the latest in safety and industrial news. The focus is on new and upcoming regulations and how they will affect your workplace.

Subscribe by sending email to listserv@labsafety.com

Send messages to the list at saftnews@labsafety.com

TOTAL-QUALITY-ISOSTDS

A forum for the exchange of ideas and information by researchers and practitioners involved in the certification, development, and management of environments to ISO standards.

Subscribe by sending email to mailbase@mailbase.ac.uk

Send messages to the list at total-quality-isostds@mailbase.ac.uk

For more details, contact

Jim McMenamin: jmcmenam@pmc1.pmc.port.ac.uk

Tony Moody: tmoody@pmc1.pmc.port.ac.uk

Christopher Seow: mcyi6cks@fs1.sm.umist.ac.uk

Safety Management/Risk Management

ALUM-L

University of Georgia Risk Management information.

Subscribe by sending email to listserv@uga.cc.uga.edu
Send messages to the list at alum-l@uga.cc.uga.edu

For more details, contact Sandra G. Gustavson:

sgustavson@cbacc.cba.uga.edu

CMTS-L

CMTS-L is a a forum for the exchange of ideas and information among organizations and corporations engaged in the establishment of chemical management and tracking systems, or concerned with improving chemical management processes in general.

Subscribe by sending email to listserv@ls.emax.com
Send messages to the list at cmts-l@ls.emax.com

EMERG-UNIV

The University and College Emergency Planning Mailing List.

Subscribe by sending email to majordomo@sfu.ca
Send messages to the list at emerg-univ@sfu.ca

For more details, contact Peter S. Anderson: anderson@sfu.ca

ERAPPA-L

ERAPPA-L is a list dedicated to the discussion of any topic of interest to those engaged in the management of campus facilities for colleges and universities. Discusssion topics might include announcements of meetings, seminars, and other events of interest to ERAPPA members, issues relating to the operations or maintenance of campus facilities, planning or construction of campus facilities, or management issues relating to physical plant personnel.

Subscribe by sending email to listserv@lists.psu.edu
Send messages to the list at erappa-l@lists.psu.edu

For more details, contact Jack L. Knee: jlk5@psu.edu
Pete Weiss: pete-weiss@psu.edu

ISO14000

This list is designed for the discussion of the ISO 14000 certification guidelines for environmental and related industries. It is currently unmoderated.

Subscribe by sending email to majordomo@quality.org
Send messages to the list at iso14000@quality.org

For more details, contact Gene Tatsch: cet@rti.org

IURISK-L

Indiana University risk management newsletter.

Subscribe by sending email to listserv@listserv.indiana.edu
Send messages to the list at iurisk-l@listserv.indiana.edu

For more details, contact Larry Stephens: STEPHENL@indiana.edu

LEPC

Hazardous materials emergency response planning.

Subscribe by sending email to listserv@list.uvm.edu
Send messages to the list at lepc@list.uvm.edu

For more details, contact Ralph Stuart: rstuart@esf.uvm.edu

MHEC-RMI

Discussion forum for risk management officers.

Subscribe by sending email to listserv@tc.umn.edu
Send messages to the list at mhec-rmi@tc.umn.edu

For more details, contact Greg Earhart: gearhart@tc.umn.edu

MTN

A mailing list for the Midwest Association of Physical Plant Administrators (MAPPA) Trainers Network.

Subscribe by sending email to listserv@listserv.indiana.edu
Send messages to the list at mtn@listserv.indiana.edu

For more details, contact Cindy Stone: stonec@indiana.edu

NCSC

Student-based campus safety programs.

Subscribe by sending email to majordomo@majordomo.srv.ualberta.ca
Send messages to the list at ncsc@majordomo.srv.ualberta.ca

For more details, contact safewalk@su.ualberta.ca

QUALITY

Total Quality Management (TQM) in manufacturing and service industries.

Subscribe by sending email to listserv@pucc.princeton.edu

Send messages to the list at quality@pucc.princeton.edu

QUEST

The emerging national and international requirement to combine quality, environment, and safety structures and systems assessments is creating the need for the development of a uniform and integrated management discipline. The purpose of this discussion is to define and develop a management system which integrates these three disciplines into an effective operational management system. Discussion will include the interrelation of support and responsibility and the identification of benefits, resources, and assessment objectives.

Subscribe by sending email to listserv@listserv.nodak.edu

Send messages to the list at quest@listserv.nodak.edu

For more details, contact jennejohn@uwstout.edu

RISK

The list is for the application of economics, social science, and management to policy problems where risk is a central feature of the decision-making process.

Subscribe by sending email to mailbase@mailbase.ac.uk

Send messages to the list at risk@mailbase.ac.uk

For more details, contact Paul Anand: panand@dmu.ac.uk

Fergus Bolger: bolger@few.eur.nl

RISKANAL

RISKANAL is a discussion list focused on risk analysis. RISKANAL is affiliated with the Society for Risk Analysis, a multidisciplinary, interdisciplinary, scholarly, international society that provides an open forum for all those who are interested in risk analysis.

Subscribe by sending email to listserv@bbs.pnl.gov

Send messages to the list at riskanal@bbs.pnl.gov

For more details, contact js_dukelow@ccmail.pnl.gov

RISKOM

The European Risk Communication Network provides a bridge between research and practice in the U.K. and Europe. This list supports the ongoing interaction between researchers and practitioners in the development of best practice for risk communication.

> Subscribe by sending email to mailbase@mailbase.ac.uk
> Send messages to the list at riskom@mailbase.ac.uk

For more details, contact Simon Gerrard: s.gerrard@uea.ac.uk

TOTAL-QUALITY-HEALTHCARE

A forum to present information, comment, and debate on quality management issues, trends, and developments which affect the healthcare industry.

> Subscribe by sending email to mailbase@mailbase.ac.uk
> Send messages to the list at total-quality-healthcare@mailbase.ac.uk

For more details, contact Christopher Seow: mcyi6cks@fs1.sm.umist.ac.uk

TQM

Total Quality Management

> Subscribe by sending email to listserv@frodo.carfax.co.uk
> Send messages to the list at tqm@frodo.carfax.co.uk

For more details, contact Sharron Lawrence: sharron.lawrence@carfax.co.uk

TQMEDU-L

Total Quality Management in Educational Institutions.

> Subscribe by sending email to listserv@admin.humberc.on.ca
> Send messages to the list at tqmedu-l@admin.humberc.on.ca

For more details, contact green@admin.humberc.on.ca

TRANSDERM

Exchange and dissemination of ideas, information and data, including unpublished material, on all aspects of skin penetration by chemicals. Seminars, conferences, and research posts can be publicized. It should interest workers in dermatology, pharmaceuticals, cosmetics, and risk assessment.

> Subscribe by sending email to mailbase@mailbase.ac.uk
> Send messages to the list at transderm@mailbase.ac.uk

For more details, contact John Pugh: pugh@cardiff.ac.uk

Standards

ISO14000

This list is designed for the discussion of the ISO 14000 certification guidelines for environmental and related industries. It is currently unmoderated.

Subscribe by sending email to majordomo@quality.org

Send messages to the list at iso14000@quality.org

For more details, contact Gene Tatsch: cet@rti.org

TOTAL-QUALITY-ISOSTDS

A forum for the exchange of ideas and information by researchers and practitioners involved in the certification, development, and management of environments to ISO standards.

Subscribe by sending email to mailbase@mailbase.ac.uk

Send messages to the list at total-quality-isostds@mailbase.ac.uk

For more details, contact

Jim McMenamin: jmcmenam@pmc1.pmc.port.ac.uk

Tony Moody: tmoody@pmc1.pmc.port.ac.uk

Christopher Seow: mcyi6cks@fs1.sm.umist.ac.uk

Toxicology

NBTOX-L

Neurobehavioral Toxicology Discussion list.

Subscribe by sending email to listserv@listserv.navy.al.wpafb.af.mil

Send messages to the list at nbtox-l@listserv.navy.al.wpafb.af.mil

For more details, contact John Rossi: jrossi@cs.bgsu.edu

NTPMAIL

A number of NTP reports are available from the NTP web site. These offerings are continually updated as new information/documents are made available. We will notify you by email of such happenings as upcoming NTP events and new publications. (U.S. National Toxicology Program)

Subscribe by sending email to rowley@niehs.nih.gov

Send messages to the list at ntpmail@niehs.nih.gov

PSST

Graduate student discussions in pharmacology and toxicology.

Subscribe by sending email to listserv@listserv.arizona.edu
Send messages to the list at psst@listserv.arizona.edu

For more details, contact Eric Wildfang: wildfang@u.arizona.edu

TOX

Integrated toxicology program list.

Subscribe by sending email to majordomo@acpub.duke.edu
Send messages to the list at tox@acpub.duke.edu

For more details, contact David Watson: dew@acpub.duke.edu

Workers Compensation, Rehabilitation, Disability

BLINDJOB

This list is intended as a discussion on any and all aspects of employment as it relates to blind people.

Subscribe by sending email to listserv@maelstrom.stjohns.edu
Send messages to the list at blindjob@maelstrom.stjohns.edu

DISABILITY-CAREERS-FORUM

This list discusses issues related to employment for people with disabilities. It is a list for university careers advisors, disabled graduates, employers, and employment professionals.

Subscribe by sending email to mailbase@mailbase.ac.uk
Send messages to the list at disability-careers-forum@mailbase.ac.uk

For more details, contact Deborah Birchall: deborah.birchall@lancaster.ac.uk
David Filmer: d.filmer@lancaster.ac.uk

DISABILITY-DIALOGUES

A forum for disseminating via the Internet the papers presented at an ongoing seminar for academics and others concerned with disability policy entitled "Dialogues in Disability Theory and Policy," together with the discussion generated by these papers.

Subscribe by sending email to mailbase@mailbase.ac.uk
Send messages to the list at disability-dialogues@mailbase.ac.uk

For more details, contact Colin Low: c.m.low@city.ac.uk

DISABILITY-INFO

Disability information, from the U.S. Centers for Disease Control and Prevention.

Subscribe by sending email to listserv@listserv.cdc.gov
Send messages to the list at disability-info@listserv.cdc.gov

For more details, contact Alex Null: axn0@cdc.gov
Angela Donaldson: amd2@cdc.gov

DISABILITY-RESEARCH

This list is intended for all those interested in research as it affects disabled people both in the U.K. and internationally. It provides a forum for the exchange of ideas, information, and news, particularly among researchers working within a social model of disablement.

Subscribe by sending email to mailbase@mailbase.ac.uk
Send messages to the list at disability-research@mailbase.ac.uk

For more details, contact Mark Priestley: splmap@lucs-01.novell.leeds.ac.uk

DISABLED

DISABLED aims to examine ways of exploring disability within a social and relational context. We take a dialectical approach to the body and psyche which attempts to resist both social or biological determinism.

Subscribe by sending email to listserv@maelstrom.stjohns.edu
Send messages to the list at disabled@maelstrom.stjohns.edu

For more details, contact Dr. Deborah Marks: d.s.marks@sheffield.ac.uk

EURAHEAD

European Association for H.E. Access and Disability.

Subscribe by sending email to listserv@listserv.heanet.ie
Send messages to the list at eurahead@listserv.heanet.ie

For more details, contact Alexis Donnelly: Alexis.Donnelly@CS.TCD.IE

FIBROM-L

FIBROM-L is a discussion forum for the disease/syndrome known as fibromyalgia/fibrositis. It is an opportunity for patients, family, and friends of patients, physicians, and researchers and other interested persons to discuss this condition. FIBROM-L is an unmoderated list open to all interested subscribers.

Subscribe by sending email to listserv@mitvma.mit.edu
Send messages to the list at fibrom-l@mitvma.mit.edu

For more details, contact Mark London: mrl@pfc.mit.edu

HR-CANADA

This list is intended for discussions of general human resource issues in Canada.

Subscribe by sending email to hr-canada-subscribe@hronline.com
Send messages to the list at hr-canada@hronline.com

For more details, contact Jeff Hill: ohs@apache.hronline.com

ICACBR-L

Community rehabilitation programs for persons with disabilities.

Subscribe by sending email to listserv@qucdn.queensu.ca
Send messages to the list at icacbr-l@qucdn.queensu.ca

For more details, contact William Boyce: boycew@post.queensu.ca

INJURY-L

This is an open discussion list for users interested in injury topics including research, cost, intervention, rehabilitation, prevention, data, and epidemiology.

Subscribe by sending email to listserv@wvnvm.wvnet.edu
Send messages to the list at injury-l@wvnvm.wvnet.edu

For more details, contact Mike Furbee: furbee@wvu.edu

LIS-SPEC

Discuss the range of experiences covering all aspects of library and information services, including access, services, facilities, equipment, and staff training for disabled staff and students.

Subscribe by sending email to mailbase@mailbase.ac.uk
Send messages to the list at lis-spec@mailbase.ac.uk

For more details, contact Carol Bevan: wl0cbe@library2.sunderland.ac.uk
Jeanette Doull: wl0jdo@library2.sunderland.ac.uk

OCCUPATIONAL-THERAPY

This list facilitates discussion and generates ideas and information about matters relating to occupational therapy (OT). Anyone with an professional interest in OT, such as lecturers and researchers, is welcome to participate.

Subscribe by sending email to mailbase@mailbase.ac.uk
Send messages to the list at occupational-therapy@mailbase.ac.uk

For more details, contact Robert Stainer: r.a.stanier@bton.ac.uk

OT-L

Occupational therapy discussions.

Subscribe by sending email to majordomo@milwaukee.tec.wi.us
Send messages to the list at ot-l@milwaukee.tec.wi.us

OT-PSYCH

Occupational therapy and mental health discussions.

Subscribe by sending email to majordomo@dartmouth.edu
Send messages to the list at ot-psych@dartmouth.edu

For more details, contact kristin.levine@valley.net

PTHER

PTHER is a forum for the exchange of ideas pertaining to treatment protocols, clinic management, and the general advancement of the field of physical therapy. Practicing physical therapists, students of physical therapy, and those interested in physical therapy and related fields are encouraged to subscribe and participate.

Subscribe by sending email to majordomo@majordomo.srv.ualberta.ca
Send messages to the list at pther@majordomo.srv.ualberta.ca

For more details, contact Jim Doree: j.d@pobox.com

SCIPIN-L
.

Spinal Cord Injury Peer Net

Subscribe by sending email to listserv@health.state.ny.us
Send messages to the list at scipin-l@health.state.ny.us

TECH4DEVCOUNTRY

Rehabilitation technology professionals for developing countries.

Subscribe by sending email to listserv@listserv.uic.edu
Send messages to the list at tech4devcountry@listserv.uic.edu

For more details, contact politano@uic.edu

VOCEVAL

Vocational evaluation for rehabilitation.

Subscribe by sending email to listserv@maelstrom.stjohns.edu
Send messages to the list at voceval@maelstrom.stjohns.edu

For more details, contact rbanks@discover-net.net

Usenet Newsgroups

Newsgroup names are hierarchical. Some of the hierarchies are international (*e.g.*, alt, misc), some originate from a particular geographic area (*e.g.*, can, de), some specialize in certain types of discussions (*e.g.*, sci, comp), and some are from specific organizations (*e.g.*, ncf, uiuc).

To access Usenet Newsgroups, you must have access to a news server (NNTP server) and news reader software. Virtually all Internet Service Providers (ISPs) have news servers for their customers' use. Many universities and colleges have their own news servers for use by their faculty, staff, and students. Not all news servers provide access to all newsgroups. "Articles" (messages) within a particular newsgroup may be kept on a particular news server for a few days, or at most, for a couple of weeks. Web-based services such as DejaNews (http://www.dejanews.com) and InReference Inc. (http://www.reference.com) maintain searchable archives of Usenet Newsgroups.

You may have access to some or all of the following safety and health-related Usenet Newsgroups.

alt.building.construction

alt.building.health-safety

alt.construction

alt.disasters.planning

alt.hvac (heating, ventilation, and air conditioning)

alt.med.ems (emergency medical services)

alt.med.fibromyalgia

alt.support.hearing-loss

alt.sustainable.agriculture

bionet.audiology

bionet.toxicology

bit.listserv.c+health (computers and health)

bit.listserv.fire-l

bit.listserv.labmgr

biz.ergonomic_sciences

can.construction

can.schoolnet.firefighters

de.etc.notfallrettung (Emergency Services - German language)

misc.emerg-services

misc.health.injuries.rsi.moderated (repetitive strain injuries)

misc.health.injuries.rsi.misc

misc.health.therapy.occupational

misc.industry.printing

misc.industry.pulp-and-paper

misc.industry.quality

misc.industry.safety.personal

misc.transport.air-industry.cargo

misc.transport.trucking

ncf.sigs.health.emergency-care

sci.chem.labware

sci.eng.heat-vent-ac

sci.engr.electrical.compliance

sci.engr.safety

sci.environment.waste

sci.med.diseases.cancer

sci.med.diseases.hepatitis

sci.med.diseases.lyme

sci.med.ems

sci.med.laboratory

sci.med.occupational

sci.med.physics

sci.med.radiology

sci.med.vision

slac.emergency-ops

uiuc.misc.safety

An Internet Glossary

Administrivia: Notices sent out by the manager of a mailing list or newsgroup that have to do with the functioning of the list, rather than the subject of the discussion group.

ASCII files, Text files: Files that contain only ASCII codes, which refer to letters, numbers, or punctuation marks. These files are the surest way to transfer text and are readable by almost any platform and software. Text enhancements, such as bold face, italics, etc., are not contained within ASCII text, although "tags" can be included within the text to indicate where these effects should be used.

Binary files: Files that contain data that is not ASCII text. These require programs which can read that format to use them.

BITNET: A precursor of the Internet; a computer network (established about 1983) connecting educational institutions via their IBM mainframes.

Bookmark: A Web address, usually stored within your Web browser, which you want to remember for future visits. Some programs refer to bookmarks as "Favorites."

CGI: Common Gateway Interface; "Glue" programs that connect HTML documents to nonweb-enabled programs running on the Web server.

Client software: Software designed to format and send requests for information and then receive a file or menu and display it on your computer.

Domain: A major group in the hierarchical system of naming nodes within computer networks. For example, **edu** is a major domain of the Internet and all node names within it end in **.edu** (*e.g.*, esf.uvm.edu).

Domain name servers: Computers that convert numeric node names (*e.g.*, 132.198.205.29) to alphabetic node names (*e.g.*, esf.uvm.edu). These computers work in concert with one another to direct email and Web files to the correct destination. When there is a problem delivering a file or making a connection on the Web, DNS errors usually figure prominently in the error message.

Downloading: The act of receiving a file on one computer that has been sent from another computer.

Emoticons: Punctuation (;:(){}[]) and acronyms used to provide clues to the tone of a message. The meaning of the punctuation symbols is usually determined by rotating the characters clockwise 90 degrees.

Sample emoticons		*Sample acronyms*	
:)	happy smile	TIA	thanks in advance
;)	ironic smile	BTW	by the way
:(	sadness	IMHO	in my humble opinion

Error messages: Messages, often cryptic, generated by computer network software indicating that a command or posting you sent created unexpected results.

FAQ (Frequently Asked Questions): Text files of questions and answers that respond to the most commonly asked questions on a mailing list or newsgroup.

File libraries: Collections of computer files made available on the Internet. These files can contain text, graphics, animation, or sound data. Some Internet tools are designed to translate these types of files directly. Others require downloading the file and then using it with appropriate software.

Flames: Emotional responses to a posting, often of a personal nature; flame wars are continuing exchanges of flames between two or more parties. These are irritating to nonparticipants and generally resolve nothing between the participants.

FTP (File Transfer Protocol): A set of commands that allows the movement of files between computers connected over the Internet.

GIF: One of two popular graphic image formats on the Web. *See also* JPEG.

Gopher: A protocol that organizes file libraries and other services into a system of menus.

Headers: The first 10 to 15 lines of an email message. These provide technical data to the mailer software about who the message is to be routed to, where it is coming from, the date sent, the subject matter, and other miscellaneous information.

Hits: The list of Web sites found when a search is conducted on a search engine's index.

Home page: The opening page of a Web site. The home page of a Web site is automatically returned to a Web browser if no other file is requested with a path suffix to the URL.

Hotlist: A group of bookmarks—URL's that you have visited and noted as containing useful information. Web browsers usually provide ways of maintaining hotlists within the software.

HTML: HyperText Mark-up Language—a system of commands inserted into an ASCII text file that tell a WWW client how to treat the text within the command.

HTTP: HyperText Transmission Protocol—the computer information required to send and receive hypertext files.

Hypertext: A text file which has codes inserted in it that point to other text file locations so that the reader can move from one point in the text to a related place with a single click.

Intranet: A computer network that uses the Internet protocols, but whose use is restricted to a certain set of users (*e.g.*, employees of a particular company or students at a certain university). The advantage of such a set-up is that a single set of software tools can be used for both internal and external resources.

The Internet: A group of interconnected computer networks that use the Internet communication protocol.

Internet Service Providers (ISPs): Commercial firms that provide access to the Internet, both to homes and businesses. The connections they provide may be made through normal phone lines over modems, or through dedicated high speed lines, such as ISDN or T1 lines.

Information providers: People who set up servers to make files on the Internet for commercial, institutional, or hobby purposes.

IP address: Internet Protocol address is a four-part sequence of numbers that uniquely names a computer on the Internet. These may be dynamically assigned so that the computer at the address can change.

Java: A programming language; Java programs can be embedded in Web pages. When someone visits the Web page using a current Web browser, the Java program is downloaded to the user's computer and runs there. Java programs are "platform independent," meaning that they can be run on any computer operating system that can run a Web browser. Javascript is a subset of the Java language that is used more easily.

JPEG: One of two popular graphic image formats on the Web. *See also* GIF.

LAN: Local area network—a system that connects computers within a local unit, for example, a specific office or building. A variety of networking protocols can operate within LANs (for example, Novell or AppleTalk). Most of these protocols can be translated to the Internet protocol once a physical Internet connection is in place.

List owner: The manager of an email list. Unlike a computer guru, the list owner is a person who is interested in the subject-matter of the list, but does not neccessarily have technical expertise. So the list owner may not be able to resolve technical computer problems. While list owners have special power over list usage, the style of managing the list will vary significantly from individual to individual.

LISTSERV: One of the earliest mailing list management software packages and the most completely developed. It was originally written for IBM mainframes of the BITNET system; it has recently been translated for Unix and other operating systems.

Link: A piece of electronic text that points to another location. By clicking on the link, you are taken immediately to the other location.

Lurker: Someone who reads a newsgroup or email list regularly without participating. Probably 50 percent of the audience for any list consists of lurkers.

Mailer: Software which is used to send and receive email.

Mailing lists: Special interest groups of email users who receive email on the subject at hand. Mailing lists are generally operated by mailing list management (MLM) software, which automates routine tasks such as adding members, deleting members, etc. Because there are many different MLM packages in use, it is important to know which package is used by a particular list in order to know what commands will be accepted.

Majordomo: A freeware Unix-based mailing list management (MLM) software package. It is widely used, but lacks many of the features found in commercial MLMs such as LISTSERV.

Mirrors: Servers that duplicate the files found at popular ftp/Web sites so that the network traffic can be distributed more evenly and access to the files at the site is more reliable.

Moderated list: A moderated list is an email list in which all the postings are first sent to one or more editors who have to approve the posting of the note to the list. These lists are uncommon for free discussion because they require a significant amount of effort to maintain. However, they are used for lists set up to provide official information and announcements to particular groups.

Netiquette: Email practices that improve a person's ability to participate productively in mailing lists and newsgroups.

Network: A communications system that links two or more computers. It can be as simple as a cable strung between two computers a few feet apart or as complex as hundreds of thousands of computers around the world linked through fiber optic cables, phone lines, and satellites.

Newbie: Somebody new to the Net. The term is often used derogatorily by Net veterans who have forgotten that they themselves were once newbies.

Newsgroups: Groups of messages on a related subject that are accessed by newsreaders.

Node: A computer on network. Its location is given by its IP address (a numeric designation) or node name (an alphabetic designation).

Operating system: The software that starts a computer and provides the basic machine functions for user-installed software. The most common operating systems are DOS/Windows and Macintosh, for personal computers, and Unix and CMS, for mainframe computers.

PDF files: Binary files which contain formatted documents. These files are read with specific software (Adobe Acrobat) which is freely available on the Web.

Platform: The hardware/software combination used to access the Internet. The platform you use determines which Internet services you can access. For example, a Macintosh computer with MacTCP, an ethernet connection, and the proper software can access nearly any Internet service.

POP: Post Office Protocol is the usual way that email software receives email. Email software often requires that you indentify the POP server, which is the computer that you will receive email from. This information should be available from your internet service provider. (Contrast with SMTP server.)

POTS: Plain Old Telephone Service, a regular phone line, which may be used for a modem connection, is sometimes referred to as a POTS line to differentiate it from a digital line specifically intended to handle computer communications.

PPP: Point to Point Protocol, which allows an Internet protocol connection to be made over a phone line-modem connection.

Posting: A message sent to a mailing list or a newsgroup.

Protocols: An agreed upon format for network transmissions; the "common language" which the various pieces of computer software speak.

RTF (Rich Text Files): ASCII files that include codes to indicate the style of the text in the document. These codes are interpreted to display text with bold, italics, and underlines rather than just plain text. This increases the readability of the text. Most word processors are able to read and write RTF files.

Server software: Software that, once installed on a particular computer connected to the Internet, waits for a request from a client. In response to a request, it sends a menu or a file.

Shell account: A computer account on a mainframe computer, usually a Unix machine, that can be used to access the Internet. These accounts do not provide graphical interfaces to files, but can be used to download graphics files from the Internet. They are often more convenient or faster to use for searching the Web or managing email.

Sig file, Signature file: A file containing personal contact information which is automatically appended to your email. Often contains bumper sticker-like slogans.

Signal-to-noise ratio: The amount of useful information to be found in a given mailing list or newsgroup. Often used derogatorily. For example: *The signal-to-noise ratio in this newsgroup is pretty low.*

SMTP: Simple Mail Transfer Protocol is the computer protocol generally used to send out email from an email client. Most email software requires you to identify a SMTP server, which is the computer that your computer will use to send your email. (Contrast with POP server.)

Sysadmin: The system administrator/system operator; the person who runs a server computer.

Terminal software: Software which allows a computer to use a modem or Telnet connection to interact with another computer. This software usually transmits simple ASCII characters from one computer to another.

TCP/IP: Terminal Control Program/Internet Protocol; the technical basis for connecting computer networks together to form the Internet.

Test messages: Messages that contain no information and are sent by someone to a newsgroup or a mailing list to see if their system is working. (This is considered poor *netiquette*.)

Thread: A sequence of messages in a discussion group on the same subject. Threads often experience "thread drift," where the subject under discussion changes without a change in the subject line of the messages.

Uniform Resource Locator (URL): A string of letters and punctuation that identify the location of a file on the Internet. They are of the form resource://node/folder/filename.

Unix: An operating system which is commonly used by computers which function as information servers on the Internet.

UseNet: A term used to refer to all newsgroups collectively. Also know as network news.

Web browser: Client software which is used to access and display computer files on the WWW.

Web site: A group of files that reside on a particular computer on the Internet and which are connected by HTML links in a logical manner.

World Wide Web (The Web, WWW): Files available for public viewing which use the HTML protocol that allows them to act as hypertext documents, pointing to other files or Internet services. These files can be graphics files, sound files, or rich text files.

Email
Reference Material

The SAFETY Welcome File

An introduction to the SAFETY mailing list, written by Ralph Stuart.

Chemical Safety Coordinator
University of Vermont
List-owner, SAFETY
rstuart@siri.org
rstuart@esf.uvm.edu

Version 5.0
Last revised: February 15, 1999

Contents

1. **Introduction to the SAFETY List**

   What is SAFETY?
   Who is SAFETY?
   Where is SAFETY?
   When is SAFETY?
   Why use SAFETY?
   How do I use SAFETY?
   Reminder
   Disclaimer

2. **Netiquette and Tips for Using SAFETY**

   Information quality
   Asking questions of the list
   Managing list traffic
   Finding the right safety list
   Attachments

3. **LISTSERV Commands and Options**

   How to send email messages to the list using LISTSERV commands
   How to search the archives

1. Introduction to the SAFETY List

What is SAFETY?

SAFETY is an electronic mailing list which started in 1989. People can send email to the list and it will be redistributed to the list subscribers. Discussions on SAFETY involve environmental and occupational health and safety issues, although a wide range of subjects and participants is encouraged. Issues discussed in the past include chemical safety issues, indoor air quality, interpretation of safety standards and regulations, proper hazardous waste disposal, safety management, and electronic resources on these topics. There are currently more than 70,000 messages in the SAFETY archives. An keyword index to these archives, daily digests of the discussions, and other useful SAFETY-related files can be found at **http://siri.org/mail** and **http://list.uvm.edu/archives/safety.html**.

Who is SAFETY?

As of February 1999, SAFETY consisted of about 3000 user IDs. (The precise number changes hourly.) The subscribers are primarily in the United States (88 percent) and Canada (5 percent), although there are subscribers in many other countries, including Australia (2 percent), New Zealand, Israel, Spain, Argentina, and the U.K. People on the list represent academia, industry, the military, and government agencies: 48 percent are .COM, .NET, and .ORG addresses (the commercial email providers and corporate addresses); 27 percent of the SAFETY IDs are academic addresses (.EDU); 7

percent are government addresses; and 3 percent are military addresses. Expertise of users ranges from Ph.D.s with many years of health and safety experience to undergraduate students.

Where is SAFETY?

The LISTSERV program which manages SAFETY is located at the University of Vermont, Burlington, Vermont, U.S.A. on the LIST.UVM.EDU machine. As noted above, people on SAFETY represent all regions of the United States and several other countries.

When is SAFETY?

Email makes nearly instantaneous communication possible. However, response from the list to a particular posting is not necessarily immediate. If a posting will elicit a response, the first response will likely come within 48 hours (except around American holidays). Note that the list is moderated by human intervention; this means that it may be several hours before your note is distributed to the list. If your need for information is more urgent than that, you may want to search the SAFETY archives at **http://list.uvm.edu/archives/safety.html** for helpful information or names.

Continuing discussion can go on for a week or longer. If your posting does not draw any response, it may be too general a question for an electronic mail reply. See Section 2 of this appendix for tips on how to improve your chances of an useful answer.

SAFETY is a moderated list. This means that each posting is reviewed before distribution to the list to assure that 1) it does not contain binary data (*i.e.*, attachments) that might disrupt the variety of email software receiving SAFETY mail, 2) it is not likely to provoke flames in response, and 3) it is not repetitive of other postings. If a posting is not deemed suitable for distribution over the list, it is either directed to an appropirate individual or a response is sent to the author describing the problem.

Why use SAFETY?

SAFETY operates as a cooperative community of people with overlapping technical interests and expertise in the wide-ranging field of occupational and environmental health and safety. Few of these people have large amounts of time to devote to answering questions via email. However, they have a genuine interest in helping other people solve safety problems and concerns and learning more about other people's situations. This is what makes SAFETY work.

How do I use SAFETY?

The specifics of using SAFETY and the LISTSERV program are given in Section 3. In general, you should use SAFETY to help determine an appropriate next step in a particular situation. No question should be considered too basic for SAFETY, but remember the limitations of the medium. Vague questions requiring long answers are unlikely to bring a response, and long essays are generally not conducive to email discussions. If a long series of questions and answers will be required to explain a situation, it is probably best to ask for help on SAFETY, and then, if any willing volunteers respond, to move the discussion off the list and to correspond privately.

Reminder

SAFETY is a public resource. Summaries and excerpts of the discussions on SAFETY are reprinted in various media, both public and private. Therefore, anything written to SAFETY should be considered in the public domain and available for unlimited reproduction, unless the author or sender expressly indicates otherwise. Please respect any copyright notice attached to a particular item by contacting the author for permission to reproduce the item.

Disclaimer

Remember that SAFETY is designed to be a forum for the sharing of information and expertise in a *general* way. It is inappropriate for specific consultations about individual safety concerns to be made over a computer network. This fact must be kept in mind while interpreting SAFETY postings; they are only general statements concerning the information which the posters have been given. Specific health and safety advice should only be given by qualified individuals who are fully aware of all the factors involved in a particular situation.

More specifically: The statements on SAFETY are strictly personal opinion and do not necessarily represent the positions of any organization the writer works for or is associated with. There is no verification of the accuracy of any messages distributed over SAFETY. Therefore, each SAFETY reader is responsible for confirming information presented on SAFETY with independent resources. Any interpretations of federal, state, or local government regulations must be confirmed with the appropriate agency before being considered authoritative. Neither the users of SAFETY nor any of the organizations providing access to SAFETY are responsible for the interpretation or application of the information contained within the discussions.

2. Netiquette and Tips for Using SAFETY

The SAFETY list has been operating for close to nine years and thus is one of the older Internet institutions. Over this time, some "rules of the road" have developed that are useful in using the list. Some of these are true of most email lists; others are specific to SAFETY. This section describes some of these rules.

Information quality

The goal of the SAFETY list is to provide an avenue for practical discussions of occupational and environmental health and safety related issues. In order to enhance these discussions, please observe these practices:

- Refrain from offering an opinion without identifying it as such.

- Whenever possible, provide references that allow the receiver of your postings to verify the information presented.

- Append a signature to your postings, preferably one indicating your professional credentials.

- Announcements of a commerical nature (product availability, upgrades, etc.) are acceptable. Their subject line should be prefixed with "COMMER-CIAL:" so that the intent of the message is clear.

Asking questions of the list

>WOW, I didn't think I would get flamed for such an honest and >sincere post?!?! Is this group always so unwelcoming to new >subscribers?

There is a developing tradition on SAFETY that for a question to be taken seriously, you need to indicate a certain level of professional expertise in formulating the question. Alternatively, you can disclaim any professional background in the issue at hand and explain enough of your situation for those with appropriate professional expertise to make suggestions or point out that further information is needed.

SAFETY is not so much a question-answering service, as a point-you-in-the-right-direction service. This makes general, open-ended questions hard to deal with. Unfortunately, people new to the list take an initial brusque reaction to a ill-formed question as meaning that the list is unfriendly. Rather, it means "We'd like to help, but you need to ask the question in a short-answer format, rather than as an essay question." BTW, we like multiple-choice tests even better. Remember, some response is better than complete silence, which usually means "Huh?"

Managing list traffic

Message traffic on SAFETY varies. It currently averages around 30 to 35 messages per day, but is often more. There is a limit of 50 messages per day on the list. Messages after the 50-message limit will be held until the next day. Because of this limit, it is important to be sure that notes that you send to the list are appropriate for broadcast to 2700+ people. Particularly during heavy traffic periods, consider whether your note is better sent to a particular individual rather than to the list as a whole. Also, you may want to consider holding a new subject for discussion on a "light traffic day."

"Heavy traffic periods" occur on a semi-predictable basis. Fridays tend to be the heaviest traffic days of the week, while Saturday and Sundays are often quiet (but not silent). The list is busiest in August, January, November, and March. June, July, and December are quieter periods. American holidays tend to be much lighter than other days.

Some people find this traffic level burdensome. Remember that not every message must be read carefully. It is possible to arrange to have all SAFETY messages sent in a single file called digests on a daily basis. (See below.)

Finding the right safety list

A variety of safety-related email lists operate on the Internet. Choosing the right list to post a particular question to can be something of a challenge.

However, it is better to pick one particular list to send the note to, rather than send it to several at once. Usually, there is a lot of overlap among list memberships, and having the same question appear on several lists at the same time can confuse the readers of the various lists and make them less likely to respond. For a list of other safety-related email lists, see **http://www.ccohs.ca/resources/listserv.htm**.

Attachments

Please do not send attachments to the SAFETY list! In order for people to be able to successfully use attachments, they must know what software produced that data. In an audience of 3000 people, it is unlikely that more than a handful will be able to use an attachment successfully. In addition, many people are concerned that unwanted binary files may contain software viruses, and many email packages cannot handle some types of attachments successfully.

If you wish to share information that cannot be carried by the ASCII text format, contact Ralph Stuart to arrange to have the file placed on the Vermont SIRI Web site.

3. LISTSERV Commands and Options

This is a description of various commands that enable you to use the many features of LISTSERV, the mailing list management program that runs SAFETY. For more info on LISTSERV features, send GET LISTSERV MEMO to LISTSERV@LIST.UVM.EDU.

How to send email messages to the list using LISTSERV commands

Send your messages to SAFETY@LIST.UVM.EDU. In order to post messages to the list, you must be signed up for the list at the address that appears in the "From:" line of your message header.

Send all the LISTSERV commands described below to

LISTSERV@LIST.UVM.EDU.

SUBscribe

To add your name to the list, send a message saying SUB SAFETY <your name>. (Example: SUB SAFETY John Smith)

The LISTSERV program will ask you to confirm your subscription, to be sure that the correct email address reached it. In order to confirm, follow the instructions included in the confirmation request message.

UNSUBcribe

To unsubscribe from SAFETY send the command UNSUB SAFETY to LISTSERV. This command also requires confirmation. Be sure that this confirmation occurs before you leave your email access for an extended period. If there is any doubt about whether you're still on the list, contact Ralph Stuart.

REVIEW

To find out who is on the list, send the command REVIEW SAFETY and you will get a list of all unconcealed subscribers. (This is a large file!) If you do not want anyone (except the owner) to be able to see that you are on SAFETY, send the command SET SAFETY CONCEAL.

If your email software does not allow you to see the name of the person posting a message to the list, you can SET SAFETY DUAL to have the header information for a posting included in the body of the message. SET SAFETY SHORT turns this feature off.

DIGESTS mail option

To receive a daily collection of SAFETY messages in one file, send SET SAFETY DIGESTS to LISTSERV@LIST.UVM.EDU. A table of contents of the collected messages comes at the beginning of the DIGEST. If your email software can read MIME encoded DIGESTs, there is an option available for these to be sent to you. Send SET SAFETY MIME to LISTSERV to invoke this option.

NOMAIL mail option

To stop the flow of mail temporarily, issue the command SET SAFETY NOMAIL to LISTSERV. No mail will be sent to you until you issue the command SET SAFETY MAIL or SET SAFETY DIGESTS This command does not require the confirmation process described above for SUB and UNSUB.

How to search the archives

There are database search facilities in LISTSERV to search past SAFETY messages for discussion of a particular subject. These allow you to search for a particular keyword, and LISTSERV will send you a list of all the messages that contain it. The syntax of this job is as follows:

SEARCH <subject> in SAFETY

You can then get selected messages individually, using the command

GETPOST SAFETY <message number(s)>

with info sent to you by the LISTSERV program.

Those with Web access can also search the SAFETY archives for keywords more quickly by using accessing

http://siri.org/mail   (or)

http://list.uvm.edu/cgi-bin/wa?S1=safety

Note that the various indexes use slightly different syntaxes and cover different time periods in the life of the list. Contact Ralph Stuart for specifics of these details.

Consult the LISTSERV documentation at **http://www.lsoft.com** or the list owner for more information about LISTSERV features.

Health and Safety Canada Mailing List

Important note

Your participation in HS-Canada indicates your acceptance of the following guidelines. If you feel that you cannot participate in HS-Canada within these guidelines, please unsubscribe now using the procedure described below. The Canadian Centre for Occupational Health and Safety (CCOHS) reserves the right to terminate the subscription of anyone who fails to comply with one or more of these guidelines, or whose participation is deemed to be inappropriate.

Purpose

HS-Canada is a means of distributing messages to a group of individuals with interests in occupational health and safety in a Canadian context. Although the list is intended primarily for Canadians, anyone with an interest in Canadian occupational health and safety issues is welcome to subscribe. Messages may be sent in English or French.

List moderators

• Andrew Cutz - cutza@northernc.on.ca (technical)

• Chris Moore - chrism@ccohs.ca (administrative)

Messages sent to HS-Canada may relate to any occupational or environmental health and safety topic specific to Canada, specific to one or more provinces or territories, or of interest to people working in health and safety in Canada.

The Canadian Centre for Occupational Health and Safety (CCOHS) provides the facilities for this mailing list, but is in no way responsible for the accuracy or content of information exchanged among list subscribers.

How to subscribe

To subscribe to HS-Canada, send email to majordomo@ccohs.ca. The body of the message should be

subscribe hs-canada

If you would like to subscribe to the digest version of the list instead (*i.e.*, receive one message a day containing all of the previous days messages), the body of the message should be

subscribe hs-canada-digest

How to unsubscribe

If you would like to be removed from the list, send email to majordomo@ccohs.ca. The body of the message should be

unsubscribe hs-canada

If you are subscribed to the digest version, the body of the message should be

unsubscribe hs-canada-digest

Please do not send unsubscribe messages to hs-canada@ccohs.ca.

How to send messages to the list

Once you have subscribed, you can send messages to all list subscribers by addressing an email message to hs-canada@ccohs.ca.

The following points of "Netiquette" and the "Technical Details" are a compilation of requests received from HS-Canada subscribers.

Netiquette

Appropriate Messages on HS-Canada

HS-Canada is for the professional exchange of ideas and information related to occupational and environmental health and safety in a Canadian context, or of interest to people working in health and safety in Canada. The list is not intended as a soapbox for promoting causes, nor is it meant for airing personal complaints against individuals and organizations. If you want information or advice about a specific situation, describe the situation in a nonjudgmental way. If you cannot do so without using defamatory statements about individuals or organizations, do not post the message to this list.

If you send an off-topic message, a message containing defamatory statements, or a message deemed by the list owner to be inappropriate, he/she will send you private email informing you that your message is inappropriate for this list and asking you not to send such messages in the future. If you send another such message, your right to send messages to the list will be revoked.

If you are unsure whether a message is appropriate for the list or not, send it privately to Chris Moore at chrism@ccohs.ca.

Signatures

At a minimum, please add your name, affiliation, geographical location, and email address at the end of messages sent to the list. Your email address alone may not identify you clearly and may not be visible to some message recipients. Subscribers are more likely to take your requests for information and your statements seriously if they know who you are. Most email software packages allow you to create "signature files" for this very purpose.

Posting and responding to job ads

OSH-related job advertisements open to people in Canada are welcome, but only from the organization with the available position or published advertisements. (No employment agencies please!) If you are posting a job advertisement, please include an email address and/or phone number where interested individuals can contact you privately. If you are replying to an advertisement, please do not send your reply to hs-canada@ccohs.ca. Reply directly to the organization that has posted the advertisement.

Upcoming health and safety-related conferenences, meetings, and courses

Announcements about upcoming events are welcome. One notice per event please!

Commercial notices

Information about commercial health and safety-related products and services is acceptable. Advertising is not. If you provide or know about a product or service that you feel might be of interest to list subscribers, provide a brief description of the product or service and details about where interested parties can get more information (*e.g.*, email and/or WWW address, phone number).

If you post a commercial message, type "COMMERCIAL:" at the beginning of the subject line.

Abbreviations

When your message includes abbreviations or acronyms, please indicate what those abbreviations or acronyms stand for. What is obvious to you may be a complete mystery to someone reading your message.

Regulations

When asking questions related to regulations and/or legislative interpretations, please indicate the jurisdiction that you want to know about. When giving an opinion or information about such a question, please indicate what jurisdiction your response refers to.

Forwarding messages from other mailing lists

Many subscribers receive messages from more than one mailing list. There may be times when you see a message on another list that you think might be appropriate for HS-Canada. If the message is an announcement of a conference, a new publication, a new Web site, etc., that might be of interest to HS-Canada subscribers, it is acceptable to "cross-post" to HS-Canada. It would be advisable to wait until the day after you see the announcement on another list, in case someone else has already cross-posted the message.

Please do not cross-post requests for information from other lists, as the responses from HS-Canada subscribers will probably not reach the person asking the question.

Technical details

When you are going to be away

If you are going to be away from your email for a period of time, and if your email software has an autoreply function (*i.e.*, the ability to automatically generate messages to the effect of "I'm going to be out of the office until such and such a date"), please unsubscribe from HS-Canada before using autoreply. Otherwise, all the HS-Canada subscribers will receive multiple copies of your "I'm going to be out of the office" message.

No "Confirm Reading" or "Confirm Delivery"

If you use Pegasus Mail or another email software package with "Confirm Reading" and/or "Confirm Delivery" options, please ensure that they are switched off when you send a message to HS-Canada. Otherwise, a number of "Confirm Reading" and/or "Confirm Delivery" messages will be send to all list members.

Duplicate messages to the list

When you reply to a message on HS-Canada, please make sure that the reply is only addressed to *either* **hs-canada@ccohs.ca** or **hs-canada@ kate.ccohs.ca**, but not both. When some people reply to HS-Canada messages, their email software automatically addresses their replies to hs-canada@ccohs.ca and copies hs-canada@kate.ccohs.ca. This causes two copies of the message to be sent to HS-Canada subscribers. It may be caused by users software having a "reply to all" option switched on.

File attachments

When sending a message to HS-Canada, please do not include any file attachments with your email message. Many users' email software will not be able to "decode" attachments. Please use plain text only within the body of the message. If you have a document that you would like to share with HS-Canada subscribers, send it to Chris Moore for inclusion on CCOHS's Web site. As a contributor, you will continue to own the copyright on documents that you contribute.

All such contributions are available at **http://www.ccohs.ca/hscanada/ hsdocs.html.**

Please do not send documents for which you do not own the copyright (*e.g.,* other people's articles), unless you have been given express permission to do so.

Quoting URLS

When pointing people to a Web site, please enter the address in the form **http://www.something.com** rather than **www.something.com**. If the first format is used, many email packages will allow the message recipient to click on the address. The recipient's Web browser will load and access the referenced site directly.

Any further questions about the list should be directed to the list owner, Chris Moore (chrism@ccohs.ca).

Last updated: December 9, 1998

Index

public health fact sheets 266
public safety 369, 492
   databases 369
   educational resources/training 369
   journals and newsletters 369
   products 370
   programs 372
   publications 370
public safety lighting 198, 317

Q

quality in manufacturing 503
quantitative risk analysis 186

R

radiation and laser safety
   conferences 372
   consulting services 372
   databases 373
   directories 374
   journals and newsletters 374
   products 376
   programs 378
   publications 376
   software 379
radiation detection and analysis 197
radiation safety 78, 172, 222, 224, 495
radioactive decay calculator 226
radiochemistry 188
radiological safety training 225
regulatory compliance products 147, 196
regulatory information 268, 496
rehabilitation 506
repetitive strain injury 201, 328, 470
repetitive strain injury prevention software 213
research 1
respirator information 155
Right-to-Know 123, 167
risk assessment 215
risk management 501
   consulting services 379
   databases 380
   directories 380
   educational resources/training 381
   policy/procedures documents 382

   programs 384
   publications 382
   software 384
risk management 183, 228
   software 210
Rocky Mountain Center 160
root cause analysis 215
rules of thumb 10

S

SAFETY 35, 94, 95.
safety and health
   professionals 3
   software 287
safety
   audits 422
   clip art 158
   cybermall 149
   directors 134
   equipment rentals 184
   eyewear 197
   gloves 197
   information 33
   list 111
   malls 135
   plans 182
   posters 133
   products 135
   seminars 164, 183
   services 130
   software 131, 179
Safety98 127
SafetyLine 80
sample software 154
scheduling program 162
search engines 30, 33, 34, 390
search strategy 33
security 204
self-assessment exam 350
sensing products 202
sharps 227
shiftwork 472
signal-to-noise ratio 25
signs 201
smoke-free environments 417
Society for Chemical Hazard Communication 81

GOVERNMENT INSTITUTES ORDER FORM

4 Research Place, Suite 200 • Rockville, MD 20850-3226
Tel (301) 921-2323 • Fax (301) 921-0264
Internet: http://www.govinst.com • E-mail: giinfo@govinst.com

3 EASY WAYS TO ORDER

1. Phone: **(301) 921-2323**
Have your credit card ready when you call.

2. Fax: **(301) 921-0264**
Fax this completed order form with your company purchase order or credit card information.

3. Mail: **Government Institutes**
4 Research Place, Suite 200
Rockville, MD 20850-3226   USA
Mail this completed order form with a check, company purchase order, or credit card information.

PAYMENT OPTIONS

❑ **Check** *(payable to Government Institutes in US dollars)*

❑ **Purchase Order** *(This order form must be attached to your company P.O. Note: All International orders must be prepaid.)*

❑ **Credit Card**   ❑ VISA   ❑ MasterCard   ❑ AMERICAN EXPRESS

Exp.___/___

Credit Card No. _____

Signature _____

(Government Institutes' Federal I.D.# is 52-0994196)

CUSTOMER INFORMATION

Ship To: (Please attach your purchase order)

Name: _____

GI Account # (*7 digits on mailing label*): _____

Company/Institution: _____

Address: _____
(Please supply street address for UPS shipping)

City: _____ State/Province: _____

Zip/Postal Code: _____ Country: _____

Tel: (_____) _____

Fax: (_____) _____

Email Address: _____

Bill To: (if different from ship-to address)

Name: _____

Title/Position: _____

Company/Institution: _____

Address: _____
(Please supply street address for UPS shipping)

City: _____ State/Province: _____

Zip/Postal Code: _____ Country: _____

Tel: (_____) _____

Fax: (_____) _____

Email Address: _____

Qty.	Product Code	Title	Price

Subtotal_____
MD Residents add 5% Sales Tax_____
Shipping and Handling (see box below)_____
Total Payment Enclosed_____

❑ **New Edition No Obligation Standing Order Program**
Please enroll me in this program for the products I have ordered. Government Institutes will notify me of new editions by sending me an invoice. I understand that there is no obligation to purchase the product. This invoice is simply my reminder that a new edition has been released.

15 DAY MONEY-BACK GUARANTEE
If you're not completely satisfied with any product, return it undamaged within 15 days for a full and immediate refund on the price of the product.

Within U.S:	**Outside U.S:**
1-4 products: $6/product	Add $15 for each item (Airmail)
5 or more: $3/product	Add $10 for each item (Surface)

SOURCE CODE: BP01

Government Institutes Mini-Catalog

PC # ENVIRONMENTAL TITLES

PC #		Title	Pub Date	Price
629		ABCs of Environmental Regulation: Understanding the Fed Regs	1998	$49
627		ABCs of Environmental Science	1998	$39
585		Book of Lists for Regulated Hazardous Substances, 8th Edition	1997	$79
579		Brownfields Redevelopment	1998	$79
4088	◉	CFR Chemical Lists on CD ROM, 1997 Edition	1997	$125
4089	💾	Chemical Data for Workplace Sampling & Analysis, Single User Disk	1997	$125
512		Clean Water Handbook, 2nd Edition	1996	$89
581		EH&S Auditing Made Easy	1997	$79
587		E H & S CFR Training Requirements, 3rd Edition	1997	$89
4082	◉	EMMI-Envl Monitoring Methods Index for Windows-Network	1997	$537
4082	◉	EMMI-Envl Monitoring Methods Index for Windows-Single User	1997	$179
525		Environmental Audits, 7th Edition	1996	$79
548		Environmental Engineering and Science: An Introduction	1997	$79
643		Environmental Guide to the Internet, 4rd Edition	1998	$59
560		Environmental Law Handbook, 14th Edition	1997	$79
353		Environmental Regulatory Glossary, 6th Edition	1993	$79
625		Environmental Statutes, 1998 Edition	1998	$69
4098	◉	Environmental Statutes Book/CD-ROM, 1998 Edition	1997	$208
4994	💾	Environmental Statutes on Disk for Windows-Network	1997	$405
4994	💾	Environmental Statutes on Disk for Windows-Single User	1997	$139
570		Environmentalism at the Crossroads	1995	$39
536		ESAs Made Easy	1996	$59
515		Industrial Environmental Management: A Practical Approach	1996	$79
510		ISO 14000: Understanding Environmental Standards	1996	$69
551		ISO 14001: An Executive Repoert	1996	$55
588		International Environmental Auditing	1998	$149
518		Lead Regulation Handbook	1996	$79
478		Principles of EH&S Management	1995	$69
554		Property Rights: Understanding Government Takings	1997	$79
582		Recycling & Waste Mgmt Guide to the Internet	1997	$49
603		Superfund Manual, 6th Edition	1997	$115
566		TSCA Handbook, 3rd Edition	1997	$95
534		Wetland Mitigation: Mitigation Banking and Other Strategies	1997	$75

PC # SAFETY and HEALTH TITLES

PC #	Title	Pub Date	Price
547	Construction Safety Handbook	1996	$79
553	Cumulative Trauma Disorders	1997	$59
559	Forklift Safety	1997	$65
539	Fundamentals of Occupational Safety & Health	1996	$49
612	HAZWOPER Incident Command	1998	$59
535	Making Sense of OSHA Compliance	1997	$59
589	Managing Fatigue in Transportation, *ATA Conference*	1997	$75
558	PPE Made Easy	1998	$79
598	Project Mgmt for E H & S Professionals	1997	$59
552	Safety & Health in Agriculture, Forestry and Fisheries	1997	$125
613	Safety & Health on the Internet, 2nd Edition	1998	$49
597	Safety Is A People Business	1997	$49
463	Safety Made Easy	1995	$49
590	Your Company Safety and Health Manual	1997	$79

Government Institutes

4 Research Place, Suite 200 • Rockville, MD 20850-3226
Tel. (301) 921-2323 • FAX (301) 921-0264
Email: giinfo@govinst.com • Internet: http://www.govinst.com

Please call our customer service department at (301) 921-2323 for a free publications catalog.

CFRs now available online. Call (301) 921-2355 for info.